FUNDAMENTALS OF CALCULUS

WITH APPLICATIONS AND COMPANION TO CALCULUS

FUNDAMENTALS OF CALCULUS

WITH APPLICATIONS AND COMPANION TO CALCULUS

William J. Adams

Pace University

with illustrations by

Ramunė B. Adams

Library of Congress Control Number: 2007902902
ISBN: Hardcover 978-1-4257-4554-7
Softcover 978-1-4257-4553-0

Fundamentals of Calculus with Applications and Companion to Calculus is available on the web at webpage.pace.edu/wadams.

This book was printed in the United States of America.

To order additional copies of this book, contact:
Xlibris Corporation
1-888-795-4274
www.Xlibris.com
Orders@Xlibris.com
38063

To
Ramunė Adams

CONTENTS

PREFACE

Still another book on calculus? Have mercy might be your initial reaction. With what seems like a ton of books on calculus now available, why another one?

First, let me note that this book is not intended for students whose studies require them to undertake a sequence of calculus courses that have a major commitment to building mathematical technique. It is intended for students who plan to pursue studies in business, economics, social sciences, and other disciplines that may, directly or indirectly, bring them into contact with calculus. My other concern is with the student who, apart from future course work and professional needs, wants to be a well-educated person. Calculus is one of the great triumphs of the human intellect and, like fine art, fine music, and fine literature, brings us together as human beings interested in partaking of the best of the human spirit.

Calculus books for this audience should, in my judgment, give high priority to the following five commandments, as I term them:

- The focus of the exposition should be on the development of basic concepts.
- While background from algebra is necessary for discussion of calculus, it is counterproductive to allow algebra technique to be given excessive attention.
- The development of calculus technique should not be allowed to take on a life of its own out of proportion to the focus stated in the first commandment.
- Ideas from other disciplines that have the structure of concepts developed in calculus should be given a prominent role in the exposition.

- A fundamental objective of the book should be development of perspective on what calculus, as a mathematical discipline, can do for us, and its limitations, in the study of real world problems.

I do not believe that these commandments are adequately addressed in the calculus literature I am aware of, and this is the rationale that underlies still another book on calculus.

Self-tests have been included to help students get blocks of material under better control.

I am greatly indebted to my daughter Ramunė for preparing the illustrations.

Fundamentals of Calculus with Applications and its accompanying *Companion to Calculus* were originally published by the Kendall/Hunt Publishing Company. In this revised edition by Xlibris they are combined to make the *Companion* easily accessible to the reader. This revised edition contains revisions in the exposition which I thought to be appropriate, refinements to make the joining of these books as smooth as possible, and corrections of errata that have come to my attention.

To make this book accessible to as wide an audience as possible I have made it available on the web at webpage.pace.edu/wadams.

W.J.A.

PERSPECTIVE ON THE CHAPTERS AND EXERCISES

As noted in the preface, the point of view underlying *Fundamentals of Calculus with Applications and Companion to Calculus* is stated in the afore five commandments. The commentary that follows is intended to provide an overview of the chapters. Exercises that I find of particular interest are noted.

Chapter 1: Miscellaneous Preliminaries

This chapter is intended as a review of topics and techniques needed for undertaking the study of calculus presented in *Fundamentals of Calculus with Applications*. It may be assigned for independent review, gone over quickly, or turned to as needed when situations requiring background review arise.

Chapter 2: Functions

The pivotal concept of function is introduced and its dimensions are indicated through discussion of a number of examples. Much attention is paid to the domain of definition of functions arising in applied settings.

Chapter 3: Limits of Functions

A fundamental chapter in the book which takes up the pivotal concept of limit of a function and its extension to continuity.

Chapter 4: An Application of Limits to Finance

Shows off an important application of limits to defining continuous compounding of interest.

Chapter 5: Derivatives of Functions

The second fundamental chapter of the book. It takes up the pivotal concept of derivative of a function and discusses its importance in terms of providing a mathematical vehicle for defining fundamental concepts in applied settings.

Chapter 6: Tools for Computing Derivatives

With the appropriate foundation having been established in Chapter 5, the natural follow up is to develop tools for computing derivatives as quickly and painlessly as possible, and then apply them to the applied settings discussed earlier and others. This is the intent of Chapter 6.

Chapter 7: Optimization Problems

Shows off a second dimension of the importance of derivatives in terms of finding maximum and minimum values of functions that arise in applied settings.

Chapter 8: Curve Sketching

Shows off a third dimension of the importance of derivatives in terms of getting a handle on graphs of functions.

Chapter 9: Integral Calculus: Indefinite Integrals

Provides an introduction to tools for finding indefinite integrals of functions, the significance of which emerges in Chapters 11-13.

Chapter 10: Introduction to Mathematical Modeling

This chapter provides an introductory discussion of the mathematical modeling perspective. It discusses what mathematical methods can do for us and their limitations by employing as a vehicle the problem of planning a vacation trip. The perspective developed here is applied in Chapter 11.

Chapter 11: Differential Equation Models

Shows off a fourth dimension of the importance of derivatives and one dimension of the importance of indefinite integrals in the construction of mathematical models for real-world phenomena and the derivation of valid conclusions.

Chapter 12: Integral Calculus: Definite Integrals

The third fundamental chapter of the book. The geometric problem of defining the area of a region is used as a vehicle for developing the concept of definite integral. This is done in two ways. A third approach to defining the definite integral is also discussed. The fundamental theorem of integral calculus for evaluating definite integrals shows another dimension to the importance of indefinite integrals.

Chapter 13: Concepts with a Definite Integral Structure

This follow up to Chapter 12 shows off one dimension of the importance of definite integrals by considering a variety of applied situations which exhibit the definite integral structure.

Chapter 14: Topics in Multivariable Calculus

Calculus developed in the preceding chapters is one dimensional in the sense that it is based on functions of one variable. This chapter introduces refinements needed for development of multivariable calculus.

Exercises

In the spirit of the show 20-questions, the following twenty questions are among my favorites in being thought provoking and insightful.

1. Ch. 3, Sec. 3.2, 101/13
2. Self-Test 1, 131/2, 3, 4
3. Self-Test 3, 135/4, 5, 6
4. Ch. 5, Sec. 5.1, 160/9
5. Ch. 5, Sec. 5.2, 163/2
6. Ch. 6, Sec. 6.8, 211/15
7. Ch. 7, Sec. 7.1, 232/23
8. Ch. 7, Sec. 7.2, 243/1
9. Ch. 7, Sec. 7.3, 249/1
10. Self-Test 3, 273/14
11. Self-Test 3, 274/17
12. Ch. 10, Sec. 10.2, 303/2
13. Ch. 11, Sec. 11.4, 324/1
14. Ch. 13, Sec. 13.4, 377/4
15. Ch. 13, Sec. 13.4, 377/5
16. Self-Test 3, 429/3
17. Self-Test 3, 429/4
18. Self-Test 3, 429/11
19. Self-Test 3, 429/13
20. Self-Test 3, 429/16

CHAPTER 1

Miscellaneous Preliminaries

1.1. OVERVIEW

Calculus rests on three fundamental building blocks: the real-number system, the concept of function, and the concept of limit of a function. The real-number system and concept of function are met in the study of algebra as well as calculus, but it is the limit concept and developments stemming from it which give calculus its own special tone, substance and spectrum of applications.

The real-number system, accompanying algebra, functions and a host of miscellaneous mathematical topics make up what has come to be called precalculus. The real-number system and needed support members of the cast are taken up in this chapter. Functions are discussed in the following chapter. The pivotal actors in the calculus cast are introduced in Chapters 3, 5, and 12.

1.2. THE REAL-NUMBER SYSTEM

Since algebraic operations and developments are dependent on the number system adopted as a foundation, let us begin our study of algebraic principles and applications by reviewing and summarizing basic data about numbers and their mathematical life. The real-number system, as it is called, underlies a wide range of developments and applications and will serve as the foundation for the topics considered in this book.

The real-number system is a collection of mathematical objects, called real-numbers, which acquire mathematical life by virtue of certain fundamental principles, or rules, that we adopt. The situation is somewhat similar to a game, like chess, for example. The chess system, or game, is a collection of objects, called chess pieces, which acquire life by virtue of the rules of the game, that is, the principles that are adopted to define allowable moves for the pieces and the way in which they may interact. A word about representation by symbols is perhaps in order at this point. The number two, for example, is represented by such symbols as 2, $1+1$, $3-1$, $\frac{4}{2}$, and so on. We should be careful not to equate the symbol used to express a number and the number itself. They are not the same, just as the signature of a person is not the same as the person, although a person's signature can be used to identify or represent him under certain circumstances. When it is necessary to be ultracareful about this distinction, the symbol used to represent a number is called a **numeral**.

Our working experience with numbers has provided us all with some familiarity with the principles that govern the real-number system. However, to establish a common ground of understanding and avoid certain errors that have become very common, we shall explicitly state and illustrate many of these principles. The real-number system includes such numbers as

$$-27,\ -\pi,\ -\sqrt{2},\ -1,\ -\frac{2}{3},\ 0,\ \frac{1}{2},\ \frac{1}{\sqrt{3}},\ \sqrt{3},\ \pi,\ 87.4$$

It is worthy of note that positive numbers, $\frac{1}{2}$, 1, $\sqrt{3}$, and 4, for example, are sometimes expressed as $+\frac{1}{2}$, $+1$, $+\sqrt{3}$, $+4$. The plus sign, $+$, used here does not express the operation of addition, but is rather part of the symbolism for the numbers themselves. Similarly, the minus sign, $-$, used in expressing such numbers as $-\frac{1}{2}$, -1, and -2, is part of the symbolism used for these numbers.

Within the real-number system, numbers of various kinds are identified and named. The numbers

$$1, 2, 3, 4, 5, \ldots$$

which are used in the counting process, are called **natural numbers**. The natural numbers, together with

$$-1, -2, -3, -4, -5, \ldots$$

and zero, are called **integers**. Since 1, 2, 3, 4, 5, . . . are greater than 0, they are also called **positive integers**; $-1, -2, -3, -4, -5, \ldots$ are less than 0, and for this reason are called **negative integers**. A real number is said to be a **rational number** if it can be expressed as the ratio of two integers, where the denominator is not zero. (For reasons we shall say more about later, division by zero is not defined, and thus denominators with zero must be excluded.) For example,

$$-10, \frac{1}{2}, -\frac{2}{3}, 0, 1, \frac{25}{7}, 5$$

are **rational numbers**. The integers are included among the rational numbers since any integer can be expressed as the ratio of the integer itself and one. For example,

$$-10 = \frac{-10}{1},\ 1 = \frac{1}{1},\ 0 = \frac{0}{1},\ 5 = \frac{5}{1}$$

A real number that cannot be expressed as the ratio of two integers is said to be an **irrational number**. $\sqrt{2}$, π, and $-\sqrt{3}$ illustrate irrational real-numbers. This classification for real-numbers is shown in Figure 1.1.

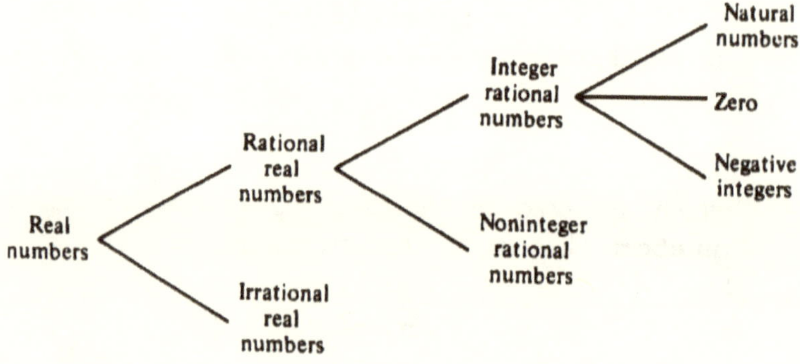

Figure 1.1

One of the basic properties of the real-number system is that any two real-numbers can be compared for size. If a and b are real numbers, we write

$$a < b$$

to signify that a is less than b. Another way of saying the same thing is to write

$$b > a$$

which is read "b is greater than a." Thus $0 < 1$, $1 < 3$, $-\frac{1}{2} < 0$, $-2 < -1$. These relations are also expressed by writing $1 > 0$, $3 > 1$, $0 > -\frac{1}{2}$, $-1 > -2$. The concepts positive number and negative number are defined by comparing the number in question with 0. If $a > 0$, then a is said to be **positive**; if $a < 0$, a is said to be **negative**. Thus $-\frac{1}{2}$ is negative since $-\frac{1}{2} < 0$; 1 is positive since $1 > 0$; 0 itself is neither positive nor negative.

We write

$$a \leq b$$

and, equivalently,

$$b \geq a$$

to mean that a is less than or equal to b, and, equivalently, b is greater than or equal to a. Thus $1 \leq 2$ is correct since $1 < 2$; $1 \leq 1$ is correct since $1 = 1$. Relations of the $\leq$ and $\geq$ type, which contain an equality component, are also referred to as inequalities.

Geometrically, real numbers are identified with points on a straight line. We choose a straight line, a unit of distance, and an initial point of reference called the **origin**. To the origin we assign the number zero. By marking off the unit of length in both directions from the origin, we assign positive integers to marked-off points in one direction (by convention, to the right of the origin) and negative integers to marked-off points in the other direction. By following through in terms of the chosen unit of length, a real number is attached to each point on the number line, and each point on the number line has attached to it one number (see Figure 1.2).

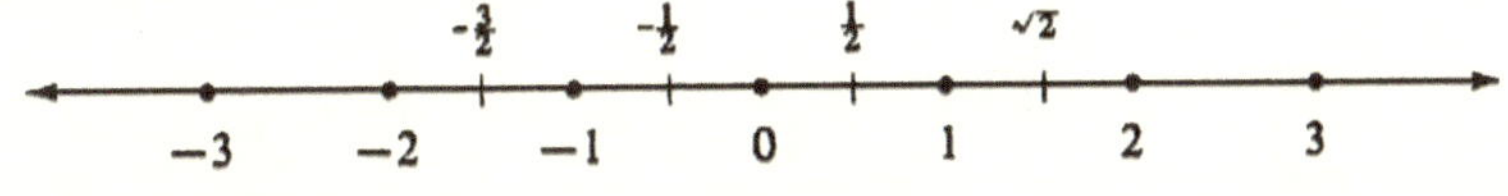

Figure 1.2

Geometrically, in terms of our number line, to say that $a < b$ is to say that a is to the left of b; $b > a$ means that b is to the right of a. Thus $-4 < -3$ since -4 is to the left of -3, and $-\frac{1}{2} > -1$ since $-\frac{1}{2}$ is to the right of -1.

Properties of Addition and Multiplication

Addition and multiplication are primary operations on real numbers. Most, if not all, of the following basic properties of these operations are familiar to us from experience.

Closure property of addition and multiplication. Whenever two real-numbers are added or multiplied, we obtain a real-number as the result. That is, performing the operations of addition and multiplication leaves us within the real-number system. We are not taken outside this system. Formally stated, we have the following:

A1. If a and b are any numbers in the real-number system R, then the sum of a and b, denoted by $a + b$, is in R.

M1. If a and b are any numbers in the real-number system R, then the product of a and b, denoted by $a \cdot b$ or ab, is in R.

Commutative property of addition and multiplication. The order in which two real-numbers are added or multiplied does not affect the result obtained. Such a property is called a commutative property. Addition and multiplication of real numbers are commutative operations. Formally stated, we have the following:

A2. If a and b are any numbers in the real-number system R, then

$$a + b = b + a.$$

M2. If a and b are any numbers in the real-number system R, then $ab = ba$.

Associative property of addition and multiplication. Parentheses, brackets, and the like, we recall, are used in algebra to group together whatever terms are within them. Thus $2 + (3 + 4)$ means that 2 is to be added to the sum of 3 and 4, yielding $2 + 7$ or 9, whereas $(2 + 3) + 4$ means that the sum of 2 and 3, or 5, is to be added to 4, yielding 9. Similarly, $2 \cdot (3 \cdot 4)$ yields $2 \cdot (12) = 24$, whereas $(2 \cdot 3) \cdot 4$ yields the same end result via the route $(6) \cdot 4 = 24$. That such is the case in general is the content of the associative property of addition and multiplication of real numbers.

A3. If a, b, and c are any numbers in the real-number system R, then

$$a + (b + c) = (a + b) + c.$$

M3. If a, b, and c are any numbers in the real-number system R, then

$$a(bc) = (ab)c.$$

The properties we have surveyed apply no matter how many terms are involved in a sum or product. In total they say that the sum, or product, of any number of real-numbers is a real-number, and the order in which they are added, or multiplied, is immaterial.

The numbers zero and one. The following are the basic properties of the numbers zero and one:

A4. There is an unique real-number, called **zero** and denoted by 0, with the property that $a + 0 = 0 + a$, where a is any real number.

M4. There is an unique real-number, different from zero, called **one** and denoted by 1, with the property that $a \cdot 1 = 1 \cdot a = a$, where a is any real-number.

Additive and multiplicative inverses of a number. Corresponding to the numbers 2 and $-\frac{1}{3}$, for example, are the numbers -2 and $\frac{1}{3}$ with the property that

$$2 + (-2) = (-2) + 2 = 0$$

$$(-\frac{1}{3}) + \frac{1}{3} = \frac{1}{3} + (-\frac{1}{3}) = 0$$

-2 is called the **additive inverse of**, or **negative of**, 2, and vice versa. $\frac{1}{3}$ is called the additive inverse of, or negative of, $\frac{1}{3}$, and vice versa.

Also, corresponding to the number 2 is the number $\frac{1}{2}$ with the property that

$$2(\frac{1}{2}) = \frac{1}{2}(2) = 1$$

Corresponding to $-\frac{1}{3}$ is -3 with the property that

$$(-\frac{1}{3})(-3) = (-3)(-\frac{1}{3}) = 1$$

$\frac{1}{2}$ is called the **multiplicative inverse**, or **reciprocal**, of 2, and vice versa; -3 is called the **multiplicative inverse, or reciprocal, of** $-\frac{1}{3}$, and vice versa. More generally, every real-number has a unique additive inverse, and every nonzero real-number has a unique multiplicative inverse. Formally stated, we have the following:

> **A5.** If a is any real-number, then there is an unique real-number x, called the additive inverse of a, or negative of a, with the property that $a + x = x + a = 0$.
>
> **M5.** If a is any **nonzero** real-number, then there is an unique real-number y, called the multiplicative inverse of a, or reciprocal of a, with the property that $ay = ya = 1$.

The negative of a (denoted by x in A5) is usually symbolized by prefixing the minus sign, $-$, before a, yielding $-a$. Thus, in terms of this notation, the negative of -3, which is 3 [since $(-3)+3=3+(-3)=0$], is denoted by $-(-3)$. We have, then, that $-(-3)=3$. More generally, if a is any real-number, then

$$-(-a) = a$$

The concept of the negative of a number should not be confused with the concept of a negative number; they are not the same. **"Negative of" means "additive inverse of." A "negative number," on the other hand, is a number that is less than zero.** The negative of -2 is 2, which is positive; the negative of 0 is 0, which is neither positive nor negative.

The multiplicative inverse of a (denoted by y in M5) is often represented by the symbol $1/a$ or a^{-1}. Note that since the product of any number y and 0 is 0, that is,

$$y \cdot 0 = 0 \cdot y = 0$$

0 cannot have a multiplicative inverse. Thus 1/0 does not exist.

Distributive property. We know that $2(3 + 4) = 2(7) = 14$, and that $2 \cdot 3 + 2 \cdot 4 = 6 + 8 = 14$. Thus $2(3 + 4) = 2 \cdot 3 + 2 \cdot 4$. That such is the case in general for all real-numbers is the content of the distributive property of multiplication over addition, more simply called the distributive property.

D. If a, b, and c are any real-numbers, then $a(b + c) = ab + ac$.

That is, the multiplier a is distributed between the terms b and c. In reading from left to right, we speak of multiplying out or removing the parentheses. In reading from right to left, we speak of taking out the common factor a from $ab + ac$, thus leaving the product of a and the other factor $b + c$.

More generally, the distributive property holds for any number of terms.

$$a(b + c + d + \ldots + z) = ab + ac + ad + \ldots + az$$

To illustrate, consider the sum

$$3x + 2x + 4x$$

By taking out the common factor x, the distributive property yields

$$\begin{aligned} 3x + 2x + 4x &= x(3 + 2 + 4) \\ &= x(9) \\ &= 9x \end{aligned}$$

From this example we see how the distributive property allows us to combine a sum of like terms into one term.

As another example consider

$$3xy + 3xz + 3x$$

By taking out the common factor $3x$ and observing what is left over in each term in our sum, we obtain

$$3xy + 3xz + 3x = 3x(y + z + 1)$$

In general, using the distributive property to take out a factor common to all members of a sum of terms and expressing the sum as a product of the term taken out and the remaining sum is called **factoring.** The role of factoring in simplifying algebraic fractions is considered in Section 1.4.

To illustrate the role of the distributive property in removing parentheses, consider $-1(x+3)$:

$$\begin{aligned} -1(x+3) &= (-1)x + (-1)3 \\ &= -x + (-3) \end{aligned}$$

As another example, consider $2a(x + 2y + 3z)$. Multiplying each term in the sum $x + 2y + 3z$ by $2a$ yields

$$\begin{aligned} 2a(x + 2y + 3z) &= 2ax + 2a(2y) + 2a(3z) \\ &= 2ax + 4ay + 6az \end{aligned}$$

From our experience with multiplication we recognize the correctness of such statements as

$$(2)(-4) = -8$$
$$(-2)(4) = -8$$
$$(-2)(-4) = 8$$

These examples illustrate a general result called the **rule of signs** for multiplication. If a and b are positive real-numbers, then

$$a(-b) = -(ab) \tag{1.1}$$
$$(-a)b = -(ab) \tag{1.2}$$
$$(-a)(-b) = ab \tag{1.3}$$

In colloquial language, the product of two numbers with unlike signs is negative; the product of two numbers with like signs is positive.

If $a = 1$, then from (1.2) we obtain

$$(-1)b = -b$$

Thus, for example,

$$\begin{aligned} -(x+3) &= (-1)(x+3) \\ &= (-1)x + (-1)3 \\ &= -x + (-3) \end{aligned}$$

We also have

$$\begin{aligned} -(a+2b+4c+d) &= (-1)(a+2b+4c+d) \\ &= (-1)a + (-1)2b + (-1)4c + (-1)d \\ &= -a + (-2b) + (-4c) + (-d) \end{aligned}$$

In colloquial language, a minus sign in front of terms grouped by parentheses changes the sign of each term in the group.

The following are two other important properties of real-number multiplication.

M6. The product of two nonzero real-numbers is nonzero. That is, if $a \neq 0$ and $b \neq 0$, then $ab \neq 0$.

This implies that if the product of a and b is zero, then a, b, or both a and b must be zero. The only way zero can be obtained from a product is for at least one of the members of the product to be zero.

M7. Cancellation Property. If $ax = ay$, and $a \neq 0$, then $x = y$.

For example, if

$$3xy = 3xz$$

and $x \neq 0$, then the common factor $3x$ can be canceled, yielding

$$y = z$$

EXERCISES

Find the following sums.

1. $-2+(-4)$
2. $3+(-2)$
3. $4+-(-2)$
4. $2+(-2)4$
5. $-3+(-1)(-2)$
6. $3(-1)+(-2)$
7. $(-2)(-3)+4$
8. $-(-3)+(-5)$
9. $-3+-(-4)$
10. $-(-1)+(-2)(-1)+(-1)3$
11. $(-1)(-4)+-(-3)+3(-2)$
12. $(-1)2+(-2)(-3)+-(-2)$

Simplify by removing parentheses or combining terms.

13. $3a+5a+7a$
14. $2xy+3xy+4xy$
15. $4wz+(-2)wz+3wz$
16. $3x^2+4x^2+2x^2$
17. $6ab+(-3)ab+4ab+(-5)ab$
18. $2ab+4ab+ab$
19. $5xyz+(-2)xyz+7xyz$
20. $3(2y+3a+4z)$
21. $-2(2x+3y+4z)$
22. $-1(4w+2v+3z)$

23. $-3(3x+3)+4x$
24. $-(2a+4b+6c)$
25. $-(2n+5m)+4n+(-3)m$
26. $-(3x+2y+3)+2x+4$
27. $-(xy+2a)+4xy+3a$
28. $-(3x^2+4y)+4x^2+3y$

Simplify by factoring.

29. $2x + 4y$
30. $9n + 9m$
31. $xy + xz$
32. $3ab + 6ac$
33. $4xyz + 8xyw$
34. $3x^2y + 6x^2z$
35. $2x + 4x^2$
36. $2ab + 4ac + 8ad$
37. $a^2y + a^2z + a^2w$
38. $2mnp + 4mnv + 10mnw$
39. $4wk^2 + (-3)k^2 + ak^2$
40. $3xmy + 2my + 5xm$
41. $3abc + 6ab + 3ac$
42. $4vw^2 + 2xw^2 + 8w^2$
43. $2xy + 3xz + x$
44. $3rst + 6rs + 12rsk$
45. Is $-x + 4$ the negative of $x-4$? Explain.
46. Is $xy + 3$ the negative of $-xy-2$? Explain.
47. Is -2 the multiplicative inverse of 2? Explain.
48. Does 0 have an additive inverse? Explain.
49. Does 1 have a multiplicative inverse? Explain.
50. What are the additive inverses of 6, -1, 1/4, $\sqrt{2}$, and π?
51. What are the multiplicative inverses of 6, -1, 1/4, $\sqrt{2}$, and π?

Subtraction and Division

Subtraction is defined in terms of addition in the following way.

If a and b are any two real-numbers, then the difference $a-b$ is defined by

$$\boldsymbol{a - b = c}$$

where c is such that $b + c = a$. Let us note that

$$c = a + (-b)$$

since $$b+c=b+a+(-b)=a$$

Thus the difference $a-b$ is defined by

$$\boldsymbol{a-b=a+(-b)}$$

That is, to *subtract* b from a means to add the negative of b (additive inverse of b) to a. For example,

$$\begin{aligned} 3-2&=3+(-2)=1 \\ -4-3&=-4+(-3)=-7 \\ -5-(-4)&=-5+-(-4) \\ &=-5+4 \\ &=-1 \end{aligned}$$

In connection with this last example, recall that $-(-4)$ expresses the additive inverse of -4, that is, the number that if added to -4 yields 0, which, of course, is 4 itself. In colloquial language such a result is often described by saying that "two minus signs make a plus sign."

The distributive property can be extended to include subtraction. Thus we have

$$\boldsymbol{a(b-c+d-e+\cdots+z)=ab-ac+ad-ae+\cdots+az}$$

For example,

$$\begin{aligned} -3(x-2y+1)&=(-3)x-(-3)2y+(-3)1 \\ &=-3x+6y-3 \end{aligned}$$

Division is defined in terms of multiplication in the following way.

If a and b are any real-numbers, where $b \neq 0$, then the quotient $a \div b$ is defined by

$$a \div b = a \times \left(\frac{1}{b}\right) = a \times b^{-1}$$

That is, to *divide a* by *b* means to multiply *a* by the multiplicative inverse (reciprocal) of *b*. The quotient $a \div b$ is also expressed by the fraction symbol $\frac{a}{b}$ or *a*/*b*. For example,

$$3 \div 2 = 3 \cdot (\frac{1}{2}) = \frac{3}{2}$$

$$5 \div (-3) = 5 \cdot (-\frac{1}{3}) = -\frac{5}{3}$$

$$6 \div \frac{1}{3} = 6 \cdot (\frac{1}{3})^{-1} = 6 \cdot 3 = 18$$

$$0 \div 4 = 0 \cdot (4^{-1}) = 0$$

Note that the last example illustrates that **division into zero is permitted**. However, **division by zero is not permitted**. For any real-number *a*, $a \div 0$ is not defined because the multiplicative inverse of 0 does not exist.

Although this analysis shows that division by zero cannot be defined in terms of the multiplicative inverse of zero, might it not be possible to define division by zero in some other way? The answer is no; **no definition of division by zero that is compatible with the structure of the real-number system as a whole is possible.** Any proposed definition for division by zero must lead to a contradiction. As an indication of this mathematical fact, let us suppose that division by zero has been defined in some way and consider the following argument (which requires some recall of elementary algebra).

Let *x* and *y* denote nonzero numbers such that

$$x = y$$

Multiplying both sides of this equation by y yields

$$xy = y^2$$

Subtracting x^2 from both sides yields

$$xy - x^2 = y^2 - x^2$$

By factoring we obtain

$$xy - x^2 = x(y - x), \qquad y^2 - x^2 = (y - x)(y + x)$$

Thus we have

$$x(y - x) = (y - x)(y + x)$$

Dividing both sides by $y - x$ (which is 0 since $x = y$) yields

$$x = y + x$$

Since $x = y$, by substituting x for y in the above we obtain

$$x = x + x$$

or

$$x = 2x$$

Dividing both sides by x yields

$$1 = 2$$

But then

$$3 = 2 + 1 = 1 + 1 = 2$$
$$4 = 3 + 1 = 2 + 1 = 3$$

and so on. That is, $1 = 2 = 3 = 4$, and so on, which is in contradiction to the structure of the real-number system. The source of this inconsistency is the step in which we divided by $y-x=0$ (under the assumption that division by zero could be defined in some way).

If 1 = 2, then we may conclude that \$1 = \$2. If a person can be found who accepts this, then he, logically speaking, should be prepared to give you two dollars for every one dollar that you give him. Whether he would be prepared to do so emotionally speaking, is another matter.

Division by zero does not usually announce itself with trumpets blaring, and when division by an algebraic expression arises it must be ascertained whether division by zero could arise under certain circumstances. A spectacular case in point arose in connection with Albert Einstein's theory of relativity. At one point in his work Einstein found that certain of his equations yielded peculiar results; as a consequence he turned in another direction, a direction that subsequently had to be abandoned. These peculiarities were later explained in 1922 when the Russian physicist Alexandre Friedman found that in his derivation Einstein had divided an equation by a quantity that, Friedman showed, could be zero under certain circumstances. This discovery and the refinement in Einstein's analysis that it brought about laid the foundation for the mathematical theory of the expanding universe.

While noting that division by zero is not defined, we should again observe that there is nothing wrong with dividing *into* zero as long as we are dividing by a nonzero number. For example,

$$0 \div 3 = 0 \cdot (\frac{1}{3}) = 0$$

$$0 \div c = 0 \cdot \left(\frac{1}{c}\right) = 0$$

where $c \neq 0$.

EXERCISES

Find the following values.

52. $2-4$
53. $4-6$
54. $-3-4$
55. $-2-5$
56. $-1-6$
57. $3-(-2)$
58. $5-(-3)$
59. $-3-(-4)$
60. $-(-2)-(-1)$
61. $-(-3)-(-7)$
62. $-3-(-5)+2$
63. $-3+2-4$
64. $2-(-3)-3$
65. $-2(-3+2-4)$
66. $-3(4-2-(-1))$
67. $-1(-3+4-(-2))$
68. $-2 \div 4$
69. $-3 \div \frac{1}{3}$
70. $4 \div -\frac{1}{2}$
71. $\frac{1}{3} \div 2$
72. $\frac{1}{4} \div -2$
73. $\frac{1}{2} \div -4$
74. $\frac{1}{5} \div 0$
75. $0 \div \frac{1}{2}$
76. $\frac{1}{8} \div -2$

Simplify by removing parentheses or combining terms.

77. $-2(3a-4b+3c)$
78. $3a-3(4a-2b-3c)$
79. $-(3a+4b-2c)$
80. $2a-4b-(a-2b+4c)$
81. $-2(x-2y+3z)-3(2x+3y-2z)$
82. $-(-3x+2y-z)-(4x-4y-2z)$

83. $-3(-2x-4y-z)-(-x+2y-4z)$
84. $3ab+2ab-4ab$
85. $-6xy+2xy-3xy$
86. $-3x^2+4x^2-x^2$
87. $-3ab+4a-2ab-2a$
88. $-3x+2xy+4y+4x-3xy+2y-x$

Simplify by factoring.

89. $3xy-4xz-5x$
90. $4a^2b-3bx-b$
91. $4mn-5mn^2+3mnp$
92. $2xy^2-4y^2+6zy^2$
93. $\frac{1}{2}a-2ab$
94. $gh-3gh^2$
95. $\frac{1}{3}xy+6xyz$
96. $3x^2y^3-2xy^2+xy$

1.3. FRACTIONS AND CANCELLATION

$$-\frac{103}{4},\ -\frac{1}{2},\ \frac{2}{3},\ \frac{\sqrt{3}}{2},\ \frac{\pi}{5}$$

The above are examples of numerical fractions;

$$\frac{x}{2},\ \frac{x-1}{y+1},\ \frac{x^2+3x+4}{3x-1}$$

illustrate algebraic fractions, which become numerical fractions when the variables involved are given specific numerical values. Of course it is understood that the variables in such expressions cannot take on values that make the denominator zero.

To begin our survey of basic properties of fractions, let us first recall that

$$\frac{a}{b}=\frac{ax}{bx}$$

That is, a common factor may be introduced into, or removed from, the numerator and denominator of a fraction without changing its value. When the need is to simplify a fraction by removing, or canceling, common factors, the process is often called cancellation. For example,

$$\frac{2}{4}=\frac{1\cdot\overset{1}{\cancel{2}}}{2\cdot\underset{1}{\cancel{2}}}=\frac{1}{2}$$

Cancellation of the common factor 2 yields $\frac{1}{2}$.

$$\frac{2\overset{1}{\cancel{(x-1)}}}{3\underset{1}{\cancel{(x-1)}}}=\frac{2}{3}$$

Cancellation of the common factor $(x-1)$ yields $\frac{2}{3}$, assuming that $x-1 \neq 0$.

> **Danger, be on the alert!**
>
> $$\frac{2\cancel{x}^{1}}{3+\cancel{x}_{1}} = \frac{2}{4}$$
>
> **is an incorrect cancellation; x is not a common factor of both numerator and denominator.**

To be sure, x is a factor of the numerator since we have the product $2x$. But x is not a factor of the denominator since we do not have the product $3x$, but rather the sum $3 + x$. Thus the cancellation principle is not applicable in this situation. Incorrect simplifications of the kind just illustrated have gained enormous popularity over the years and one must be on continuous alert against them.

To simplify algebraic fractions by cancellation, the numerator and denominator components must be expressed as products of factors. To express these components as products of factors, we often employ the distributive property. The following examples illustrate.

EXAMPLE 1

$$\frac{5x-10}{3x-6}$$

By factoring the numerator and denominator, we obtain

$$\frac{5x-10}{3x-6} = \frac{5(x-2)}{3(x-2)}$$

Canceling the common factor $(x-2)$, where $x-2 \neq 0$, yields

$$\frac{5x-10}{3x-6}=\frac{5\overset{1}{\cancel{(x-2)}}}{3\underset{1}{\cancel{(x-2)}}}=\frac{5}{3}$$

EXAMPLE 2

$$\frac{2xy+3x}{6y+9}$$

By factoring numerator and denominator, we obtain

$$\frac{2xy+3x}{6y+9}=\frac{x(2y+3)}{3(2y+3)}$$

Canceling the common factor $(2y + 3)$, where $2y + 3 \neq 0$, yields

$$\frac{2xy+3x}{6y+9}=\frac{x\overset{1}{\cancel{(2y+3)}}}{3\underset{1}{\cancel{(2y+3)}}}=\frac{x}{3}$$

EXAMPLE 3

$$\frac{x-2}{2-x}$$

At first sight it seems that no simplification is possible. But let us observe that $2-x$ is the negative of $x-2$ (since the sum of $2-x$ and $x-2$ is zero). The quotient of two quantities that are negatives of each other is -1. This we establish by factoring -1 from the numerator and following through as shown.

$$\frac{x-2}{2-x}=\frac{-1(-x+2)}{2-x}=\frac{-1(2-x)}{2-x}=\frac{-1\overset{1}{\cancel{(2-x)}}}{\underset{1}{\cancel{2-x}}}=-1$$

It is assumed that $2-x\neq 0$; that is, $x\neq 2$.

EXERCISES

Specify whether the cancellation is correct. Explain the basis for your conclusion.

1. $\dfrac{x+\overset{1}{\cancel{2}}}{3y+\underset{1}{\cancel{2}}}=\dfrac{x}{3y}$

2. $\dfrac{\overset{1}{\cancel{a}}-3}{\underset{1}{\cancel{a}}+4}=-\dfrac{3}{4}$

3. $\dfrac{\overset{1}{\cancel{2}}}{x+\underset{1}{\cancel{2}}}=\dfrac{1}{x+1}$

4. $\dfrac{2\overset{1}{\cancel{x}}(x+y)}{3\underset{1}{\cancel{x}}(x-y)}=\dfrac{2(x+y)}{3(x-y)}$

5. $\dfrac{4\overset{1}{\cancel{(x+y)}}}{\underset{1}{\cancel{(x+y)}}}=4$

6. $\dfrac{\cancel{2}^{1}+(x+y)}{\cancel{2}_{1}}=1+x+y$

7. $\dfrac{\overset{1}{\cancel{a}}-\overset{1}{\cancel{b}}}{\underset{1}{\cancel{a}}+\underset{1}{\cancel{b}}}=\dfrac{1-1}{1+1}=0$

8. $\dfrac{2\overset{1}{\cancel{a+3}}}{\underset{1}{\cancel{a+3}}}=\dfrac{5}{4}$

Simplify, if possible, by canceling common factors.

9. $\dfrac{9}{27}$

10. $\dfrac{8}{24}$

11. $\dfrac{-9}{-33}$

12. $\dfrac{2ax}{3ay}$

13. $\dfrac{3abc}{4b}$

14. $\dfrac{3x+1}{x+1}$

15. $\dfrac{4(a+b)}{a+b}$

16. $\dfrac{3}{x+3}$

17. $\dfrac{2+(x+y)}{x+y}$

18. $\dfrac{3a(x+y)}{4x(x+y)}$

19. $\dfrac{3(x-y)(x+y)}{2(x-y)}$

20. $\dfrac{-4(a+2b)}{-2(a+2b)}$

21. $\dfrac{2a+b}{a+b}$

22. $\dfrac{x+y+1}{y+1}$

23. $\dfrac{3-x}{x-3}$

24. $\frac{x-5}{x-5}$ 25. $\frac{2x+1}{2x}$

Simplify, if possible, by factoring and canceling common factors.

26. $\frac{2x+10}{x+5}$ 27. $\frac{2x-4}{x-2}$ 28. $\frac{2a-10}{5-a}$

29. $\frac{xy+y}{y(x+1)}$ 30. $\frac{2a+3ab}{2a(2+3ab)}$ 31. $\frac{4x-16y}{x-4y}$

32. $\frac{3xy+4xyz}{xy}$ 33. $\frac{3x-15}{10-2x}$ 34. $\frac{3xy+x}{-3y-1}$

35. $\frac{4m+3n}{4mx^2+3nx^2}$ 36. $\frac{yz-3ys}{3yz-9ys}$ 37. $\frac{4cd-c}{4de-e}$

38. $\frac{5x-5y}{x^2-xy}$ 39. $\frac{3ab-abc}{9-3c}$

40. For what values of x are the fractions stated in 23-27 undefined? Explain.

Rules of Sign

In Section 1.2 we noted that $(-1)b=-b$. For fractions this implies that

$$\frac{-a}{-b}=\frac{\cancel{(-1)}a}{\cancel{(-1)}b}=\frac{a}{b}$$

By multiplying numerator and denominator of $-a/b$ by -1, we obtain

$$\frac{-a}{b}=\frac{(-1)(-a)}{(-1)b}=\frac{a}{-b}$$

Thus we have the following rules of sign:

$$\frac{-a}{-b}=\frac{a}{b}$$

$$\frac{-a}{b}=\frac{a}{-b}$$

$-a/b$ and $a/-b$ are equal to the negative (additive inverse) of a/b, as we now show. The negative of a/b, denoted by $-(a/b)$ (with the minus sign in front of the fraction), is that number which yields zero when added to a/b. We must show that

$$-\frac{a}{b}=\frac{-a}{b}$$

(In colloquial language, the minus sign in front of a fraction can be moved to the numerator of the fraction.) To establish this result we must show that the sum of a/b and $-a/b$ is zero. No request could be simpler.

$$\frac{a}{b}+\frac{-a}{b}=\frac{1}{b}\cdot a+\frac{1}{b}\cdot(-a)=\frac{1}{b}(a+(-a))=\frac{1}{b}\cdot 0=0$$

Since $$-\frac{a}{b}=\frac{-a}{b} \text{ and } \frac{-a}{b}=\frac{a}{-b}$$

we have the following rule of signs:

$$\frac{-a}{b}=\frac{-a}{b}=\frac{a}{-b}$$

That is, the minus sign in front of a fraction can be shifted from the front of the fraction to the numerator of the fraction or to the denominator of the fraction, and vice versa.

Test for Equality

$a/b = c/d$ if and only if $ad = bc$. Passing from $a/b = c/d$ to $ad = bc$ is sometimes called cross multiplication. Thus, for example, $\frac{7}{9} \neq \frac{13}{15}$ since $15(7) = 105$, whereas $13(9) = 117$; $\frac{6}{12} = \frac{14}{28}$ since $28(6) = 168$ and $14(12) = 168$.

Multiplication of Fractions

As we recall, to multiply fractions, multiply numerators and denominators. That is,

$$\frac{a}{c} \cdot \frac{c}{d} = \frac{ac}{bd}$$

EXAMPLE 4

$$\frac{1}{2} \cdot \frac{3}{4} = \frac{3}{8}$$

EXAMPLE 5

$$\frac{2x}{3y} \cdot \frac{y}{z} = \frac{2x\cancel{y}^{1}}{3\cancel{y}_{1}z} = \frac{2x}{3z}$$

Alternatively,

$$\frac{2x}{3\cancel{y}_{1}} \cdot \frac{\cancel{y}^{1}}{z} = \frac{2x}{3z}, \text{ where } y \neq 0$$

In multiplying fractions, common factors can be canceled after numerators and denominators are multiplied, as was first done, or before numerators and denominators are multiplied, as is illustrated by the second method.

EXAMPLE 6

$$\frac{a-1}{b}\cdot\frac{2}{b(a-1)}=\frac{\overset{1}{\cancel{(a-1)}}2}{b^2\underset{1}{\cancel{(a-1)}}}=\frac{2}{b^2}$$

where $a-1\neq 0$.

Division of Fractions

$$\frac{a}{b}\div\frac{c}{d}=\frac{a}{b}\cdot\frac{d}{c}=\frac{ad}{bc}$$

That is, to divide fractions we invert the divisor and multiply. For example,

$$\frac{1}{2}\div\frac{2}{3}=\frac{1}{2}\cdot\frac{3}{2}=\frac{3}{4}$$

This procedure for finding the quotient of two fractions follows from the definition of division of real numbers. By definition,

$$\frac{a}{b}\div\frac{c}{d}=\frac{a}{b}\cdot\left(\text{reciprocal of }\frac{c}{d}\right)$$

The reciprocal of $\frac{c}{d}$ is $\frac{d}{c}$ $\left(\text{since }\frac{c}{d}\cdot\frac{d}{c}=1\right)$. Thus we have

$$\frac{a}{b}\div\frac{c}{d}=\frac{a}{b}\cdot\frac{d}{c}=\frac{ad}{bc}$$

We should also note that sometimes the quotient $\frac{a}{b} \div \frac{c}{d}$ is written

$$\frac{\frac{a}{b}}{\frac{c}{d}} \text{ or } \frac{a}{b} \Big/ \frac{c}{d}$$

To avoid ambiguity, we make the division line longer or heavier than the fraction lines of the component fractions. To illustrate the ambiguity that arises when we are not careful, let us note that $\frac{\frac{1}{3}}{2}$ could mean $\frac{1}{3}$ divided by 2, which equals $\frac{1}{6}$, or 1 divided by $\frac{3}{2}$, which equals $\frac{2}{3}$.

EXAMPLE 7

Find $\frac{1}{6} \div \frac{2}{3}$ and simplify.

$$\frac{1}{6} \div \frac{2}{3} = \frac{1}{6} \cdot \frac{3}{2} = \frac{1}{4}$$

EXAMPLE 8

Find $\frac{3a}{b-1} \div \frac{2a}{b}$ and simplify.

$$\frac{3a}{b-1} \div \frac{2a}{b} = \frac{3\cancel{a}^{1}}{b-1} \cdot \frac{b}{2\cancel{a}_{1}} = \frac{3b}{2(b-1)},$$

where $a \neq 0$, $b \neq 0$, $b \neq 1$

EXAMPLE 9

Find $\dfrac{4a}{a-1} \div 3a$ and simplify.

$$\frac{4a}{a-1} \div 3a = \frac{4\cancel{a}^{1}}{a-1} \cdot \frac{1}{3\cancel{a}_{1}} = \frac{4}{3(a-1)},$$

where $a \neq 0, 1$

EXERCISES

Determine if the given fractions are equal.

41. $\dfrac{6}{21}, \dfrac{8}{24}$
42. $\dfrac{7}{12}, \dfrac{9}{15}$
43. $\dfrac{14}{23}, \dfrac{17}{25}$
44. $\dfrac{3}{18}, \dfrac{5}{30}$
45. $\dfrac{6}{21}, \dfrac{14}{49}$
46. $\dfrac{6}{16}, \dfrac{8}{25}$

Express the product in simplest terms.

47. $\dfrac{2}{3} \cdot \dfrac{6}{7}$
48. $\dfrac{1}{4} \cdot \dfrac{8}{7}$
49. $\dfrac{2}{7} \cdot \dfrac{14}{5}$
50. $\dfrac{2a}{3b} \cdot \dfrac{5b}{3c}$
51. $\dfrac{3x}{4y} \cdot \dfrac{8y-1}{x}$
52. $\dfrac{xy}{z} \cdot \dfrac{x}{y}$
53. $\dfrac{a}{4} \cdot \dfrac{16}{ab}$
54. $\dfrac{x-1}{3} \cdot \dfrac{x}{x-1}$
55. $\dfrac{3y}{x+1} \cdot \dfrac{x}{6y-1}$
56. $\dfrac{x-1}{x+1} \cdot \dfrac{x}{x-1}$
57. $\dfrac{4x}{y} \cdot \dfrac{3y}{x^2}$
58. $\dfrac{1}{x-5} \cdot \dfrac{5-x}{3}$
59. $\dfrac{3x-2}{x} \cdot \dfrac{1}{6x-4}$
60. $\dfrac{5a}{a-1} \cdot \dfrac{2a-2}{5}$
61. $\dfrac{3}{2-x} \cdot \dfrac{x-2}{4}$

62. $\frac{4a+1}{b}\cdot\frac{2a}{4ab+b}$ 63. $\frac{4}{3-n}\cdot\frac{2n-6}{n}$ 64. $\frac{3}{v-3}\cdot\frac{vx-3x}{x}$

Divide and express the quotient in simplest terms.

65. $\frac{1}{2}\div\frac{3}{4}$ 66. $-\frac{1}{6}\div\frac{2}{3}$ 67. $\frac{3}{5}\div-\frac{1}{4}$

68. $3\div\frac{1}{2}$ 69. $\frac{1}{4}\div-2$ 70. $\frac{2}{3}\div-6$

71. $\frac{2a}{b}\div\frac{3}{ab}$ 72. $\frac{3x}{y}\div\frac{x}{y-1}$ 73. $\frac{4}{a}\div\frac{2b}{a}$

74. $\frac{x-1}{2y}\div\frac{3}{x-1}$ 75. $\frac{4}{x-2}\div\frac{2x}{x-2}$ 76. $\frac{a}{a-b}\div\frac{a+b}{a-b}$

77. $\frac{1}{x-2}\div\frac{2}{2-x}$ 78. $\frac{2}{a+4}\div\frac{1}{2a+8}$ 79. $\frac{2}{3-a}\div\frac{2a}{a-3}$

1.4. POLYNOMIALS AND EQUATIONS

Algebraic expressions such as

$$2x+1, \qquad 2x^2+\frac{1}{2}x-1$$

$$x^3-x^2+4, \qquad 2x^8-\frac{1}{3}x^6-4$$

illustrate the nature of polynomials in one variable. The **general polynomial in *x* of degree *n***, where *n* is a positive integer, is an expression that is, or can, be put into the form

$$\boldsymbol{ax^n+bx^{n-1}+cx^{n-2}+\cdots+hx+k}$$

where a, b, c, . . . , h and k are constants, and $a \neq 0$. Polynomials of degree 1, illustrated by

$$2x+1, \qquad \frac{1}{4}x-2, \qquad 3x$$

are called linear, and polynomials of degree 2, illustrated by

$$2x^2+\frac{1}{2}x-1, \qquad 3x^2-x, \qquad x^2-9$$

are called **quadratic**. Our main concern is with linear and quadratic polynomials.

Addition and Subtraction of Polynomials

Addition and subtraction of polynomials is no different from addition and subtraction of algebraic expressions in general; but since we shall have occasion to employ such operations, we pause to refresh our memories at this point. To add two polynomials, we write like terms under each other and then add.

EXAMPLE 1

Add $2x^3+4x^2+2$ and $-2x^2+x-6$.

$$\begin{array}{r} 2x^3+4x^2+2 \\ -2x^2+x-6 \\ \hline 2x^3+2x^2+x-4 \end{array}$$

Thus the sum is $2x^3+2x^2+x-4$.

To subtract a first polynomial from a second, write the first under the second, placing like terms under each other, change the signs of the terms in the lower polynomial, and add.

EXAMPLE 2

Subtract $4x^3 - 2x + 1$ from $3x^2 - 3x + 4$.

We obtain

$$\begin{array}{lll} & 3x^2 - 3x + 4 & \\ 4x^3 & \quad - 2x + 1 \xrightarrow[\text{signs}]{\text{change}} & \end{array} \qquad \begin{array}{r} 3x^2 - 3x + 4 \\ -4x^3 \qquad\quad + 2x - 1 \\ \hline -4x^3 + 3x^2 - x + 3 \end{array}$$

Thus the difference is $-4x^3 + 3x^2 - x + 3$.

The basis for this procedure is the use of the distributive property for removing parentheses. The problem is to find $(3x^2 - 3x + 4) - (4x^3 - 2x + 1)$. Removing parentheses yields $3x^2 - 3x + 4 - 4x^3 + 2x - 1$. The follow-through involves combining like terms.

EXERCISES

1. Add $3x^2 + 4x - 3$ and $x^3 + 5x^2 - 1$.
2. Add $2x^4 - 4x^3 + 6$ and $5x^3 - x^2 + 3$.
3. Add $-4x^2h + 2xh - 1$ and $-4x^2h + 3xh + h^2$.
4. Add $-x^3h - 3x^2h + 4h$ and $2x^3 - 4x^2h - xh^2$.
5. Add $3x^2y - 3y + 2$ and $x^2y - 3xy + 5$.
6. Subtract $2x^2 + x - 1$ from $4x^2 - 3x + 5$.
7. Subtract $-x^2 + 4x - 3$ from $x^2 + 3x - 8$.
8. Subtract $x^2h - 3xh + h$ from $3x^2h + 4xh + 7$.
9. Subtract $4x^3h + 2x^2h + 3xh$ from $4x^3h - 3x^2h + 2x + 1$.
10. Subtract $3x^2h^2 + 2xh - 4$ from $4x^2h^2 + 3x^2h + 7xh$.

Multiplication of Polynomials

The distributive property for the real-number system states

$$a(b - c + d - e + \cdots + z) = ab - ac + ad - ae + \cdots + az$$

For example,

$$\begin{aligned} -5(3x^2 + x - 1) &= -15x^2 - 5x + 5 \\ 3x^2(4x^3 - 3x^2 - 2) &= 3x^2(4x^3) + 3x^2(-3x^2) + 3x^2(-2) \\ &= 12x^5 - 9x^4 - 6x^2 \end{aligned}$$

More generally, to multiply a first polynomial by a second polynomial, multiply each term of the first polynomial by each term of the second, and take the algebraic sum of these products.

EXAMPLE 3

Multiply $(x-5)(2x^2+x-1)$.

By the distributive property, we have

$$\begin{aligned} (x-5)(2x^2+x-1) &= (x-5)2x^2 + (x-5)x + (x-5)(-1) \\ &= (2x^3 - 10x^2) + (x^2 - 5x) + (-x + 5) \\ &= 2x^3 - 10x^2 + x^2 - 5x - x + 5 \\ &= 2x^3 - 9x^2 - 6x + 5 \end{aligned}$$

This work may be more conveniently carried out by means of the following vertical arrangement.

$$\begin{array}{rrr} 2x^2 + & x - 1 & \\ x \; - & 5 & \\ \hline 2x^3 + & x^2 - x & \rightarrow \text{ product } x(2x^2+x-1) \end{array}$$

$$-10x^2 - 5x + 5 \rightarrow \text{ product } -5(2x^2 + x - 1)$$

$$2x^3 - 9x^2 - 6x + 5 \rightarrow \text{ sum of like terms}$$

EXAMPLE 4

Find $(x+h)^3$.

Multiplying $(x + h)$ by itself yields

$$\begin{array}{r} x + h \\ x + h \\ \hline x^2 + hx \qquad \\ + hx + h^2 \\ \hline x^2 + 2hx + h^2 \end{array}$$

Multiplying $(x+h)^2 = x^2 + 2hx + h^2$ by $x + h$ yields

$$\begin{array}{r} x^2 + 2hx + h^2 \qquad \\ x + h \qquad \\ \hline x^3 + 2hx^2 + h^2x \qquad \\ + hx^2 + 2h^2x + h^3 \\ \hline x^3 + 3hx^2 + 3h^2x + h^3 \end{array}$$

Thus $(x+h)^3 = x^3 + 3hx^2 + 3h^2x + h^3$.

EXERCISES

Find the following products.

11. $3x(x+4)$
12. $2x(2x^2+1)$
13. $-x(x^2-x+1)$
14. $2x(3x^2-3x+4)$
15. $-3x(x^3-4x^2+7)$
16. $-2x^2(3x^2+4x-9)$

17. $2xh(3x^2h+4h-1)$
18. $3xh(5x^2-3xh+4)$
19. $(x-1)(x+2)$
20. $(x-3(x+3)$
21. $(x-h)(x+h)$
22. $(x-h)^2$
23. $(2x-3)(x+1)$
24. $(5x+1)(x-3)$
25. $(2x-1)(3x+2)$
26. $(5x-3)(2x-4)$
27. $(x-h)^3$
28. $(x-1)(x^2+3x+4)$
29. $(x-2)(x^3+3x^2+4)$
30. $(x+h)(2hx^2+4hx-3)$
31. $(x+2)(x^4-3x^2+2x+1)$
32. $2(x-3)(4x+5)$
33. $(3x-2)(3x^3+4x^2+7)$
34. $(2x^2+h)(x^2h+3hx-4)$
35. $(4x+3)(4x^3+3x+7)$
36. $4x(x-2)(x+5)$
37. $5x^2(x-3)(2x^2+1)$
38. $(x-h)^4$
39. $(x-3)(x+4)(x-6)$
40. $(3x+1)(x+3(2x+6)$
41. $(x+h)^5$
42. $(x-h)^5$

Factoring

The distributive property lies at the center of basic factoring, since it justifies taking out a factor common to all members of a sum of terms and expressing the sum as a product of the term taken out and the remaining sum. Thus, for example, from

$$2x^3-2x^2-12x$$

we may factor out $2x$ and write

$$2x^3-2x^2-12x=2x(x^2-x-6)$$

Can we do better? That is, can x^2-x-6 be expressed as a product? To explore this question, let us consider how the factors, if there are such, would multiply back to yield x^2-x-6. Consider the factors

$$x + a \text{ and } x + b$$

Multiplying these factors yields

$$\begin{array}{r} x+a \\ x+b \\ \hline x^2+ax \qquad \\ +bx+ab \\ \hline x^2+ax+bx+ab \\ =x^2+(a+b)x+ab \end{array}$$

For $x^2-x-6, a+b=-1,$ and $ab=-6$. Thus to yield a negative product, -6, a and b must be of opposite sign and add up to -1. This suggests $a=-3$ and $b=2$. To verify that $x^2-x-6=(x-3)(x+2)$, we check by multiplying $(x-3)$ and $(x+2)$.

$$\begin{array}{r} x-3 \\ x+2 \\ \hline x^2-3x \qquad \\ +2x-6 \\ \hline x^2 \;-\; x-6 \end{array}$$

Thus we have

$$2x^3-2x^2-12x=2x(x^2-x-6)=2x(x-3)(x+2)$$

Factoring is useful because it often permits simplifications through application of the cancellation principle and is useful in solving certain kinds of equations. Much can be said about factoring a variety of special forms, but for our purposes it suffices to focus on factoring quadratic polynomials of the form

$$ax^2+bx+c$$

where $a \neq 0$.

EXAMPLE 5

Factor, if possible, $2x^2 - 4x - 30$.

The first order of business is to see if any terms that are common to all members of the sum can be factored out. From $2x^2 - 4x - 30$ we can factor out 2.

$$2x^2 - 4x - 30 = 2(x^2 - 2x - 15)$$

To factor $x^2 - 2x - 15$, we need factors that must be of the form

$$(x + a)(x + b)$$

to obtain the term x^2. The numbers a and b must yield a product of -15 (and thus be of opposite sign) and a sum of -2. If we try $a = 5$ and $b = -3$, we obtain a product of -15, but a sum of 2, so that this possibility washes out. On the other hand, $a = -5$ and $b = 3$ seem to work. Verification via multiplication confirms this.

$$\begin{array}{r} x - 5 \\ x + 3 \\ \hline x^2 - 5x \qquad \\ +3x - 15 \\ \hline x^2 - 2x - 15 \end{array}$$

Thus we have

$$2x^2 - 4x - 30 = 2(x^2 - 2x - 15) = 2(x - 5)(x + 3)$$

EXAMPLE 6

Factor, if possible, $x^2 + 4$.

To factor $x^2 + 4$, we consider factors of the form

$$(x+a)(x+b)$$

which yield the term x^2. The numbers a and b must yield a product of 4 and a sum of 0. To yield a product of 4, a and b must be of the same sign (both positive or both negative), and therefore cannot yield a sum of 0. Thus it is not possible to factor $x^2 + 4$.

EXAMPLE 7

Factor, if possible, $6x^2 + x - 15$.

This problem is a bit more troublesome because the number of possibilities to be considered is larger; but our basic approach still pertains. We need linear factors that yield a product of $6x^2$ rather than x^2. To begin, consider factors of the form

$$(6x+a)(x+b)$$

The product ab must be -15, and thus a and b must be of opposite sign. The following possibilities are suggested, and we only stop to determine the middle term since the other requirements are met.

$$\begin{array}{r} 6x+15 \\ x-\ 1 \\ \hline 15x \\ -6x \\ \hline 9x \end{array} \qquad \begin{array}{r} 6x+5 \\ x-3 \\ \hline 5x \\ -18x \\ \hline -13x \end{array} \qquad \begin{array}{r} 6x-5 \\ x+3 \\ \hline -5x \\ 18x \\ \hline 13x \end{array}$$

None of these products work, so let's consider factors of the form

$$(2x+a)(3x+b)$$

Since we must have $ab = -15$, the following possibilities are suggested.

$$\begin{array}{r} 2x+5 \\ \underline{3x-3} \\ 15x \\ \underline{-6x} \\ 9x \end{array} \qquad \begin{array}{r} 2x-3 \\ \underline{3x+5} \\ -9x \\ \underline{10x} \\ x \end{array}$$

Thus $(2x-3)(3x+5)=6x^2+x-15$.

There is often a fair amount of trial and error in this kind of analysis, and in the end the factors we are searching for may not exist. One may well ask, what kind of factors are we searching for? As the examples considered suggest, we shall restrict our attention to factors with integer values. Thus, for example, x^2-5 cannot be factored further from this point of view. If irrational numbers are permitted in the factors, then x^2-5 can be expressed as the product $(x-\sqrt{5})(x+\sqrt{5})$.

EXAMPLE 8

Simplify, if possible, $\dfrac{2x^2-4x-30}{x-5}$.

One possibility that should be explored is that of factoring the numerator. From Example 5 we see that this works out nicely.

$$\frac{2x^2-4x-30}{x-5}=\frac{2\overset{1}{\cancel{(x-5)}}(x+3)}{\underset{1}{\cancel{(x-5)}}}=2(x+3)$$

where $x \neq 5$.

EXERCISES

Factor each of the following as completely as possible and check by multiplication.

43. $5x+3xy+xz$
44. $3a^2+2ab+a$
45. $6ax-4a^2y$
46. $3b+6b^2+12b^3$
47. $2xy+x^2y-xy^2$
48. $2ab-3a^2b^2+a^3b^2$
49. x^2+5x-6
50. x^2+4x+3
51. $x^2+12x+11$
52. x^2-9
53. t^2-6t+5
54. y^2+16
55. $x^2-10x+21$
56. $2x^2-18x+16$
57. x^2+4x+5
58. $2y^3-12y^2+16y$
59. $3x^3-27x^2+42x$
60. $x^2-18x+72$
61. t^3-16t^2+60t
62. $x^2-16x+64$
63. $4x^3-8x^2-32x$

Simplify to the extent possible.

64. $\dfrac{x^2-9}{x-3}$
65. $\dfrac{x^2+9}{x+3}$
66. $\dfrac{t^2-6t+5}{t-1}$
67. $\dfrac{2x^2-18x+16}{x-1}$
68. $\dfrac{x^2-4}{x+2}$
69. $\dfrac{3x^3-27x^2+42x}{x-7}$
70. $\dfrac{x^2-16x+64}{x-8}$
71. $\dfrac{2x^2-2x-12}{x-3}$

Equations

Equations which can be written in the form of a polynomial set equal to zero arise in many situations of interest. The quadratic equation with the form

$$\boldsymbol{ax^2+bx+c=0},$$

where a, b, and c are constants and $a \neq 0$ is perhaps the most generally known of these.

In undertaking to solve such equations factoring can be an invaluable ally due to the real-number property that the product of two numbers, and more generally two algebraic expressions, cannot be zero unless at least one of them is zero.

EXAMPLE 9

Solve $2x^2 - 8 = 0$.

By factoring the left side of this equation we obtain:

$$2(x-2)(x+2) = 0$$

This reduces the problem to consideration of two cases.

Case 1: $x - 2 = 0$	Case 2: $x + 2 = 0$
$x = 2$	$x = -2$

By substituting 2 and -2 for x in $2x^2 - 8$ we obtain 0, thus verifying that 2 and -2 are the solutions of the given equation.

EXAMPLE 10

Solve $2x^3 - 2x^2 - 12x = 0$.

We begin by factoring out $2x$. This yields

$$2x(x^2 - x - 6) = 0$$

Since $x^2 - x - 6$ factors into $(x-3)(x+2)$, we obtain

$$2x(x-3)(x+2) = 0$$

Case 1: $2x = 0$	Case 2: $x - 3 = 0$	Case 3: $x + 2 = 0$
$x = 0$	$x = 3$	$x = -2$

By substitution, it is easily verified that 0, 3, and -2 are solutions of $2x^3 - 2x^2 - 12x = 0$.

Since many quadratic equations cannot be easily factored, and, indeed, most cannot be factored at all, a more general approach to solving quadratic equations is needed.

The Quadratic Formula

> The solutions of the quadratic equation $ax^2+bx+c=0$ are given by
>
> $$x=\frac{-b+\sqrt{b^2-4ac}}{2a} \text{ and } x=\frac{-b-\sqrt{b^2-4ac}}{2a}.$$

These solutions require some knowledge of square roots which, along with radicals in general, are discussed in the next section.

EXAMPLE 11

Solve $6x^2+x-15=0$.

This equation can be solved by factoring the left side (the factors are $2x-3$ and $3x+5$), but it may happen that they are not readily seen. Rather than spending what may turn out to be an inordinate amount of time trying to track them down, we can appeal to the quadratic formula.

In this situation $a = 6$, $b = 1$, and $c=-15$. We have:

$$x=\frac{-1\pm\sqrt{(1)^2-4(6)(-15)}}{2(6)}=\frac{-1\pm\sqrt{1+360}}{12}$$
$$=\frac{-1\pm\sqrt{361}}{12}=\frac{-1\pm 19}{12}=\frac{3}{2}\text{ and }-\frac{5}{3}$$

EXERCISES

Solve the following equations by factoring, if possible, or by use of the quadratic formula.

72. $x^2+5x-6=0$	73. $x^2-25=0$	74. $x^2+4x+3=0$
75. $x^2+4x=0$	76. $x^2+12x+11=0$	77. $2x^2-18x+16=0$
78. $x^2-18x+72=0$	79. $3x^3-27x^2+42x=0$	80. $4x^2-12x+5=0$
81. $6x^2+5x-6=0$	82. $x^2+7x+2=0$	83. $(x-3)^2=64$
84. $2x^2-8x+1=0$	85. $x^2-x-5=0$	86. $x^2-2x-7=0$

1.5. EXPONENTS, ROOTS, AND RADICALS

If n is a positive integer, then x^n stands for the product $x \cdot x \cdot \cdots \cdot x$ with n factors. n, the number of factors in $x \cdot x \cdot \cdots \cdot x = x^n$, is called the **exponent** of x^n and x^n itself is called the ***n*th power of x**. Thus

$$x^1 = x \qquad \text{the exponent of } x^1 \text{ is } 1$$
$$x^2 = x \cdot x, \qquad \text{the exponent of } x^2 \text{ is } 2$$
$$x^3 = x \cdot x \cdot x, \qquad \text{the exponent of } x^3 \text{ is } 3$$
$$(-2)^4 = (-2)(-2)(-2)(-2) = 16$$

In particular, x^2 is called the square of x, and x^3 is called the cube of x. Exponents are extended to include zero and negative integers as follows.

> If $x \neq 0$, then x^0 is **defined by**
>
> $$x^0 = 1$$
>
> If $x \neq 0$ and n is a positive integer, then x^{-n} is defined by
>
> $$x^{-n} = \frac{1}{x^n}$$

For example,

$$8^0 = 1, \qquad x^{-2} = \frac{1}{x^2}$$

$$x^{-3}=\frac{1}{x^3}, \qquad 3^{-2}=\frac{1}{3^2}=\frac{1}{9}$$

$3^{-2}=\frac{1}{9}$ serves to show that a number raised to a negative exponent need not necessarily be negative.

Let us also note that

$$\frac{1}{x^{-n}}=\frac{1}{\left(1/x^n\right)}=1\cdot\frac{x^n}{1}=x^n$$

Thus a negative exponent of a term in the numerator throws the term with corresponding positive exponent into the denominator, whereas a negative exponent of a term in the denominator throws the term with corresponding positive exponent into the numerator. As further illustrations, we have

$$\frac{1}{x^{-3}}=x^3, \quad \frac{y^2}{x^{-4}}=y^2x^4, \quad \frac{2^{x-3}}{3^{y-2}}=\frac{2y^2}{3x^3}$$

The definitions of zero and negative exponents, which perhaps seem strange at first glance, are natural in terms of the properties that positive, negative, and zero exponents have. The following examples illustrate one of these basic properties.

$$x^2\cdot x^3=x^{2+3}=x^5, \qquad x^2\cdot x^{-4}=x^{2-4}=x^{-2} \qquad x^3\cdot x^0=x^{3+0}=x^3$$

We verify these relations as follows:

$$x^2\cdot x^3=\underbrace{(x\cdot x)(x\cdot x\cdot x)}_{\text{5 factors}}=x^5$$

$$x^2\cdot x^{-4}=\frac{x^2}{x^4}=\frac{x\cdot x}{x\cdot x\cdot x\cdot x}=\frac{1}{x^2}=x^{-2}$$

$$x^3 \cdot x^0 = x^3 \cdot 1 = x^3$$

More generally,

$$x^n \cdot x^m = x^{n+m}$$

where n and m are positive, negative, or zero. Thus we have

$$x^4 \cdot x = x^4 \cdot x^1 = x^5 \qquad (1+v)(1+v) = (1+v)^2$$

$$x^{-3} \cdot x^{-2} = x^{-5} \qquad (1+v)^2(1+v) = (1+v)^3$$

$$x^6 \cdot x^{-2} = x^4 \qquad (1+v)^3(1+v)^{-1} = (1+v)^2$$

The division counterpart of the above result is the following. If x is not zero, then

$$\boxed{\frac{x^n}{x^m} = x^{n-m}}$$

where n and m are positive, negative, or zero. For example,

$$\frac{x^4}{x^2} = x^{4-2} = x^2 \qquad \frac{x^3}{x^{-2}} = x^{3-(-2)} = x^{3+2} = x^5$$

$$\frac{x^{-2}}{x^3} = x^{-2-3} = x^{-5} \qquad \frac{x^{-3}}{x^{-4}} = x^{-3-(-4)} = x^1 = x$$

The following are three other basic properties of exponents:

$$(x^n)^m = x^{nm}$$

$$(xy)^n = x^n y^n$$

$$\left(\frac{x}{y}\right)^n = \frac{x^n}{y^n}$$

For example,

$$(x^2)^3 = x^{2(3)} = x^6 \qquad (x^{-3})^2 = x^{(-3)2} = x^{-6}$$

$$(xy)^5 = x^5y^5 \qquad (xy)^{-3} = x^{-3}y^{-3}$$

$$\left(\frac{x}{y}\right)^2 = \frac{x^2}{y^2} \qquad \left(\frac{x}{y}\right)^{-3} = \frac{x^{-3}}{y^{-3}}$$

In summary, then, we have the following basic definitions and properties of exponents.

Definition. If n is a positive integer, then

$$x^n = \underbrace{x \cdot x \cdot x \cdot \cdots \cdot x)}_{n \text{ factors}}$$

$$x^{-n} = \frac{1}{x^n}$$

$$x^0 = 1, \text{ where } x \neq 0$$

Properties.

$$\frac{1}{x^{-n}} = x^n \qquad x^n x^m = x^{n+m}$$

$$\frac{x^n}{x^m} = x^{n-m} \qquad (x^n)^m = x^{nm}$$

$$(xy)^n = x^n y^n \qquad \left(\frac{x}{y}\right)^n = \frac{x^n}{y^n}$$

The following examples further illustrate how these properties are used to simplify expressions involving exponents.

$$(x^2)^0 = 1$$

$$(2x^{-3})^2 = 2^2(x^{-3})^2 = 4x^{-6} = \frac{4}{x^6}$$

$$(x^2y^{-3})^3 = (x^2)^3(y^{-3})^3 = x^6y^{-9} = \frac{x^6}{y^9}$$

$$\frac{3x^2y^{-3}}{2x^{-3}y^2} = \frac{3}{2}x^{2-(-3)}y^{-3-2} = \frac{3}{2}x^5y^{-5} = \frac{3x^5}{2y^5}$$

$$\left(\frac{x}{y}\right)^{-3} = \frac{x^{-3}}{y^{-3}} = \frac{y^3}{x^3}$$

$$\left(\frac{x^{-1}}{y^2}\right)^3 = \frac{(x^{-1})^3}{(y^2)^3} = \frac{x^{-3}}{y^6} = \frac{1}{x^3y^6}$$

EXERCISES

Evaluate each of the following.

1. $(-2)^5$
2. $(-3)^{-2}$
3. $(-2)^{-3}$
4. $(\frac{1}{2})^0$
5. $3(4^{-2})$
6. $(4^0)(3^{-2})$
7. $\frac{(-2)^{-3}}{3^{-2}}$
8. $\frac{(3^2)^{-2}}{2^{-3}}$
9. $(5^0)(10^3)$
10. $\frac{(-2)^{-4}}{5^{-2}}$
11. $(-6)^{-2}(3^{-2})$
12. $[(-2)^{-3}]^{-2}$
13. $[(-4)^0]^{-3}$
14. $\frac{4^{-2}}{[(-3)^{-1}]^{-3}}$
15. $\frac{(-4)^{-3}}{(-5)^{-2}}$

Simplify each of the following by using properties of exponents. Remove zero and negative exponents.

16. $3x^{-2}y$
17. $4x^{-1}y^{-2}$
18. $\frac{3a^{-1}}{b^{-2}}$
19. $(x^{-2})^{-3}$
20. $\frac{x^{-2}}{(y^{-2})^3}$
21. $(4m^{-3})^0$
22. $(1+a)^3(1+a)$
23. $(a+b)^2(a+b)^{-2}$
24. $(m+2n)^3(m+2n)^{-1}$

25. $(2x^3y^4)(3x^2y^{-2})$ 26. $\dfrac{(2x^2)x^{-3}y}{4x^2y^2}$ 27. $(5x^{-2})^2(3x^4)^{-1}$

28. $2x^2(3x^4+4x^{-2})$ 29. $x^{-3}(2x^3+x^2+1)$ 30. $\dfrac{2y^4z^3}{3y^{-2}z^4}$

31. $(2ab^2)^{-3}$ 32. $(4a^2b^3)^{-2}$ 33. $(2a^{-1}b^2)^{-3}$

34. $(2xy)^3(3x^2y^{-1})^2$ 35. $\dfrac{4x^2y^{-3}}{3x^{-2}y^2}$ 36. $\left(\dfrac{x^2}{y^3}\right)^3\cdot\dfrac{x^{-2}}{y^{-1}}$

Roots and Radicals

Let us recall that if $x^2 = c$, then x is called a **square root of c.** Thus -3 and 3 are square roots of 9, since $(-3)^2 = 9$ and $3^2 = 9$. More generally, if n is a positive integer greater than 1 and $x^n = c$, then x is called an **nth root of c.** In particular, if $x^3 = c$, then x is called a **cube root of c.** Thus 4 is a cube root of 64, since $4^3 = 64$; -2 and 2 are fourth roots of 16, since $(-2)^4 = 16$ and $2^4 = 16$.

From these examples it is clear that a number c may have more than one nth root. One of these, the principal nth root, is singled out in the following way.

Definition. If c is positive, the **principal nth root of c** is the positive nth root of c; if c is negative and n is odd (that is, $n = 3$, 5, 7, 9, etc.), the principal n*th* root of c is the negative nth root of c.

Thus, for example, the principal square root of 25 is 5, since 5 is positive and $5^2 = 25$; the principal cube root of -8 is -2 since -2 is negative, $n = 3$ is odd, and $(-2)^3 = -8$.

The principal *n*th root of *c* is denoted by the radical sign $\sqrt[n]{c}$ The number c under the radical sign is called the radicand, and the number n, which indicates the root to be taken and is written above the radical sign, is called the index of the radical.

> **By definition**, then, $\sqrt[n]{c}$ designates the principal *n*th root of *c* and has the property that
>
> $$(\sqrt[n]{c^n}) = c$$

When the principal square root of *c*, denoted by $\sqrt[2]{c}$, is involved, it is customary to omit the index 2 on the radical sign and write $\sqrt{c}$. In terms of radical notation, we have

$$\sqrt{25} = 5, \quad \sqrt[3]{-8} = -2, \sqrt[4]{16} = 2$$

Note $\sqrt{25} \neq -5$ since $\sqrt{25}$ designates the principal square root of 25, which is positive 5. The negative square root of 25, -5, is designated by $-\sqrt{25}$.

EXERCISES

Determine the value of each of the following.

37. $\sqrt{64}$
38. $\sqrt[3]{-27}$
39. $\sqrt[3]{-64}$
40. $\sqrt{49}$
41. $\sqrt[5]{-32}$
42. $\sqrt[6]{64}$
43. $\sqrt[4]{81}$
44. $\sqrt[3]{-125}$
45. $(\sqrt{5})^2$
46. $(\sqrt[3]{-9})^3$
47. $(\sqrt[5]{-32})^2$
48. $\dfrac{1}{(\sqrt{25})^{-3}}$
49. $(\sqrt[4]{6})^4$
50. $\dfrac{1}{\sqrt[4]{81}}$
51. $\dfrac{1}{\sqrt[4]{16}} \cdot \dfrac{1}{(\sqrt[3]{125})^{-2}}$
52. Does $\sqrt{(-2)^2} = -2$? Explain.

Computation with Radicals

Only a comparatively small number of radicals can be determined directly from the definition of radical itself. Thus we have such results as $\sqrt{4} = 2, \sqrt[3]{-8} = -2, \sqrt[4]{16} = 2$. But how are such radicals as $\sqrt{2}, \sqrt{3}$ and $\sqrt[3]{10}$ to be determined? How does one work with numbers expressed in terms of radicals? "Simple, use a calculator," you might say. Indeed the calculator is a most valuable computational ally, but we should be careful not to allow ourselves to become its slave so that without its use the simplest of calculations becomes an impossible task. Therefore, basic principles that aid calculation never become obsolete.

The calculation of radicals (with or without a calculator) and the simplification of expressions involving radicals are facilitated by certain basic properties of radicals, two of which we now state and illustrate. In many situations it suffices to express a numerical result in radical form and leave the explicit calculation of the radical to be undertaken when necessary.

Product and Quotient Theorems for Radicals. If a and b are positive real numbers, then

$$\sqrt[n]{ab} = \sqrt[n]{a} \cdot \sqrt[n]{b}, \qquad \sqrt[n]{\frac{a}{b}} = \frac{\sqrt[n]{a}}{\sqrt[n]{b}}$$

That is, the nth root of a product (or quotient) of positive numbers is equal to the product (or quotient) of the nth roots of the component parts. For example,

$$\sqrt{2} \cdot \sqrt{8} = \sqrt{2(8)} = \sqrt{16} = 4$$
$$\sqrt{200} = \sqrt{100(2)} = \sqrt{100}\sqrt{2} = 10\sqrt{2}$$
$$\sqrt[3]{\frac{1}{8}} = \frac{\sqrt[3]{1}}{\sqrt[3]{8}} = \frac{1}{2}$$
$$3\sqrt{50} = 3[\sqrt{25(2)}] = 3(\sqrt{25}\sqrt{2}) = 3(5\sqrt{2}) = 15\sqrt{2}$$
$$4\sqrt[3]{16} = 4[\sqrt[3]{8(2)}] = 4(\sqrt[3]{8}\sqrt[3]{2}) = 4(2\sqrt[3]{2}) = 8\sqrt[3]{2}$$
$$3\sqrt{4x^2y^5} = 3\sqrt{4}\sqrt{x^2}\sqrt{y^4y} = 3(2)x\sqrt{y^4}\sqrt{y} = 6xy^2\sqrt{y}$$

Danger, be on the alert! The product and quotient theorems for radicals cannot be extended to sums and differences. In general,

$$\sqrt[n]{a+b} \neq \sqrt[n]{a} + \sqrt[n]{b}, \quad \sqrt[n]{a-b} \neq \sqrt[n]{a} - \sqrt[n]{b}$$

To illustrate, compare $\sqrt{2} = \sqrt{1+1}$ with $\sqrt{1} + \sqrt{1}$.

$$\sqrt{1+1} = \sqrt{2} \simeq 1.414$$
$$\sqrt{1} + \sqrt{1} = 1 + 1 = 2$$

EXAMPLE 1. TEMPTATION

Simplify, to the extent possible, $\sqrt{9x^2 + 81y^2}$.

For some reason this sort of situation brings out the mathematical Mr. Hyde in many of us. Suggested simplifications include $3x + 9y$, which may be obtained by overlooking the aforenoted danger sign and arguing thusly:

$$\begin{aligned}\sqrt{9x^2+81y^2} &= \sqrt{9x^2}+\sqrt{81y^2}\\ &= 3x+9y\end{aligned}$$

If this were correct, then squaring $3x+9y$ should give us $9x^2+81y^2$. It does not; we obtain $9x^2+54xy+81y^2$.

The only possibility for simplification rests on expressing $9x^2+81y^2$ as a product and using the product theorem for radicals. The pickings are thin.

$$\begin{aligned}\sqrt{9x^2+81y^2} &= \sqrt{9(x^2+9y^2)}\\ &= \sqrt{9}\sqrt{x^2+9y^2}\\ &= 3\sqrt{x^2+9y^2}\end{aligned}$$

EXERCISES

Simplify the following to the extent possible.

53. $\sqrt{48}$
54. $\sqrt{72}$
55. $\sqrt{54}$
56. $\sqrt{300}$
57. $\sqrt{126}$
58. $4\sqrt{50}$
59. $3\sqrt{175}$
60. $\frac{1}{2}\sqrt{180}$
61. $-\frac{1}{2}\sqrt{28}$
62. $\sqrt[3]{250}$
63. $\sqrt[4]{48}$
64. $\sqrt[3]{108}$
65. $\sqrt{4a^2b}$
66. $\sqrt{9a^4c}$
67. $\frac{1}{2}\sqrt{16x^3y}$
68. $\sqrt{x^2-4}$
69. $\sqrt{4x^2+16y^2}$
70. $\sqrt{x^2+2x+1}$

Radicals and Fractional Exponents

Earlier in this section we defined a concept of exponent for integer values and observed that the following properties hold:

1. $\dfrac{1}{x^{-n}} = x^n$.
2. $\dfrac{x^n}{x^m} = x^{n-m}$.
3. $(xy)^n = x^n y^n$
4. $x^n x^m = x^{n+m}$.
5. $(x^n)^m = x^{nm}$.
6. $\left(\dfrac{x}{y}\right)^n = \dfrac{x^n}{y^n}$

We now turn our attention to an extension of the concept of exponent to rational numbers. In making such an extension it is natural to preserve properties 1 through 6, which hold for integral exponents, since it is these very properties that make the exponent concept useful.

To obtain a clue about how to proceed, consider $x^{1/2}$, which has not been defined and is therefore a candidate for definition. If property 5 is to hold for fractional exponents, then $(x^{1/2})^2$ must yield $x^{(1/2)\cdot 2} = x^1 = x$. Thus $x^{1/2}$ will have to stand for a quantity which if squared yields x. This leads us to define

$$x^{1/2} = \sqrt{x}$$

where $\sqrt{x}$ is the principal square root of x. This definition also makes sense from the point of view of property 4 since we obtain

$$x^{1/2} \cdot x^{1/2} = x^{1/2+1/2} = x^1$$

which corresonds to the result $\sqrt{x} \cdot \sqrt{x} = x$.

More generally, **the definition**

$$x^{1/q} = \sqrt[q]{x},$$

ensures that property 5 holds for rational exponents of the form $1/q$. We obtain

$$(x^{1/q})^q = x^{(1/q) \cdot q} = x$$

which corresponds to the meaning of $\sqrt[q]{x}$, that $(\sqrt[q]{x})^q = x$. Furthermore, we define $x^{-1/q}$ by

$$x^{-1/q} = \frac{1}{x^{1/q}}$$

Thus, for example,

$$9^{1/2} = \sqrt{9} = 3 \qquad (-8)^{1/3} = \sqrt[3]{-8} = -2$$

$$(81)^{-1/4} = \frac{1}{(81)^{1/4}} = \frac{1}{3} \qquad (-32)^{-1/5} = \frac{1}{(-32)^{1/5}} = -\frac{1}{2}$$

Finally, if property 5 is to hold when $x^{1/q}$ is raised to the power p and when x^p is raised to the power $1/q$, we must have

$$(\sqrt[q]{x})^p = (x^{1/q})^p = x^{(1/q) \cdot p} = x^{p/q}$$

and
$$\sqrt[q]{x^p} = (x^p)^{1/q} = x^{p \cdot (1/q)} = x^{p/q}$$

This leads us to the following definition of $x^{p/q}$: If p and q are nonnegative integers and $q \neq 0$, $x^{p/q}$ and $x^{-p/q}$ are defined by

$$x^{p/q} = (\sqrt[q]{x})^p = \sqrt[q]{x^p}, \qquad x^{-p/q} = \frac{1}{x^{p/q}}$$

To avoid unpleasant surprises we also require that x be nonnegative in defining $x^{p/q}$ and positive for $x^{-p/q}$.

For example, we have

$$9^{3/2} = (\sqrt{9})^3 = 3^3 = 27$$

$$8^{2/3} = (\sqrt[3]{8})^2 = 2^2 = 4$$

$$(81)^{-3/4} = \frac{1}{81^{3/4}} = \frac{1}{(\sqrt[4]{81})^3} = \frac{1}{3^3} = \frac{1}{27}$$

It can be proved that rational exponents satisfy conditions 1 through 6, and that any rational exponent can be replaced by an equivalent rational number. Thus, for example, $x^{2/4} = x^{1/2}$. The following examples further illustrate the application of properties 1 through 6 to the simplification of expressions involving radicals or rational exponents.

$$\sqrt{x} \cdot \sqrt[3]{x} = x^{1/2} \cdot x^{1/3} = x^{1/2+1/3} = x^{5/6} = \sqrt[6]{x^5}$$

$$(3^{1/2})^4 = 3^{(1/2)\cdot 4} = 3^2 = 9 \qquad (8x^6)^{1/3} = 8^{1/3}(x^6)^{1/3} = 2x^2$$

EXERCISES

Evaluate each of the following.

71. $25^{1/2}$
72. $49^{1/2}$
73. $125^{1/3}$
74. $64^{-1/2}$
75. $216^{-1/3}$
76. $100^{-3/2}$
77. $8^{-1/3}$
78. $36^{-2/4}$
79. $27^{2/3}$
80. $8^{-4/3}$
81. $3^{5/2} \cdot 3^{1/2}$
82. $(4^3)^{-1/3}$
83. $(2^8)^{-1/2}$
84. $2^{1/4} \cdot 2^{-3/4}$
85. $6^{1/3} \cdot 6^{5/3}$
86. $\dfrac{32^{1/2}}{32^{1/3}}$
87. $\dfrac{4^{-1/2}}{2^{-6}}$
88. $3^{-1/2} \cdot 3^{3/2}$

Simplify each of the following.

89. $(9x^4)^{1/2}$

90. $(x^2x^{1/2})^4$

91. $\dfrac{x^{1/2}}{x^{-1/2}}$

92. $\dfrac{x}{x^{1/2}}$

93. $\left(\dfrac{x^{1/2}}{x^3}\right)^2$

94. $(x^{-1/4}y^{1/2})^4$

95. $\sqrt{x}(\sqrt{x})^5$

96. $\dfrac{(\sqrt[3]{x})^2}{\sqrt[3]{x^5}}$

97. $\dfrac{(\sqrt[3]{x})^5}{\sqrt[10]{x}}$

1.6. LINEAR EQUATIONS AND LINES

Let us recall that the graph of every linear equation in two variables is a line, and that every line is the graph of a linear equation in two variables. The problem we turn to now is that of determining the equation of a given line. Let L denote a nonvertical line (see Figure 1.3). L is described by some linear equation that, upon solving for y in terms of x, can be written as

$$y = mx + b$$

where m and b are constants. Let $R(x_1, y_1)$ and $P(x_2, y_2)$ denote two points on L. Then we have

$$y_2 = mx_2 + b$$

and

$$y_1 = mx_1 + b$$

Subtracting and simplifying yields

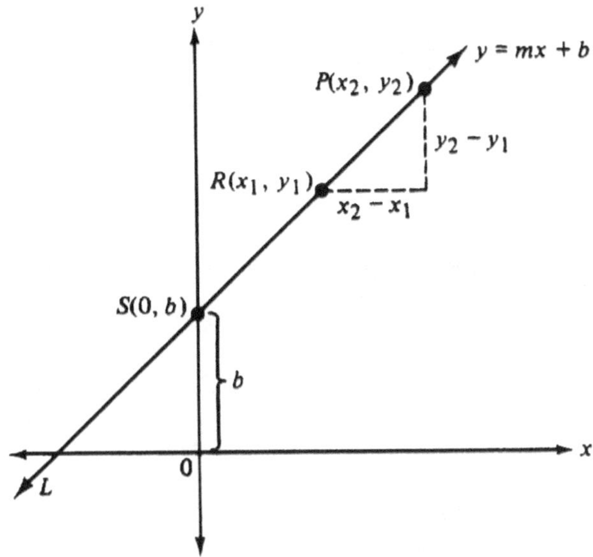

Figure 1.3

$$y_2 - y_1 = mx_2 + b - (mx_1 + b)$$
$$y_2 - y_1 = mx_2 + b - mx_1 - b$$
$$y_2 - y_1 = mx_2 - mx_1$$
$$y_2 - y_1 = m(x_2 - x_1)$$

By dividing both sides by $x_2 - x_1$, we obtain

$$\frac{y_2 - y_1}{x_2 - x_1} = m$$

The constant m is called the **slope** of line L. Geometrically, m expresses the ratio of the vertical climb to the horizontal run in going from any one point $R(x_1, y_1)$ on L to any other point $P(x_2, y_2)$ on L. The slope m is a measure of the inclination of line L to the x-axis and is a fundamental characteristic of the line. To determine the equation of L, it suffices to know the slope of L and a point on L, or two points on L.

Let us note that if we substitute 0 for x in the equation $y = mx + b$, $y = b$ is obtained. The constant b is called the **y-intercept** of line L, since it is the y-value of the point of intersection of L with the y-axis (see Figure 1.3).

EXAMPLE 1

Find the equation of the line L that passes through $R(1, 3)$ and has slope 2.

If $P(x, y)$ is any point on L, then

$$y = 2x + b$$

Since $R(1, 3)$ is on L, we have

$$3 = 2(1) + b$$
$$1 = b$$

Thus

$$y = 2x + 1$$

is an equation of line L (see Figure 1.4).

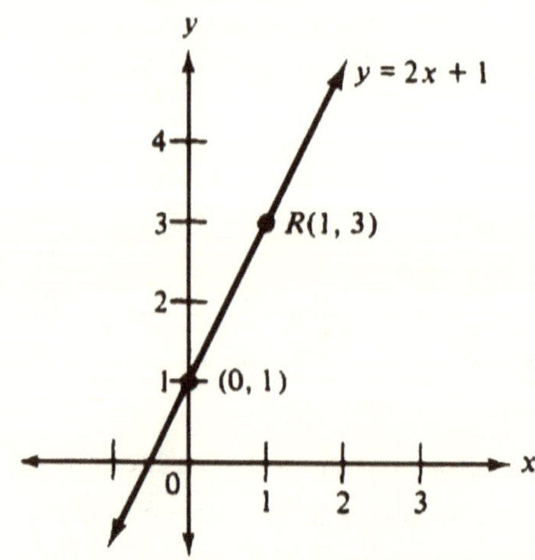

Figure 1.4

EXAMPLE 2

Find the equation of the line L that passes through $R(1, 1)$ and $S(3, 2)$.

We first determine the slope of L from the points $R(1, 1)$ and $S(3, 2)$.

$$m = \frac{2-1}{3-1} = \frac{1}{2}$$

Thus

$$y = \frac{1}{2}x + b$$

Since $R(1, 1)$ is on L, we have

$$1 = \frac{1}{2}(1) + b$$
$$\frac{1}{2} = b$$

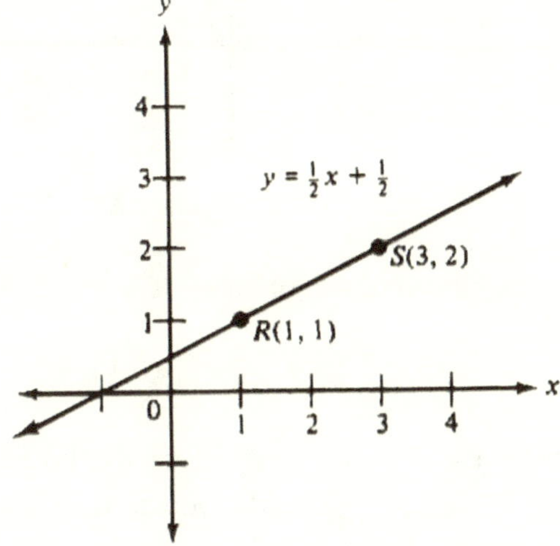

Figure 1.5

Thus

$$y = \frac{1}{2}x + \frac{1}{2}$$

is an equation for L (see Figure 1.5).

Horizontal and Vertical Lines

If L is an horizontal line that passes through the point $S(0, b)$ (see Figure 1.6), then all points on L have y-value b. Thus $R(1, b)$ is another point on L, and the slope of L is

$$m = \frac{b - b}{1 - 0} = \frac{0}{1} = 0$$

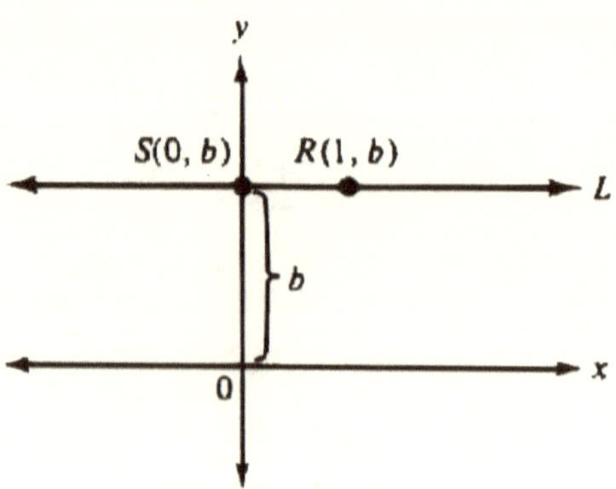

Figure 1.6

As the equation of L, we obtain

$$y = mx + b = 0 + b = b$$

Therefore, $y = 2$ is the equation of the horizontal line that passes through (0, 2), $y = -3$ is the equation of the horizontal line that passes through $(0, -3)$, and $y = 0$ is the equation of the horizontal line that passes through the origin (0, 0), the x-axis (see Figure 1.7).

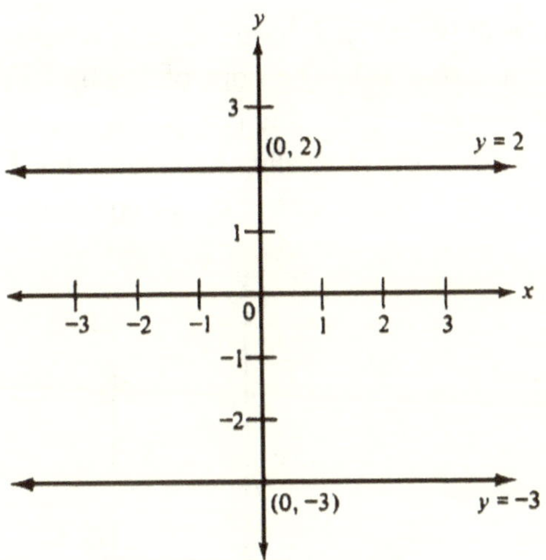

Figure 1.7

The slope of a vertical line, on the other hand, is undefined. To see why this is the case, consider the vertical line L that passes through the point (1, 0) (see Figure 1.8).

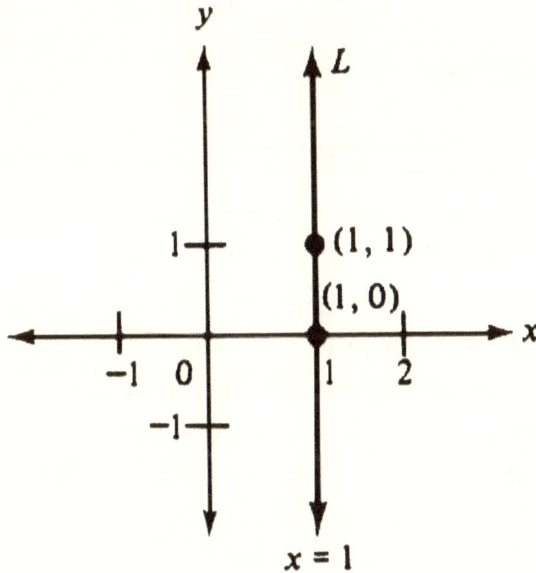

Figure 1.8

Since all points on L have the same x-value, 1, this line is described by the equation $x = 1$. The slope of L is undefined, since the difference between the x-values for any two points is $1-1=0$, and division by 0 is undefined. More generally, the vertical line L that passes through $(b, 0)$ (see Figure 1.9) is described by the equation $x = b$.

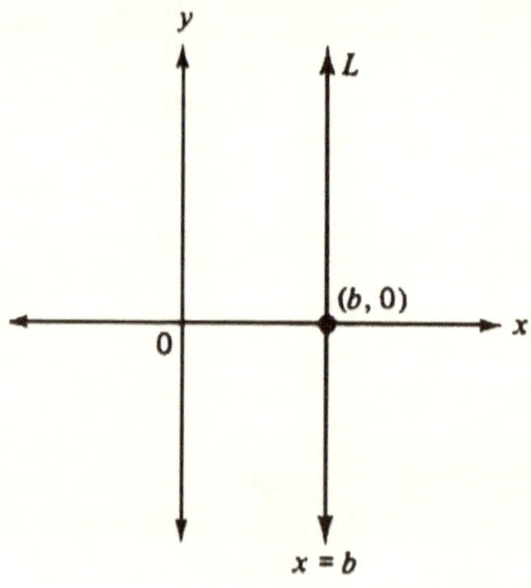

Figure 1.9

EXERCISES

Write an equation of the line satisfying the following conditions.

1. Passing through (1, 4) with slope 2.
2. Passing through $(-1, 3)$ with slope 3.
3. Passing through (2, 1) with slope -1.
4. Passing through $(4, -1)$ with slope $-\frac{1}{2}$.
5. Passing through $(1, -3)$ with slope -2.
6. Passing through $(-1, 2)$ and (3, 4).
7. Passing through (1, 1) and (4, 2).
8. Passing through $(-2, 1)$ and (4, 1).
9. Passing through (1, 4) and (1, 5).
10. Passing through $(-1, -1)$ and (2, 5).

CHAPTER 2

Functions

2.1. FUNCTIONS OF ONE VARIABLE

The idea of function is suggested by essentially the following background. Let us assume that two factors connected with some process under study have been singled out for consideration. Moreover, these factors can be quantified and expressed in terms of variables x and y. Also, the process gives rise to a rule of association such that, whenever a value for x is specified, exactly one value for y is determined by means of this rule. Such a structure is called a function.

In summary, then, a **function**, or **function of one variable**, has the following structure. There are two variables x and y, say, and a rule with the property that for each value of x the rule determines exactly one value for y. Here x is envisioned as taking on values with corresponding values of y being determined. To distinguish these different roles, x is called an **independent variable** and y is called a **dependent variable**. If their roles were to be reversed, with y taking on values and corresponding values of x being determined, then y would be the independent variable and x would be the dependent variable. If x is the independent variable and y is the dependent variable, then we say that **y is a function of x**. The term "function of one variable" is due to the presence of one independent variable. The set of values that can be taken on by the independent variable is called the **domain of definition**, or **domain**, of the function and the independent variable. When the domain of definition is not explicitly stated, it is understood to be the set of all real numbers for which the rule makes mathematical sense. If the function arises from some applied setting, economics, for example, then the domain, when not explicitly stated, is understood to be all numbers for which the rule makes economic as well as mathematical sense. Often, but not always, the rule of the function can be expressed in terms of a single algebraic expression. For example, the rule

$$y = x^3 - x + 10$$

defines y as a function of x. The domain, which is not explicitly stated, is understood to be the set of all real numbers, since this rule can be applied to all real numbers.

$$\text{For } x = 2, \qquad y = (2)^3 - 2 + 10 = 16$$
$$\text{For } x = -1, \qquad y = (-1)^3 - (-1) + 10 = 10$$

The rule

$$y = \sqrt{x}$$

defines y as a function of x. Recall that $\sqrt{x}$ expresses the principal, or nonnegative, square root of x. The domain, which is not explicitly stated, is understood to be all nonnegative real numbers, since we are operating within the real-number system and $\sqrt{x}$ does not yield a real-number for negative x.

The practice of using letters such as f, g, h, F, G, H, and so on, to designate either the rule of the function or the function itself has become widespread in recent years. In terms of this practice, the notation $f(x)$, read "f of x" or "f at x," is used to express the value assigned by the function f to x. $f(x)$, in other words, is another symbol for the dependent variable and, in fact, we often see the statement $y = f(x)$ written as an expression of this fact. The notation $f(x)$ is also often used to denote the function itself, as well as the value assigned by the function to x. Thus, for example, the rule of the function

$$y = x^3 - x + 10$$

is expressed in terms of the $f(x)$ notation by

$$f(x) = x^3 - x + 10$$

The value assigned by this function to 2, for example, is expressed by $f(2)$; the value assigned to -1 is expressed by $f(-1)$. We have

$$f(2) = (2)^3 - 2 + 10 = 16$$
$$f(-1) = (-1)^3 - (-1) + 10 = 10$$

Two functions f and g are said to be **equal** if (1) they have the same domain of definition, and (2) the rules of f and g assign the same value to each number in their common domain of definition. Thus, for example, the functions f and g defined by

$$f(x) = x^3 - x + 10$$

and $$g(x) = x^3 - x + 10, \qquad \text{where } x \geq 0$$

are not equal since they have different domains of definition. $f(-1) = 10$, whereas $g(-1)$ is not defined.

The following examples further illustrate the nature of a function.

EXAMPLE 1. *THE CONSTANT FUNCTION*

Let f denote the function that assigns the value 3 to each real number. That is,

$$f(x) = 3, \qquad \text{for all real-numbers } x$$

Thus $f(-4) = 3, f(0) = 3, f(\frac{1}{2}) = 3, f(\sqrt{2}) = 3, f(10) = 3$, and so on. Since the dependent variable takes on one value, the value 3 in this case, such a function is called a **constant function**.

EXAMPLE 2. *A TWO-PIECE FUNCTION*

Let f denote the function defined on the reals by

$$f(x)=\begin{cases}2x+1, \text{ for } x\le 1\\ x-2, \text{ for } x>1.\end{cases}$$

This is one function with one rule and one domain of definition. The unusual feature is that the rule behaves differently on different parts of the domain of definition according to whether $x\le 1$ or $x>1$. Thus $f(1)=2(1)+1=3$, $f(-2)=2(-2)+1=-3$, but $f(4)=4-2=2$.

An analogy which is helpful to a point is to think of a function in terms of a production process which converts raw materials into finished products.

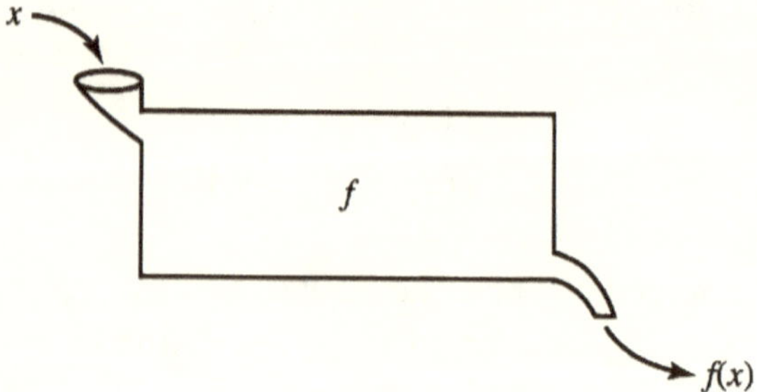

The x values in the domain of definition correspond to raw materials to be used. Only these raw materials may be fed to the machine, which corresponds to the rule of the function, for the production of finished products—the $f(x)$ values. The machine, or rule f, may be a very simple one, as in the case of a constant function which yields the same finished product, $f(x) = 3$, for example, for all raw materials x given to it. The machine may be a fairly complex one which behaves differently on different parts of the domain of definition, as in the case of Example 2.

EXERCISES

1. For $f(x)=x^2-x-4$, find $f(1), f(3), f(-1), f(-2)$, and $f(0)$.

2. For $f(x) = x^3 + 2x^2 - x + 1$, find $f(0), f(1), f(2), f(3), f(-1)$, and $f(-2)$.
3. For $f(x) = 3x^2 - 2x - 4$, find $f(-3), f(\frac{1}{2}), f(3)$, and $f(10)$.
4. What is the domain of definition of the function $f(x) = \sqrt{x+1}$? Explain. What is $f(-2), f(-1), f(3)$, and $f(2)$?
5. For $f(x) = -1$, find $f(0), f(1), f(-1), f(10)$, and $f(-2)$.
6. What is the domain of definition of the function $f(x) = 1/(x-2)$? Explain.
7. For

$$f(x) = \begin{cases} 3x-2, & \text{for } x \geq 1 \\ 4x, & \text{for } x < 1 \end{cases}$$

   find $f(1), f(2), f(3)$, and $f(-1)$.

8. For

$$f(x) = \begin{cases} 1, & \text{for } x \geq 2 \\ -1, & \text{for } x < 2 \end{cases}$$

   find $f(1), f(2), f(\frac{1}{2}), f(0)$, and $f(-1)$.

2.2. GRAPHS OF FUNCTIONS

The **graph of a function f** is the graph of the equation $y = f(x)$, where x takes on only those values in its domain of definition. A crude, but often satisfactory, approach to sketching the graph of a function f is to let the independent variable x take on a number of values, calculate the corresponding values of $y = f(x)$, plot the points $(x, f(x))$, and join these points with a smooth curve. It is an approach that must be used with caution since the graph of a function, or equation, may exhibit gaps and breaks. Particular attention must be paid to the domain of definition of the function, and care must be taken not to inadvertently allow the independent variable to take on "illegal" values.

EXAMPLE 1

Sketch the graph of the constant function $f(x) = 3$.

The domain of definition of $f(x) = 3$, or $y = 3$, is the collection of all real-numbers. For $x = -2, y = 3$; for $x = 0$, $y = 3$; for $x = 10$, $y = 3$; and so on. The graph of $f(x) = 3$ is the horizontal line three units above the x-axis and parallel to the x-axis (see Figure 2.1). More generally, the graph of the constant function $f(x) = b$ is a horizontal line b units above the x-axis if b is positive, b units below the x-axis if b is negative, and the x-axis itself if b is zero.

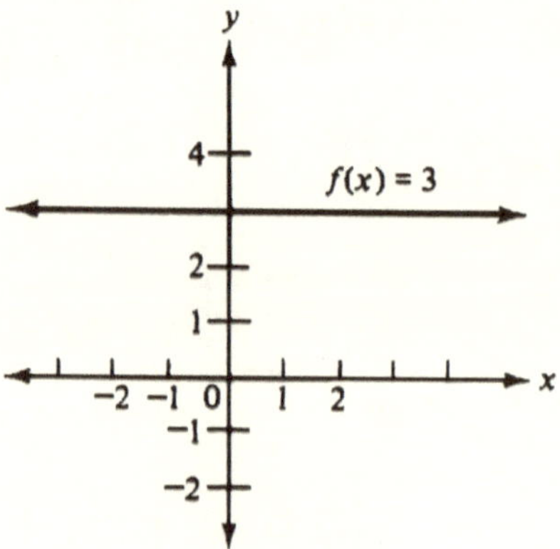

Figure 2.1

The function f with the general form

$$f(x) = mx + b$$

where m and b are constants, is said to be a **linear function**, or **linear function of one variable**. If the domain of definition is the set of all real numbers, then the graph of f is a line. If the domain of definition of f is some subcollection of real-numbers, then the graph of f is part of a line.

EXAMPLE 2

Sketch the graph of $f(x) = 3x - 2$.

Since the domain of definition consists of all real-numbers and the function is linear it suffices to plot two points to determine its graph. For $x = 1, f(1) = 1$; for $x = 2, f(2) = 4$. Thus (1, 1) and (2, 4) on the graph, which is shown in Figure 2.2.

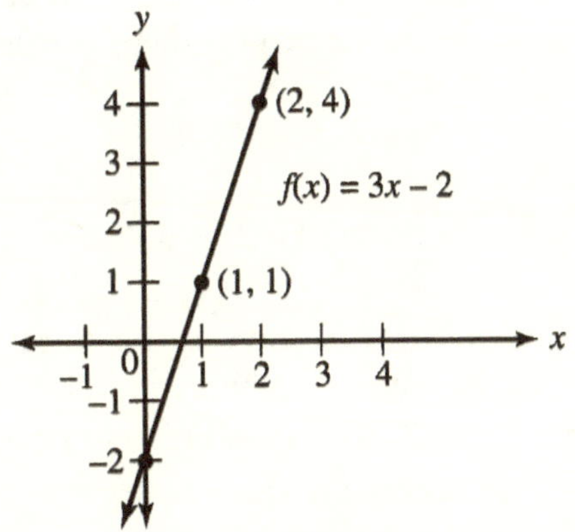

Figure 2.2

EXAMPLE 3

Sketch the graph of $f(x)=3x-2$, where $1 \leq x \leq 2$.

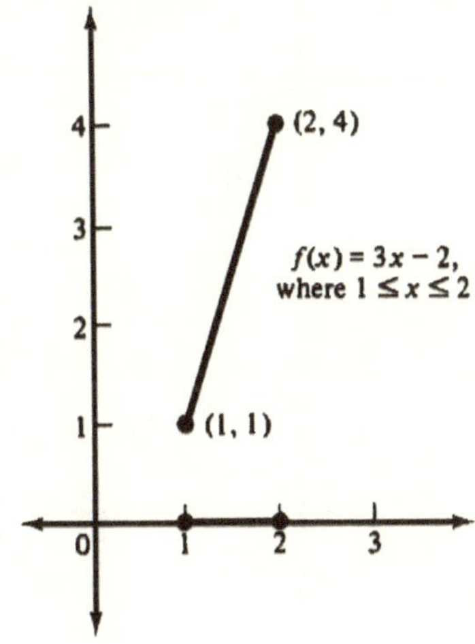

Figure 2.3

The domain of definition of this linear function consists of all real-numbers in the interval with end points 1 and 2, including 1 and 2. Thus the graph of $f(x) = 3x - 2$, where $1 \leq x \leq 2$, consists of all points on the line segment with end points $(1, f(1)) = (1, 1)$ and $(2, f(2)) = (2, 4)$ including these end points (see Figure 2.3). To emphasize that these end points are part of the graph, we display them as solid dots.

EXAMPLE 4

Sketch the graph of $f(x) = 3x - 2$, where $1 < x < 2$.

The domain of definition of this linear function consists of all real-numbers in the interval with end points 1 and 2, excluding the end points 1 and 2. The graph of $f(x) = 3x - 2$, where $1 < x < 2$, consists of all points on the line segment with end points (1, 1) and (2, 4), excluding these end points (see Figure 2.4). To properly sketch the graph, these end points must be shown, but at the same time it must be made clear that they are not on the graph. We do this by using open circles, as opposed to solid dots, to display them.

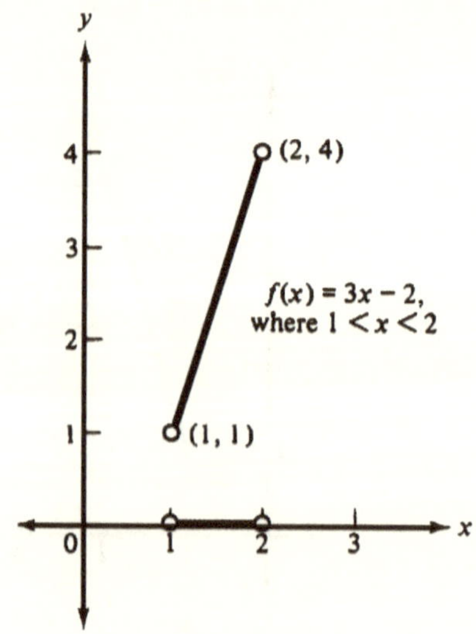

Figure 2.4

In connection with number intervals, a comment on notation is in order. The collection of all real-numbers between *a* and *b*, including the end points *a* and *b*, is often denoted by $[a, b]$. The collection of all real-numbers between *a* and *b*, excluding *a* and *b*, is denoted by (a, b), which should not be confused with the coordinates of a point (a, b). The context of the discussion makes clear which meaning is intended. Thus, to include an end point of an interval use a square bracket; to exclude an end point use a parenthesis. $[a, b)$, therefore, denotes the collection of all real-numbers between *a* and *b*, including *a* but excluding *b*.

EXAMPLE 5

Sketch the graph of $f(x) = 3x - 2$, where the domain of definition consists of the values 1, 2, and 3.

Since the domain of definition consists of three values, the graph of $f(x)$ consists of three points on the line $y = 3x - 2$, that is, (1, 1), (2, 4), and (3, 7) (see Figure 2.5).

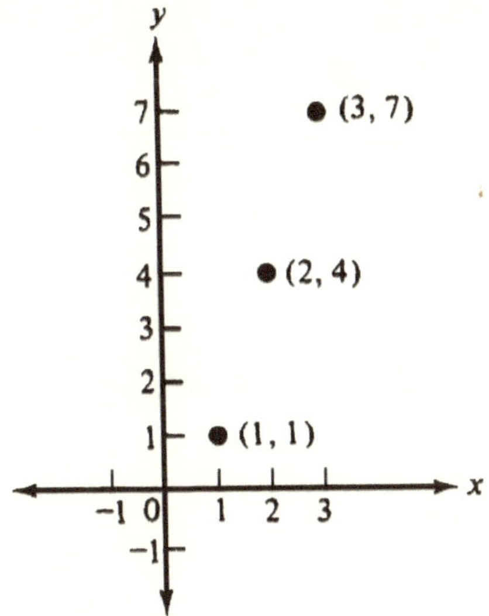

Figure 2.5

EXAMPLE 6

Sketch the graph of

$$g(x)=\begin{cases}3x-2, & \text{for } x\neq 2\\ 5, & \text{for } x=2\end{cases}$$

For $x = 2$ we obtain the point (2, 5); for $x \neq 2$ we obtain points on the line $y = 3x - 2$. The point (2, 4) on this line is not on the graph, and we therefore indicate (2, 4) by an open circle. Since (2, 5) is on the graph, we indicate it by a solid dot. The graph, shown in Figure 2.6, consists of the point (2, 5) together with all points on the line $y = 3x - 2$, except (2, 4).

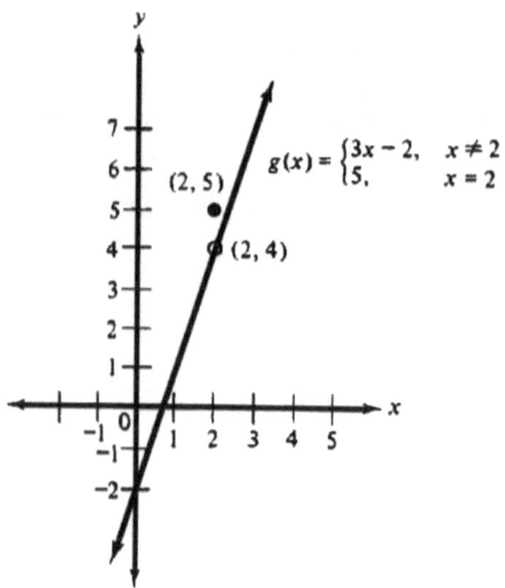

Figure 2.6

EXAMPLE 7

Sketch the graph of

$$f(x)=\begin{cases}1, & \text{for } x\geq 2\\ -1, & \text{for } x<2\end{cases}$$

For $x \geq 2$, $f(x) = 1$, so that we obtain all points on the ray one unit above the x-axis with end point (2, 1). The end point (2, 1) is on the graph and is thus indicated by a solid dot. For $x < 2, f(x) = -1$, so that we obtain all points on the ray one unit below the x-axis with end point $(2, -1)$. The end point $(2, -1)$ is not on the graph and is thus indicated by an open circle. The graph is shown in Figure 2.7.

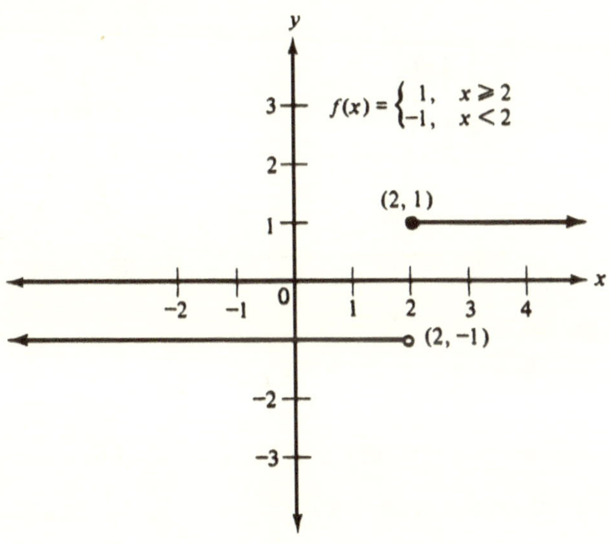

Figure 2.7

EXAMPLE 8

Sketch the graph of $f(x) = 1/(x-1)^2$.

Let us first note that $f(1)$ is not defined, since division by zero, which is not defined, arises if substitution of 1 for x is attempted. To sketch the graph, we examine the behavior of $f(x)$ for values greater than 1 and for values less than 1. For x = 2, 5, 11, and 101, we have $f(2) = 1, f(5) = \frac{1}{16}, f(11) = \frac{1}{100}$, and $f(101) = \frac{1}{10{,}000}$. In general, as x takes on larger and larger values, $f(x)$ remains positive but gets closer and closer to 0, as shown in Figure 2.8(a).

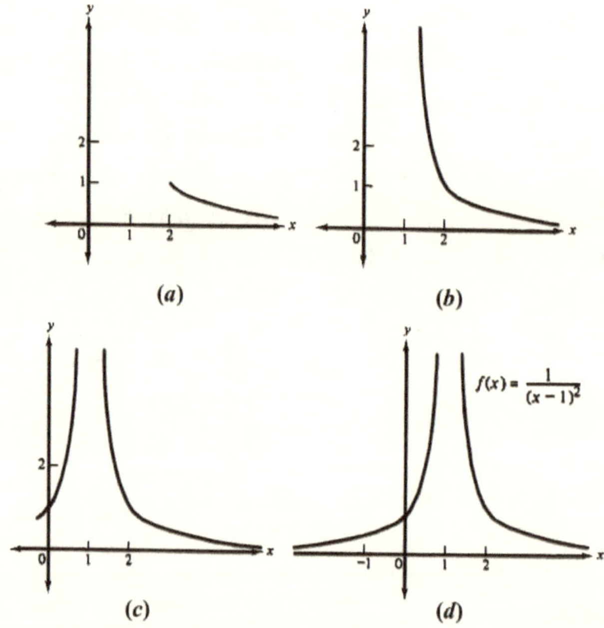

Figure 2.8

If we give x values that are getting closer and closer to 1, but that are greater than 1, such as $1+\frac{1}{10}, 1+\frac{1}{50}$, and $1+\frac{1}{100}$, we obtain $f(1+\frac{1}{10})=100, f(1+\frac{1}{50})=2500$, and $f(1+\frac{1}{100})=10,000$, values that are getting larger and larger, as shown in Figure 2.8(b).

The behavior of $f(x)$ to the right of 1 is reproduced on the left of 1. If we give x values that are getting closer and closer to 1, but that are less than 1, such as $1-\frac{1}{10}, 1-\frac{1}{50}$, and $1-\frac{1}{100}$, we obtain $f(1-\frac{1}{10})=100, f(1-\frac{1}{50})=2500$, and $f(1-\frac{1}{100})=10,000$, values that are getting larger and larger, as shown in Figure 2.8(c).

For $x=0,-9$, and -99, we obtain $f(0)=1, f(-9)=\frac{1}{100}$, and $f(-99)=\frac{1}{10,000}$, values that remain positive but get closer and closer to 0, as shown in Figure 2.8(d).

EXERCISES

Sketch the graphs of the following functions.

1. $f(x)=2$
2. $f(x)=3x+2$
3. $f(x)=x^2$
4. $f(x)=-x^2+2$
5. $f(x)=\dfrac{1}{x^2}$
6. $f(x)=2x^2-1$
7. $f(x)=2x+1, \text{for } 1\le x\le 3$
8. $f(x)=2x+1, \text{for } 1<x<3$
9. $f(x)=2x+1, \text{for } x=1,2,3$
10. $f(x)=x^2, \text{for } -1\le x\le 1$
11. $f(x)=\dfrac{1}{x^2}, \text{for } -1<x<1$
12. $h(x)=\dfrac{x^2-x}{x}$
13. $h(x)=\begin{cases} 3, & \text{for } x\ge 1 \\ -1, & \text{for } x<1 \end{cases}$
14. $f(x)=\begin{cases} x+1, & \text{for } x\ge 0 \\ -x+1, & \text{for } x<0 \end{cases}$
15. $f(x)=\begin{cases} 2x+1, & \text{for } x\ge 1 \\ -x+3, & \text{for } x<1 \end{cases}$
16. $f(x)=\begin{cases} 1+x, & \text{for } x\ge 1 \\ 1-x, & \text{for } x<1 \end{cases}$
17. $f(x)=\begin{cases} 3x+2, & \text{for } x\ge 1 \\ 5x, & \text{for } x<1 \end{cases}$
18. $f(x)=\begin{cases} 2x-3, & \text{for } x\ge 2 \\ 3x-1, & \text{for } x<2 \end{cases}$
19. $f(x)=\dfrac{1}{x-1}$
20. $f(x)=\dfrac{1}{(x-2)^2}$
21. $f(x)=\dfrac{1}{x-3}$
22. $f(x)=\dfrac{1}{(x-3)^2}$
23. $f(x)=\begin{cases} x^2, & \text{for } x\ne 2 \\ 1, & \text{for } x=2 \end{cases}$
24. $f(x)=\begin{cases} 2x-1, & \text{for } x\ne 1 \\ 5, & \text{for } x=1 \end{cases}$
25. $f(x)=\begin{cases} 2x-1, & \text{for } x<2 \\ 5, & \text{for } x=2 \\ -x+5, & \text{for } x>2 \end{cases}$
26. $f(x)=\begin{cases} 3, & \text{for } x<-1 \\ 2, & \text{for } -1\le x\le 1 \\ 3, & \text{for } x>1 \end{cases}$

2.3. FUNCTIONS FROM APPLIED SETTINGS

Functions that arise from applied settings are more complex than those with no applied background in the sense that their domains of definition are more complex. We must now be sensitive to values that not only make mathematical sense, but sense in terms of the application.

EXAMPLE 1. *COST FUNCTIONS*

Consider a firm that is producing a single, uniform commodity. The function

$$c = c(x)$$

which describes the total cost c for the production of x units of output, is called the **cost function of the firm**. Thus, for example, suppose that

$$c(x) = \frac{1}{2}x^2 + 20x + 900$$

is the cost function of a coffee producer, where x is output in tons per day and $c(x)$ is the cost in dollars per ton. Then the cost of producing 10 tons per day is

$$c(10) = \frac{1}{2}(10)^2 + 20(10) + 900 = \$1150$$

Values of $c(x)$ for selected values of x are shown in Table 2.1, and the resulting points are plotted in Figure 2.9(a). Different scales are required for the x and y axes because of the difference in magnitude between the output values x and cost values $c(x)$. While necessary in this case, these different scales have a distortion effect that we should be aware of and sensitive to. The points plotted in Figure 2.9(a) almost seem to lie on a line; in fact, they lie on a parabolic arc.

Table 2.1

x	0	4	8	12	16	20
$c(x)$	900	988	1092	1212	1348	150

The smallest value of $c(x)$ is 900. Much space would be wasted if we were to indicate the full range of values between 0 and 900 on the y-axis. It is common practice to omit such a range and to indicate this omission by employing the symbol shown on the y-axis in Figure 2.9(a).

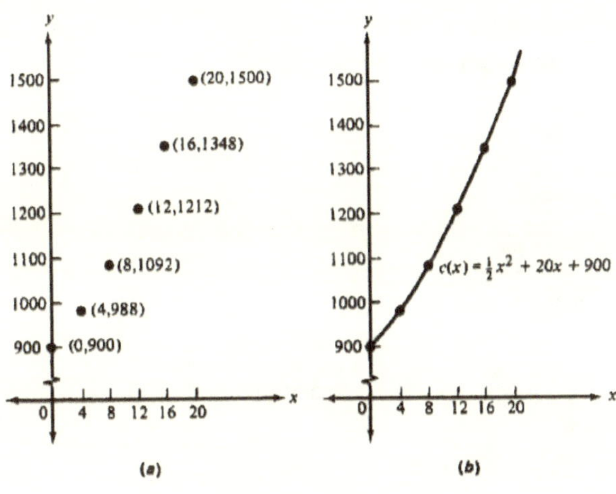

Figure 2.9

Also to be considered is the question of what values for x are appropriate, given the background of the function, and what values should be allowed for mathematical simplicity. In this situation, output, being expressed in tons of coffee per day, is divisible up to a point, so that it makes sense from an economic point of view for x to assume fractional values. Still, in every such situation some limitation on divisibility is imposed by practical or technical considerations. We should also note that x, representing output, can neither assume negative values nor values beyond a certain output level. From a mathematical point of view, however, it is often convenient in such situations to allow x to assume all real values or all real values in some interval. Opting for an interval of nonnegative values in this situation leads to the graph shown in Figure 2.9(b). It must be kept in mind, however, that, although all values of x in this interval are mathematically admissible, not all values of x are economically feasible.

EXAMPLE 2. *COST, REVENUE, AND PROFIT FUNCTIONS*

The function

$$R = R(x)$$

which describes total revenue as a function of output, is called the **revenue function** of the firm. If

$$R(x) = 306x - 5x^2$$

is the revenue function of the coffee producer, where x is output in tons per day and $R(x)$ is revenue in dollars per ton, the revenue obtained for an output of 10 tons per day is

$$R(10) = 306(10) - 5(10)^2 = \$2560$$

The domains of definition of these functions consist of all $x \geq 0$ for which $c(x)$ and $R(x)$ make economic sense.

The **profit function** $P(x)$ of the firm is the difference between the revenue and cost functions:

$$P(x) = R(x) - c(x)$$

If $c(x) = \frac{1}{2}x^2 + 20x + 900$ and $R(x) = 306x - 5x^2$, the coffee producer's profit function is:

$$P(x) = (306x - 5x^2) - (\frac{1}{2}x^2 + 20x + 900)$$
$$= -\frac{11}{2}x^2 + 286x - 900$$

EXAMPLE 3. *THE THOMAS COMPANY'S PROBLEM*

The Thomas fast-food chain has 60 restaurants in New England, each doing an average of \$20,000 worth of business per day. Studies conducted on the impact of opening new restaurants in the region indicate that the average amount of business done by each restaurant will drop by \$200 per day for each new restaurant opened. Express the total daily average income of all restaurants in the chain as a function of the number of new restaurants opened.

Let x denote the number of new restaurant to be opened. Then $60 + x$ expresses the total number of restaurants that will be in operation, and $20,000-200x$ expresses the average daily income of each restaurant. The total daily average income $I(x)$ is the product of the average income of one restaurant, $20,000-200x$, and the number of restaurants, $60 + x$. Thus

$$I(x) = (20,000-200x)(60+x).$$

What is the domain of definition of $I(x)$? First of all, $x \geq 0$ since we will either add a certain number of restaurants to the region or not. Negative values for x make no sense because we are not considering selling off existing restaurants. Fraction values, such as $\frac{1}{2}$ and $\frac{3}{4}$, clearly make no sense in this setting. And then there is the question of how large x can realistically be. It is clear that $I(x)$ must exceed zero. This leads us to consider the inequality

$$20,000-200x > 0,$$

from which we obtain $x < 100$. Thus, at the very least, x must be less than 100.

In summary, we have the non-negative integer values 0, 1, 2, . . . , perhaps extending as far as 99, as comprising the domain of definition of $I(x)$.

EXAMPLE 4. ***GAS CHARGE***

The monthly charge for natural gas in Bell City is \$2 for any amount up to and including 5 units (1 unit = 100 cubic feet), \$0.40 per unit for any amount exceeding 5 units up to and including 1000 units, and \$0.50 per unit for any amount in excess of 1000 units. Express costs as a function of the amount used.

Let x denote the number of units used and $c(x)$ the cost of x units. For x between 0 and 5 inclusive $(0 \le x \le 5), c(x) = 2.00$; for x between 5 and 1000, including $1000\,(5 < x \le 1000), c(x) = (0.40)x$; for x greater than $1000\,(x > 1000), c(x) = 400$ (for the first 1000 units) plus $0.50(x-1000)$ (for the amount by which x exceeds 1000 units). By multiplying and collecting terms, we obtain

$$400 + 0.50(x-1000) = (0.50)x - 100$$

Thus we have

$$c(x) = \begin{cases} 2.00, & \text{for } 0 \le \text{x} \le 5 \\ (0.40)x, & \text{for } 5 < x \le 1000 \\ (0.50)x - 100, & \text{for } x > 1000 \end{cases}$$

This example illustrates a function whose rule cannot be stated in terms of one algebraic expression. Three algebraic expressions are needed, with the expression used depending on where in the domain of definition the number in question falls. For example, $c(0)$ = 2.00; if no gas is used, you are still charged \$2. $c(500)$ = (0.40)500 = 200; if 500 units are used, you are charged \$200.

It is a common error to equate function with algebraic expression. Although three algebraic expressions appear, we have one cost function, not three cost functions, that expresses cost as a function of the amount of gas used.

For mathematical simplicity we take the domain of definition of $c(x)$ to be the non-negative real-numbers, but keeping in mind that not all non-negative real-numbers make sense in terms of this setting.

EXAMPLE 5. *TAXES*

Consider the tax-rate schedule for the Bell City income tax given in Table 2.2.

Table 2.2

If taxable income is		*The tax to be paid is*				
not over $1,000		0.7% (0.007) of taxable income				
over	but not over					
$1,000	$3,000	$ 7	plus	1.1% (0.011)	of excess over	$ 1,000
3,000	6,000	29	plus	1.4% (0.014)	of excess over	3,000
6,000	10,000	71	plus	1.8% (0.018)	of excess over	6,000
10,000	15,000	143	plus	2.1% (0.021)	of excess over	10,000
15,000	20,000	248	plus	2.5% (0.025)	of excess over	15,000
20,000	25,000	373	plus	2.8% (0.028)	of excess over	20,000
25,000	30,000	513	plus	3.2% (0.032)	of excess over	25,000
30,000		673	plus	3.5% (0.035)	of excess over	30,000

In the language of functions this tax-rate schedule translates to the following income-tax function $T(x)$.

$$T(x)=\begin{cases} (0.007)x, & \text{for} \quad 0<x\le 1{,}000 \\ 7+0.011(x-1{,}000), & \text{for} \quad 1{,}000<x\le 3{,}000 \\ 29+0.014(x-3{,}000), & \text{for} \quad 3{,}000<x\le 6{,}000 \\ 71+0.018(x-6{,}000), & \text{for} \quad 6{,}000<x\le 10{,}000 \\ 143+0.021(x-10{,}000), & \text{for} \quad 10{,}000<x\le 15{,}000 \\ 248+0.025(x-15{,}000), & \text{for} \quad 15{,}000<x\le 20{,}000 \\ 373+0.028(x-20{,}000), & \text{for} \quad 20{,}000<x\le 25{,}000 \\ 513+0.032(x-25{,}000), & \text{for} \quad 25{,}000<x\le 30{,}000 \\ 673+0.035(x-30{,}000), & \text{for} \quad 30{,}000<x \end{cases}$$

Thus if taxable income is $9000, the tax to be paid is $T(9000)$ = 71 + 0.018(3000) = $125; if taxable income is $12,000, the tax to be paid is $T(12{,}000)$ = 143 + 0.021(2000) = $185.

EXAMPLE 6. *GROWTH*

Each bacterium in a culture consisting of three *E. coli* organisms divides at the end of each second. Determine the function that describes the number of bacteria in the culture at the end of t seconds. Find the number of bacteria in the culture at the end of 5 seconds.

At time $t = 0$, the starting point of our study, the culture contains 3 bacteria. At time $t = 1$, each bacterium gives rise to 1 additional bacterium, so that the number of bacteria at time $t = 1$ is

$$3+3=3(2)$$

At time $t = 2$ each bacterium in the culture gives rise to 1 additional member, so that the number of bacteria in the culture at time $t = 2$ is the number present at time $t = 1$, 3(2), plus the size of the increase, 3(2). Thus at time $t = 2$ there are

$$3(2)+3(2)=3(2)[2]=3(2)^2$$

organisms. At time $t = 3$ each bacterium in the culture gives rise to 1 additional member, so that the number of bacteria in the culture at time $t = 3$ is the number present at time $t = 2$, $3(2)^2$, plus the size of the increase, $3(2)^2$. Thus at time $t = 3$ there are bacteria.

$$3(2)^2+3(2)^2=3(2)^2[2]=3(2)^3$$

More generally, this analysis leads to the function

$$y=3(2)^t$$

where y is the number of bacteria in the culture at the end of t, seconds. The domain of definition consists of nonnegative integer values ($t = 0, 1, 2$, etc.) up to a certain point. As t becomes larger and larger the accuracy of this growth function dimishes, since it does not reflect such factors as the limited ability of the environment to support life and loss through death.

The predicted number of bacteria in the culture at the end of 5 seconds is

$$y = 3(2)^5 = 96$$

EXERCISES

1. The monthly charge for telephone service in a certain region is \$8 for any number of message units up to and including 50 and \$0.09 for each additional message unit beyond 50. Express cost as a function of the number units accumulated.
2. A taxi fleet uses 100 taxis to service Johnson City. Each taxi brings in an average of \$200 per day in fares. If additional taxis are added to the fleet, it is estimated that the amount in fares brought in by each taxi will drop by \$10 per day for each additional taxi added. Express the total daily average income of the taxi fleet as a function of the number of taxis added to the fleet.
3. An author signed a contract with a publishing company that stipulated a royalty rate of 15 percent of the list price of the book (estimated at \$10 per copy) for the first 5000 copies sold, and 20 percent of list price for each copy over 5000 that is sold. Express royalty income as a function of the number of books sold.
4. A new government agency, formed to study the growth of bureaucracy in government, is beginning its operations with a staff of two persons. At the end of each month each person hires two assistants. By using an approach analogous to the one employed in Example 5, find a function that describes the number of staff in this agency at the end of t months. At this rate, what would the size of the staff be 1 year later?
5. The income tax paid in a certain region is \$0.00 if taxable income is less than or equal to \$5000, \$20 plus 1 percent of the excess over \$5000 if taxable income is over \$5000 but not over \$25,000, and \$220 plus 3 percent of the excess over \$25,000

if taxable income is over \$25,000. Express tax to be paid as a function of taxable income.

6. A first-class letter whose weight does not exceed 3 ounces is to be sent from one point in the United States to another. Express postage costs as a function of weight.
7. Some property loses value over time. This decrease in value is called **depreciation**. Thus, if an item is purchased for \$1000 and 1 year later its resale value is \$600, its depreciation after 1 year is \$400. One of several methods used for depreciating property is linear depreciation, which assumes that the loss in value over a specified time is a fixed percentage of the original value.

   Suppose that the cost of a machine is \$2000 and that it depreciates linearly at the rate of 10 percent per year. Express the value of the machine as a function of time. Find the value of the machine after 3 years. More generally, suppose that an item costs C dollars and depreciates linearly at the rate r per annum. Express the value of the item as a function of time.

CHAPTER 3

Limits of Functions

3.1. LIMITS OF FUNCTIONS: PART 1

The concept of limit of a function and the developments that stem from it give calculus the scope, tone, and substance that distinguish it from algebra.

As we have seen in Chapter 2, functions state relationships between variables. If, for example, we know that

$$y = 3x + 2$$

then we can determine the corresponding value of the dependent variable y whenever a value for the independent variable x is prescribed. But this, as far as it goes, is static, so to speak. It does not tell us about the dynamics of the underlying process, the behavior of variables as change occurs. How does the dependent variable y of a function $y = f(x)$ behave as the independent variable x takes on values closer and closer to some specific number, 2, for example? This sort of question is the basic concern of the limit of a function idea. To illustrate this thrust, we turn to some examples.

EXAMPLE 1

$$f(x) = 3x - 2.$$

Values of $f(x) = 3x - 2$ for values of x that are getting closer and closer to 2 are shown in Table 3.1, from which we see that, as x

takes on values closer and closer to 2 (values that are both less than 2 and greater than 2), $f(x)=3x-2$ reacts by taking on values closer and closer to 4. In geometric terms, as x approaches 2 on the x-axis, $f(x)=3x-2$ approaches 4 on the y-axis. Figure 3.1 shows a closeup view of the situation.

Table 3.1

x	1.9	2.1	1.99	2.01	1.999	2.001	1.9999	2.0001
$3x-2$	3.7	4.3	3.97	4.03	3.997	4.003	3.9997	4.0003

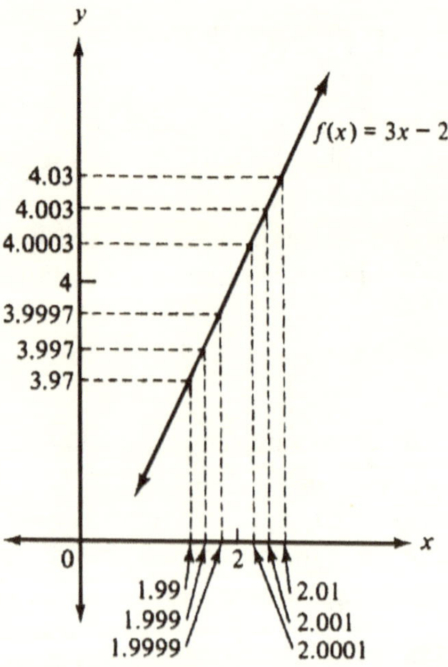

Figure 3.1

In symbols we write

$$f(x)=3x-2 \to 4 \text{ as } x \to 2$$

to express the result that $f(x)=3x-2$ approaches 4 as x approaches 2. We also use the symbolic statements

$$\underset{x\to 2}{\text{limit}}\, f(x) = 4, \qquad \text{where } f(x) = 3x - 2$$

and

$$\underset{x\to 2}{\text{limit}}\,(3x - 2) = 4$$

read, "the limit of $f(x) = 3x - 2$ is 4 as x approaches 2," to express this behavior.

Let us also observe the fact that $f(2) = 4$ did not come up in our discussion of the limit behavior of $f(x) = 3x - 2$ as x approaches 2. The reason for this is that $f(2)$ describes the behavior of $f(x)$ at 2, and this behavior is irrelevant to our concern—the behavior of $f(x)$ as x approaches 2. The fact that these different concerns lead to the same numerical value in this situation should not confuse us into equating the concerns.

That these concerns sometimes lead to different conclusions is illustrated by Example 2 and 3.

EXAMPLE 2

$$g(x) = \begin{cases} 3x - 2, & \text{for } x \neq 2 \\ 5, & \text{for } x = 2 \end{cases}$$

$g(x) = f(x)$ defined in Example 1 for all x, except 2.

$$\underset{x\to 2}{\text{limit}}\, g(x) = 4$$

since it is still true that as x approaches 2, $g(x)$ approaches 4. At the same time

$$g(2) = 5$$

EXAMPLE 3

$$h(x) = \frac{x^2 - 4}{x - 2}$$

$h(2)$ is not defined, since substitution of 2 for x leads to division by 0, which is undefined. At the same time

$$\underset{x \to 2}{\text{limit}}\ h(x) = 4$$

To see this, we first simplify $h(x)$ by factoring $x^2 - 4$.

$$h(x) = \frac{x^2 - 4}{x - 2} = \frac{(x-2)(x+2)}{(x-2)} = x + 2$$

where $x \neq 2$. As x approaches 2, $h(x) = x + 2$ approaches 4. Thus

$$\underset{x \to 2}{\text{limit}}\ h(x) = 4$$

The graph of $h(x)$ is shown in Figure 3.2.

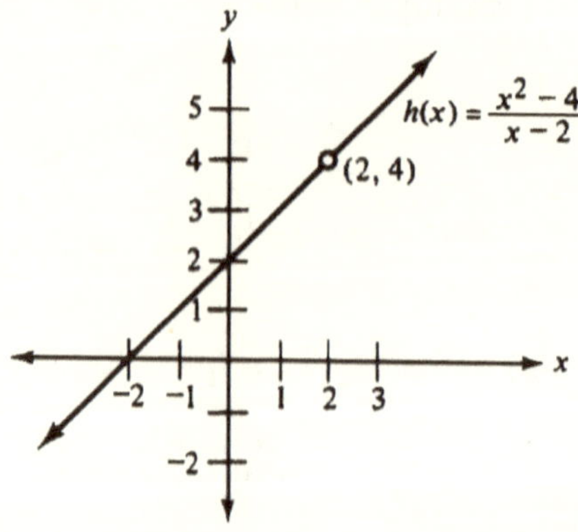

Figure 3.2

EXAMPLE 4

$$f(x) = x^2 - 1.$$

Values of $f(x) = x^2 - 1$ for values of x that are getting closer and closer to 2 are shown in Table 3.2, from which we see that, as

x takes on values closer and closer to 2, $f(x) = x^2 - 1$ gets closer and closer to 3. Thus

$$\operatorname*{limit}_{x \to 2} (x^2 - 1) = 3$$

A geometric closeup of the situation is given in Figure 3.3.

Table 3.2

x	1.9	2.1	1.99	2.01	1.999	2.001	1.9999	2.0001
$x^2 - 1$	2.61	3.41	2.9601	3.0401	2.9960	3.0040	2.9996	3.004

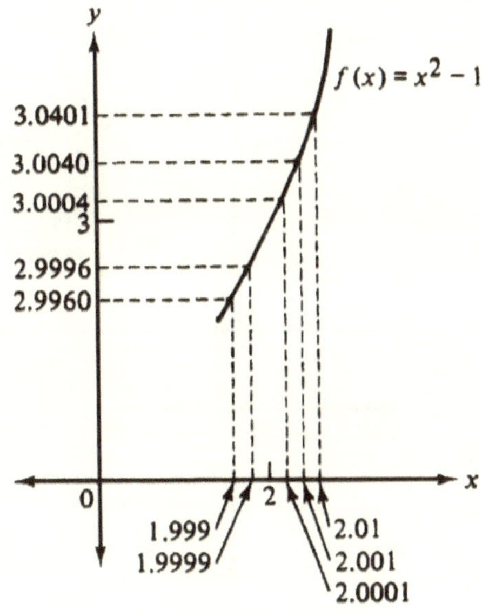

Figure 3.3

EXAMPLE 5

$$f(x) = \begin{cases} 1, & \text{for } x \geq 2 \\ -1, & \text{for } x < 2 \end{cases}$$

Values of $f(x)$ for values of x that are getting closer and closer to 2 are shown in Table 3.3, from which we see that, as x takes on values closer and closer to 2, $f(x)$ is unstable; $f(x)$ clusters around two values, −1 and 1, rather than approaching one value or behaving in some other stable manner. For this reason we say that

$$\lim_{x \to 2} f(x) \text{ does not exist}$$

Figure 3.4 shows the situation in geometric terms.

Table 3.3

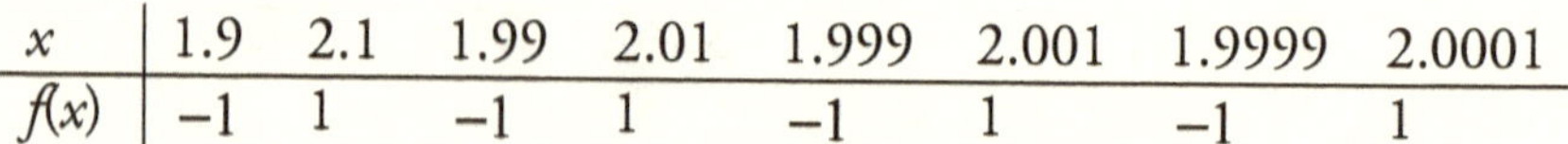

x	1.9	2.1	1.99	2.01	1.999	2.001	1.9999	2.0001
$f(x)$	−1	1	−1	1	−1	1	−1	1

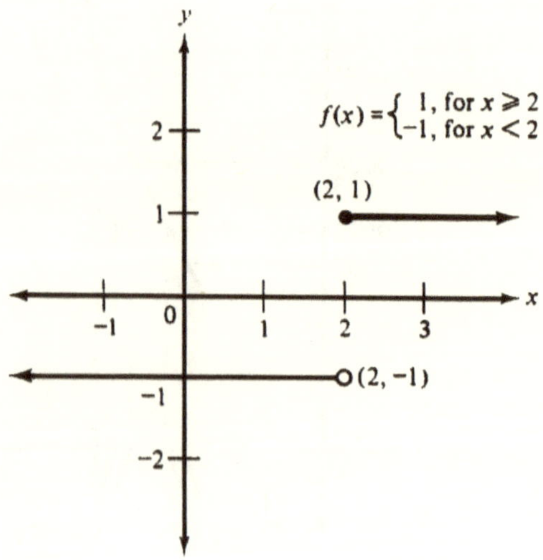

Figure 3.4

EXAMPLE 6

$$f(x) = 1/x$$

Values of $f(x) = 1/x$ for values of x that are getting closer and closer to 0 are shown in Tables 3.4, from which we see that, as x takes on values closer and closer to 0, $f(x) = 1/x$ is unstable. For values of x less than 0 but closer and closer to 0, the values of $f(x)$ get more and more negative −10, −100, −1000, and so on). For values

Table 3.4

x	$\frac{1}{10}$	$-\frac{1}{10}$	$\frac{1}{100}$	$-\frac{1}{100}$	$\frac{1}{1000}$	$-\frac{1}{1000}$
$1/x$	10	−10	100	−100	1000	−1000

of x greater than 0 but closer and closer to 0, the values of $f(x)$ get larger and larger (10, 100, 1000, and so on). For this reason we say that

$$\operatorname*{limit}_{x \to 0} \frac{1}{x} \text{ does not exist}$$

Figure 3.5 shows what is happening in geometric terms.

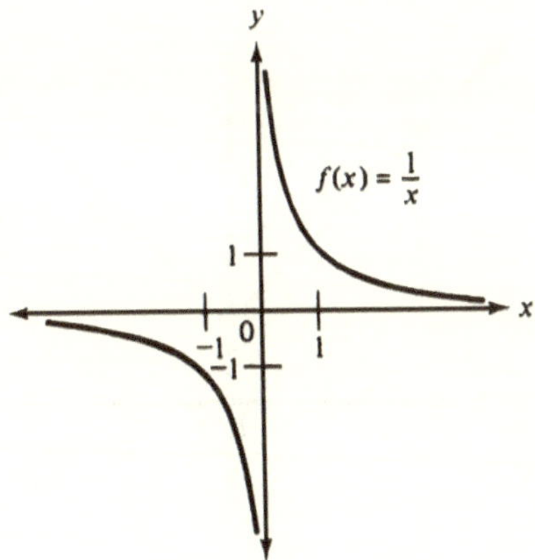

Figure 3.5

EXAMPLE 7

$$f(x) = \begin{cases} x+1, & \text{for } x \geq 0 \\ -x+1, & \text{for } x < 0 \end{cases}$$

Values of $f(x)$ for values of x that are getting closer and closer to 0 are shown in Table 3.5, from which we see that, as x takes on values closer and closer to 0, $f(x)$ gets closer and closer to 1. Thus

$$\lim_{x \to 0} f(x) = 1,$$

where

$$f(x) = \begin{cases} x+1, & \text{for } x \geq 0 \\ -x+1, & \text{for } x < 0 \end{cases}$$

The graph of this function is shown in Figure 3.6.

Table 3.5

x	0.1	−0.1	0.01	−0.01	0.001	−0.00	0.0001	−0.0001
$f(x)$	1.1	1.1	1.01	1.01	1.001	1.001	1.0001	1.0001

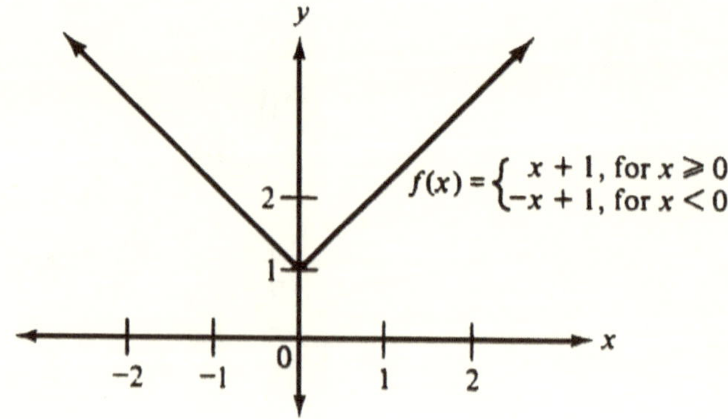

Figure 3.6

The examples considered point the way to the following general definition of limit of a function.

> **Definition**. If, as x takes on values closer and closer to a value a, $f(x)$ gets closer and closer to a single value L, or remains constant at L, then L is said to be the **limit of the function $y = f(x)$ as x approaches a.** This is expressed by writing
>
> $$f(x) \to L \text{ as } x \to a \text{ or } \lim_{x \to a} f(x) = L$$

EXERCISES

Determine if the limits of the following functions exist. When a limit exists, find its value. When a limit does not exist, explain why.

1. $\lim_{x\to 3} f(x)$, where $f(x)=2x+1$
2. $\lim_{x\to 2} f(x)$, where $f(x)=4x-3$
3. $\lim_{x\to 3} f(x)$, where $f(x)=5x+1$
4. $\lim_{x\to 3} f(x)$, where $f(x)=x^2$
5. $\lim_{x\to 2} f(x)$, where $f(x)=x^3$
6. $\lim_{x\to 1} f(x)$, where $f(x)=\dfrac{x^2-1}{x-1}$
7. $\lim_{x\to 2} f(x)$, where $f(x)=x^2+2$
8. $\lim_{x\to 1} f(x)$, where $f(x)=\begin{cases}\dfrac{x^2-1}{x-1}, & \text{for } x\neq 1\\ 4, & \text{for } x=1\end{cases}$
9. $\lim_{x\to 0} f(x)$, where $f(x)=\begin{cases}1, & \text{for } x\geq 0\\ -1, & \text{for } x<0\end{cases}$
10. $\lim_{x\to 1} f(x)$, where $f(x)=\begin{cases}1, & \text{for } x\geq 0\\ -1, & \text{for } x<0\end{cases}$
11. $\lim_{x\to 3} f(x)$, where $f(x)=\dfrac{x^2-9}{x-3}$
12. $\lim_{x\to 2} f(x)$, where $f(x)=\begin{cases}x^2, & \text{for } x\neq 2\\ 5, & \text{for } x=2\end{cases}$
13. $\lim_{x\to 4} f(x)$, where $f(x)=2$
14. $\lim_{x\to 1} f(x)$, where $f(x)=\begin{cases}3, & \text{for } x\geq 1\\ 2, & \text{for } x<1\end{cases}$
15. $\lim_{x\to 2} f(x)$, where $f(x)=\begin{cases}x+1, & \text{for } x\geq 2\\ 2x-3, & \text{for } x<2\end{cases}$

16. $\underset{x\to 1}{\text{limit}}\, f(x)$, where $f(x)=\begin{cases} 2x+1, & \text{for } x \geq 1 \\ 5x-3, & \text{for } x<1 \end{cases}$

17. $\underset{x\to 2}{\text{limit}}\, f(x)$, where $f(x)=\dfrac{1}{x-2}$

18. $\underset{x\to 3}{\text{limit}}\, f(x)$, where $f(x)=\begin{cases} 4x-2, & \text{for } x \geq 3 \\ 3x+1, & \text{for } x<3 \end{cases}$

19. $\underset{x\to 1}{\text{limit}}\, f(x)$, where $f(x)=\begin{cases} 4x+2, & \text{for } x>1 \\ 5, & \text{for } x\leq 1 \end{cases}$

20. $\underset{x\to 1}{\text{limit}}\, f(x)$, where $f(x)=\dfrac{1}{x-1}$

3.2. TOOLS FOR COMPUTING LIMITS

The calculation of limits by use of the definition of limit is feasible only in the simplest situations. This being the case, one is led to develop tools to facilitate the calculation of limits. A natural approach to the development of such tools is to view a function as being made up of component parts held together by the operations of addition, subtraction, multiplication, and division, and then to examine the limit behavior of the component parts.

To illustrate the spirit of this approach, let us return to the limit behavior of

$$f(x)=3x-2$$

as x approaches 2, considered in Example 1 of Section 3.1. As x approaches 2, we would expect the component $3x$ to approach 6, since 3 in $3x$ magnifies the value that x takes on uniformly by a factor of 3. Subtracting 2 drops values that are approaching 6 two units. Looking at the components of this function leads us to the conclusion

$$f(x)=3x-2\to 6-2=4 \quad \text{as } x\to 2$$

Such an approach via components is productive in general. The following theorems, which can be established on the basis of the definition of limit, make precise the spirit of the approach to limits via components.

Limit of a sum. Suppose that $h(x)$ can be expressed as the sum of functions $f(x)$ and $g(x)$; that is, $h(x) = f(x) + g(x)$, and $\underset{x \to a}{\text{limit}}\, f(x) = L$ and $\underset{x \to a}{\text{limit}}\, g(x) = M$; then

$$\underset{x \to a}{\text{limit}}\, h(x) = L + M$$

In colloquial language we say that the limit of a sum is the sum of the limits (provided of course, that the component parts have limits). More generally, this theorem holds for a sum of any number of component parts.

Limit of a difference. Suppose that $h(x)$ can be expressed as the difference of functions $f(x)$ and $g(x)$; that is, $h(x) = f(x) - g(x)$, and $\underset{x \to a}{\text{limit}}\, f(x) = L$ and $\underset{x \to a}{\text{limit}}\, g(x) = M$; then

$$\underset{x \to a}{\text{limit}}\, h(x) = L - M$$

In colloquial language, the limit of a difference is the difference of the limits. This theorem holds for differences formed from any number of functions.

Limit of a product. Suppose that $h(x)$ can be expressed as the product of functions $f(x)$ and $g(x)$; that is, $h(x) = f(x) \cdot g(x)$, and $\underset{x \to a}{\text{limit}}\, f(x) = L$ and $\underset{x \to a}{\text{limit}}\, g(x) = M$; then

$$\underset{x \to a}{\text{limit}}\, h(x) = L \cdot M$$

In colloquial language, the limit of a product is the product of the limits. This theorem holds for a product of any number of functions.

Limit of a quotient. Suppose that $h(x)$ can be expressed as the quotient of functions $f(x)$ and $g(x)$; that is, $h(x) = f(x)/g(x)$, and $\lim_{x \to a} f(x) = L$ and $\lim_{x \to a} g(x) = M$, where $M \neq 0$; then

$$\lim_{x \to a} h(x) = \frac{L}{M}$$

In colloquial language, the limit of a quotient is the quotient of the limits.

EXAMPLE 1

Determine, if they exist,

$$\lim_{x \to 2} 5x^3, \quad \lim_{x \to 3} 4x^2, \quad \text{and} \quad \lim_{x \to -2} 2x^4.$$

$h(x) = 5x^3$ can be expressed as the product of the constant function 5 and the function x. The limit of these components exists as x approaches 2. We have

$$\begin{aligned} \lim_{x \to 2} 5x^3 &= [\lim_{x \to 2} 5][\lim_{x \to 2} x][\lim_{x \to 2} x][\lim_{x \to 2} x] \\ &= 5 \cdot 2 \cdot 2 \cdot 2 = 40 \end{aligned}$$

The same sort of reasoning yields the following analogous results.

$$\lim_{x \to 3} 4x^2 = [\lim_{x \to 3} 4][\lim_{x \to 3} x][\lim_{x \to 3} x] = 4 \cdot 3 \cdot 3 = 36$$

$$\lim_{x \to 2} 2x^4 = [\lim_{x \to 2} 2][\lim_{x \to 2} x][\lim_{x \to 2} x][\lim_{x \to 2} x][\lim_{x \to 2} x]$$

$$= 2 \cdot (-2) \cdot (-2) \cdot (-2) \cdot (-2) = 32$$

EXAMPLE 2

Determine, if it exists,

$$\lim_{x \to 2} \frac{5x^3}{3x - 2}$$

As we have just observed, both the numerator and denominator components have limits as x approaches 2, and the limit of the denominator component is not zero. Thus, from the quotient theorem, we have

$$\operatorname*{limit}_{x\to 2} \frac{5x^3}{3x-2} = \frac{\operatorname*{limit}_{x\to 2} 5x^3}{\operatorname*{limit}_{x\to 2}(3x-2)} = \frac{40}{4} = 10$$

EXAMPLE 3

Determine, if it exists,

$$\operatorname*{limit}_{x\to 3} (2x^2 - 2x - 12).$$

All three components of $f(x) = 2x^2 - 2x - 12$ have limits as x approaches 3. We have

$$\begin{aligned}\operatorname*{limit}_{x\to 3} (2x^2 - 2x - 12) &= \operatorname*{limit}_{x\to 3} 2x^2 - \operatorname*{limit}_{x\to 3} 2x - \operatorname*{limit}_{x\to 3} 12 \\ &= 18 - 6 - 12 = 0\end{aligned}$$

Very Important: Watch the Hypothesis

If the hypothesis of a theorem is not satisfied, no conclusion can be drawn. This is particularly important for us with the quotient theorem. Example 4 is a good illustration of this point.

EXAMPLE 4

Determine, if it exists,

$$\operatorname*{limit}_{x\to 3} \frac{2x^2 - 2x - 12}{x-3}$$

Both the numerator and denominator functions have limits as x approaches 3. But the limit of the denominator function, $g(x) = x - 3$, is 0, which makes the quotient theorem inapplicable.

A theorem is inapplicable when its hypothesis is not satisfied. When this happens, the watchword is "caution," since there are a number of possible erroneous steps that one might take. When a theorem is inapplicable, we must turn elsewhere, in this case back to the definition of limit, to analyze the situation. No hasty conclusions should be drawn. Here the problem can be settled by simplifying the quotient as follows:

$$\frac{2x^2 - 2x - 12}{x - 3} = \frac{(x-3)(2x+4)}{x-3} = 2x + 4$$

where $x \neq 3$. Thus we have

$$\underset{x \to 3}{\text{limit}} \frac{2x^2 - 2x - 12}{x - 3} = \underset{x \to 3}{\text{limit}}(2x + 4) = 10$$

One popular brand of nonsense is based on the following reasoning:

$$\underset{x \to 3}{\text{limit}} \frac{2x^2 - 2x - 12}{x - 3} = \frac{\underset{x \to 3}{\text{limit}}(2x^2 - 2x - 12)}{\underset{x \to 3}{\text{limit}}(x - 3)} = \frac{0}{0}$$

which is undefined since division by zero is undefined. This argument recognizes that division by zero is undefined, but is attempting to apply the quotient theorem where it is inapplicable, ignoring the hypothesis which requires that the limit of the denominator component as x approaches 3 exist and be nonzero.

Another school of nonsense reasons as follows:

$$\underset{x \to 3}{\text{limit}} \frac{2x^2 - 2x - 12}{x - 3} = \frac{\underset{x \to 3}{\text{limit}}(2x^2 - 2x - 12)}{\underset{x \to 3}{\text{limit}}(x - 3)} = \frac{0}{0} = 1$$

This brand of nonsense not only ignores hypotheses, but also believes in division by zero.

EXAMPLE 5

If $f(x)=x^2$ and $g(h)=\dfrac{f(2+h)-f(2)}{h}$, determine $\operatorname*{limit}_{h\to 0} g(h)$.

Our problem has essentially two parts. Part 1 is algebraic: determine and simplify $g(h)$. Note that $g(h)$ has a feature that makes the quotient theorem inapplicable: the denominator component goes to 0 as h approaches 0. Thus the simplification of $g(h)$ is a must. Part 2 involves taking the limit of our simplified $g(h)$ as h approaches 0.

Part 1. Construct and simplify $g(h)$.

$$f(2+h)=(2+h)^2=4+4h+h^2$$
$$f(2)=2^2=4$$

Subtracting yields

$$f(2+h)-f(2=(4+4h+h^2)-4=4h+h^2$$

Dividing by h yields $g(h)$:

$$g(h)=\frac{f(2+h)-f(2)}{h}=\frac{4h+h^2}{h}=\frac{h(4+h)}{h}=4+h$$

where $h\neq 0$.

Part 2. Find $\operatorname*{limit}_{h\to 0} g(h)$.

$$\operatorname*{limit}_{h\to 0} g(h)=\operatorname*{limit}_{h\to 0}(4+h)=4$$

The initial reaction of some people to this sort of limit problem is, "what a peculiar structure," and of course they are quite right. But this structure is of fundamental importance, and we shall have much to say about it in Chapter 5. For us at this point, it serves as a source of examples and exercises on limits.

EXAMPLE 6

If $f(x)=2x^2+1$ and $g(h)=\dfrac{f(x+h)-f(x)}{h}$, determine $\operatorname*{limit}_{h\to 0} g(h)$.

Part 1. Construct and simplify $g(h)$.

$$\begin{aligned} f(x+h) &= 2(x+h)^2+1 \\ &= 2(x^2+2xh+h^2)+1 \\ &= 2x^2+4xh+2h^2+1 \\ f(x) &= 2x^2+1 \end{aligned}$$

Subtracting yields

$$\begin{aligned} f(x+h)-f(x) &= (2x^2+4xh+2h^2+1)-(2x^2+1) \\ &= 4xh+2h^2 \end{aligned}$$

Dividing by h yields $g(h)$:

$$\begin{aligned} g(h)=\frac{f(x+h)-f(x)}{h} &= \frac{4xh+2h^2}{h} \\ &= \frac{h(4x+2h)}{h} \\ &= 4x+2h, \end{aligned}$$

where $h \neq 0$.

Part 2. Find $\operatorname*{limit}_{h\to 0} g(h)$.

$$\operatorname*{limit}_{h\to 0} g(h) = \operatorname*{limit}_{h\to 0}(4x+2h) = 4x$$

EXERCISES

Determine the limits of the following functions.

1. $\operatorname*{limit}_{x\to 4} 5x^2$
2. $\operatorname*{limit}_{x\to -2} 3x^4$
3. $\operatorname*{limit}_{x\to 2}(3x^2+2)$
4. $\operatorname*{limit}_{x\to -1}(4x^3+3x)$
5. $\operatorname*{limit}_{x\to 3}(5x^3-3x^2)$
6. $\operatorname*{limit}_{x\to 3}\left(2x^3+\frac{2}{x}\right)$
7. $\operatorname*{limit}_{x\to 3}(3x^2+2x-4)$
8. $\operatorname*{limit}_{x\to 2}(5x^3-3x^2+2)$
9. $\operatorname*{limit}_{x\to 4}(x^2+2)(3x-1)$
10. $\operatorname*{limit}_{x\to -2}(4x^2-3)(5x-4)$
11. $\operatorname*{limit}_{x\to 2}\frac{3x^2-3}{3x+2}$
12. $\operatorname*{limit}_{x\to -1}\frac{4x^3-3x}{2x^2+1}$
13. $\operatorname*{limit}_{x\to 3}\frac{x^2+x-12}{x-3}$
14. $\operatorname*{limit}_{x\to -3}\frac{2x^3+4x-3}{2x^2-3x+4}$
15. $\operatorname*{limit}_{x\to -1}(3x^4+2x)(5x+1)$
16. $\operatorname*{limit}_{x\to -4}\frac{3x^2+10x-8}{x+4}$
17. $\operatorname*{limit}_{x\to -2}(3x^4-2x)(3x^2-5)$
18. $\operatorname*{limit}_{x\to -3}\frac{3x^2+2x-7}{4x^2-3x+2}$

In Exercises 19-24, determine $\operatorname*{limit}_{h\to 0} g(h)$, *where g(h) is defined as indicated.*

19. $g(h)=\frac{f(3+h)-f(3)}{h}$ and $f(x)=x^2$

20. $g(h)=\dfrac{f(4+h)-f(4)}{h}$ and $f(x)=2x+1$

21. $g(h)=\dfrac{f(1+h)-f(1)}{h}$ and $f(x)=x^2+3x$

22. $g(h)=\dfrac{f(x+h)-f(x)}{h}$ and $f(x)=x^2$

23. $g(h)=\dfrac{f(x+h)-f(x)}{h}$ and $f(x)=3x+6$

24. $g(h)=\dfrac{f(x+h)-f(x)}{h}$ and $f(x)=2x^2-1$

3.3. CONTINUITY

We have seen that the limit behavior of a function $y = f(x)$ as x approaches a is an entirely different concern from $f(a)$. $f(a)$ describes the behavior of $y = f(x)$ at a, whereas $\underset{x\to a}{\text{limit}}\, f(x)$ is concerned with the behavior of $f(x)$ as x gets closer and closer to a. Sometimes these two different concerns lead to the same numerical value, and this situation forms the basis for the concept of continuity of a function.

> **Definition.** If $y = f(x)$ is a function such that
>
> 1. $f(a)$ is defined,
> 2. $\underset{x\to a}{\text{limit}}\, f(x)$ exists,
> 3. $\underset{x\to a}{\text{limit}}\, f(x) = f(a)$,
>
> **then $y = f(x)$ is said to be continuous at a.**

EXAMPLE 1

Determine if $f(x) = 3x - 2$ is continuous at 2.

All we have to do is examine one by one the requirements for continuity at 2.

1. $f(2) = 3(2) - 2 = 4$, so that condition (1) is satisfied.
2. $\lim_{x \to 2} (3x - 2) = 4$. As x approaches 2, $3x - 2$ approaches 4, so that condition (2) is satisfied.
3. Since the values obtained from conditions (1) and (2) are the same, condition (3) is satisfied and $f(x) = 3x - 2$ is continuous at 2.

The graph of $f(x) = 3x - 2$ is shown in Figure 3.7.

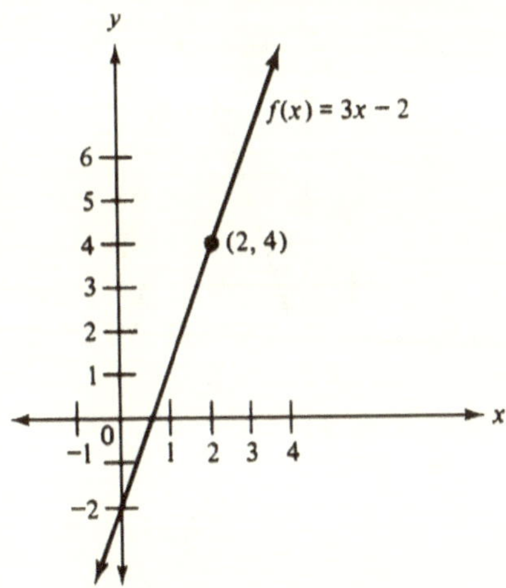

Figure 3.7

EXAMPLE 2

Determine if

$$g(x) = \begin{cases} 3x - 2, & \text{for } x \neq 2 \\ 5, & \text{for } x = 2 \end{cases}$$

is continuous at 2.

1. $g(2) = 5$, so that condition (1) is satisfied.
2. $\operatorname*{limit}_{x \to 2} g(x) = 4$. As x approaches 2, $g(x)$ approaches 4 (see Section 3.1, Example 2, p. 87), so condition (2) is satisfied.
3. $\operatorname*{limit}_{x \to 2} g(x) \neq g(2)$ since conditions (1) and (2) yield different values, 5 and 4, respectively. Thus $g(x)$ is not continuous at 2.

The graph of $g(x)$ is shown in Figure 3.8 (also see Section 2.2, Example 6). Since (2, 4) is not on the graph, it is shown by an open circle. (2, 5) is on the graph and is shown by a heavy dot.

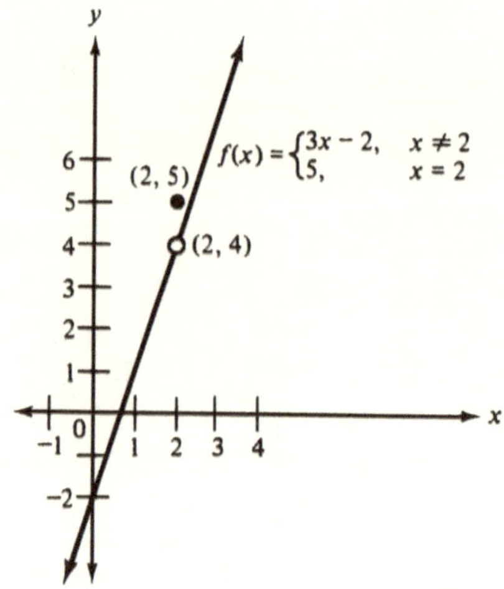

Figure 3.8

EXAMPLE 3

Determine if

$$f(x) = \begin{cases} 1, & \text{for } x \geq 2 \\ -1, & \text{for } x < 2 \end{cases}$$

is continuous at 2.

1. $f(2) = 1$, so that condition (1) is satisfied.
2. $\lim\limits_{x \to 2} f(x)$ does not exist. As x approaches 2, $f(x)$ is unstable since it clusters about two values, -1 and 1 (see Section 3.1, Example 5, p. 89). Since condition (2) is not satisfied, $f(x)$ is not continuous at 2.

The graph of $f(x)$ is shown in Figure 3.9 (also see Section 2.2, Example 7).

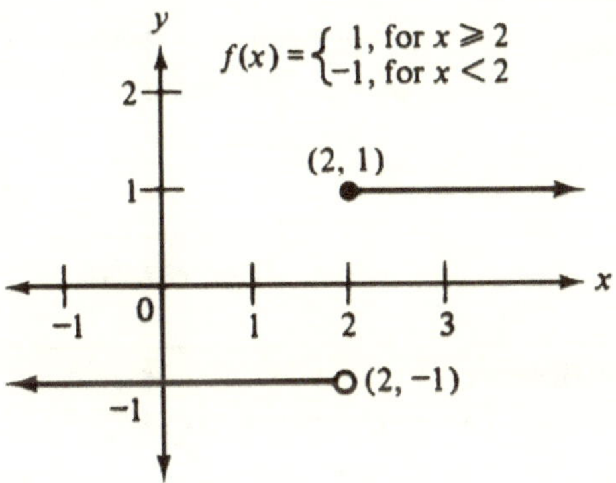

Figure 3.9

EXAMPLE 4

Determine if

$$f(x) = \begin{cases} x+1, & \text{for } x \geq 0 \\ -x+1, & \text{for } x < 0 \end{cases}$$

is continuous at 0.

1. $f(0) = 1$, so that condition (1) is satisfied.
2. $\lim\limits_{x \to 0} f(x) = 1$. As x approaches 0, $f(x)$ approaches 1 (see Section 3.1, Example 7, p. 91), so that condition (2) is satisfied.
3. Since the values obtained from conditions (1) and (2) are the same, condition (3) is satisfied and $f(x)$ is continuous at 0.

The graph of $f(x)$ is shown in Figure 3.10.

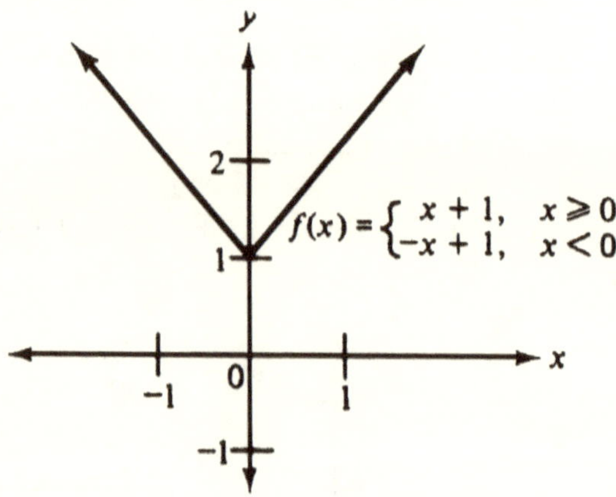

Figure 3.10

> **Definition.** A function $y = f(x)$ is said to be **continuous over an interval** I if it is continuous at each value a of I; $y = f(x)$ is said to be **continuous over the set of real numbers if it is continuous at each real-number a.**

As the term continuity itself suggests and the graphs of the functions considered indicate, the continuity of a function is intimately connected with the nature of the graph of the function. The graph of a function that is continuous over an interval I, or the set of all real-numbers R, is in one piece over I, or R; the graph has no gaps, jumps, punctures, or breaks.

EXERCISES

Determine if the following functions are continuous at the given values. Explain the basis for your conclusion in each case.

1. $f(x) = x^2 + 3$ at 2
2. $f(x) = 3x^2 + 2x - 1$ at 3
3. $f(x) = \dfrac{x^2+1}{x-1}$ at 2
4. $f(x) = \dfrac{x^2-4}{x-2}$ at 2

5. $f(x)=\begin{cases} 1+x, & \text{for } x \geq 0 \\ 1-x, & \text{for } x < 0 \end{cases}$ at 0

6. $f(x)=\begin{cases} 1, & \text{for } x \geq 0 \\ -1, & \text{for } x < 0 \end{cases}$ at 0

7. $f(x)=\begin{cases} 3x+2, & \text{for } x \geq 1 \\ 5x, & \text{for } x < 1 \end{cases}$ at 1

8. $f(x)=(3x-2(4x^2+1)$ at 3

9. $f(x)=\begin{cases} \frac{1}{x}, & \text{for } x \neq 0 \\ 2, & \text{for } x = 0 \end{cases}$ at 0

10. $f(x)=\begin{cases} 2x-3, & \text{for } x \geq 2 \\ 3x-1, & \text{for } x < 2 \end{cases}$

3.4. LIMITS OF FUNCTIONS: PART 2

Our initial contact with limits in Section 3.1 focused on the behavior of a function $y = f(x)$ as x takes on values closer and closer to a prescribed value a. If $f(x)$ in turn approaches one value L, then we say that the limit of $f(x)$ as x approaches a as L, and write

$$\operatorname*{limit}_{x \to a} f(x) = L$$

But x can be manipulated in other ways of interest, and $f(x)$ can exhibit other consistent forms of behavior. It is to such considerations that we now turn our attention.

First let us turn to the function $f(x)=1/(x-1)^2$ and examine its behavior as x approaches 1. Values of $f(x)=1/(x-1)^2$ for values of x that are getting closer and closer to 1 are shown in Table 3.6, from which we see that, as x takes on values closer and closer to 1, $f(x)=1/(x-1)^2$ reacts by becoming larger and larger without bound.

Table 3.6

x	$1+\frac{1}{10}$	$1-\frac{1}{10}$	$1+\frac{1}{100}$	$1-\frac{1}{100}$	$1+\frac{1}{1000}$	$1-\frac{1}{1000}$
$f(x)$	100	100	10,000	10,000	1,000,000	1,000,000

Figure 3.11 shows the situation in geometric terms (also see Section 2.2, Example 8). We describe this consistency in behavior by saying $f(x)=1/(x-1)^2$ approaches infinity (written ∞) as x approaches 1, and write

$$\frac{1}{(x-1)^2}\to\infty \text{ as } x\to 1 \text{ or } \underset{x\to 1}{\text{limit}}\frac{1}{(x-1)^2}=\infty$$

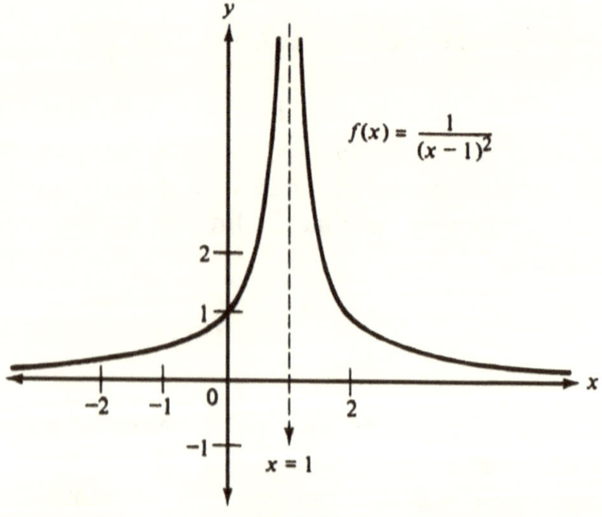

Figure 3.11

More generally, if as x takes on values closer and closer to a value a, $f(x)$ gets larger and larger without bound, we say that $f(x)$ approaches infinity as x approaches a and write

$$f(x)\to\infty \text{ as } x\to a \text{ or } \underset{x\to a}{\text{limit}}\, f(x)=\infty$$

Infinity is not a number, and thus the symbol ∞ does not stand for a number in the real-number system. It is a term that has no meaning when taken out of context. A statement such as "$f(x)$ approaches infinity" [expressed in symbols by $f(x) \to \infty$] must be viewed "in total"; it is another way of saying that $f(x)$ is increasing without bound.

We next turn to the function $f(x) = 1/x$, whose graph is shown in Figure 3.12, and examine its behavior as x increases without bound $(x \to \infty)$. As $x \to \infty$ (x = 10, 100, 1000, 10,000, and so on), $f(x) = 1/x$ gets closer and closer to 0 $[f(x) = \frac{1}{10}, \frac{1}{100}, \frac{1}{1000}, \frac{1}{10{,}000}$, and so on]. We say that $f(x) = 1/x$ approaches 0 as x approaches ∞, and write

$$\frac{1}{x} \to 0 \text{ as } x \to \infty \quad \text{or} \quad \operatorname*{limit}_{x \to \infty} \frac{1}{x} = 0$$

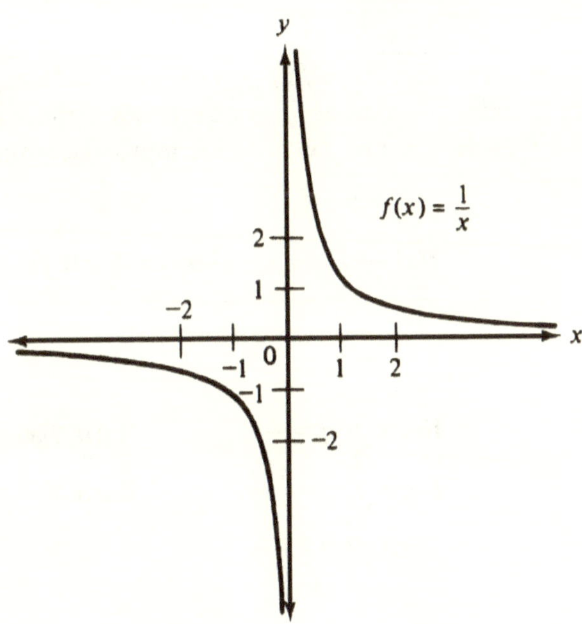

Figure 3.12

More generally, if $y = f(x)$ gets closer and closer to a number L, or remains constant at L, as x increases without bound, then we say that the limits of $f(x)$ is L as x approaches infinity. We write

$$\boldsymbol{f(x) \to L \text{ as } x \to \infty \text{ or } \lim_{x\to\infty} f(x) = L}$$

If x decreases without bound ($x = -10, -100, -1000, -10,000$, and so on), we say that x approaches minus infinity, and write $x \to -\infty$. It is often useful to study the behavior of a function as $x \to -\infty$. Let us observe, for example, that as $x \to -\infty$ ($x = -10, -100, -1000$, and so on), $f(x) = 1/x$ approaches 0 [$f(x) = -\frac{1}{10}, -\frac{1}{100}, -\frac{1}{1000}$, and so on]. We write

$$\frac{1}{x} \to 0 \text{ as } x \to -\infty \text{ or } \lim_{x\to-\infty} \frac{1}{x} = 0$$

More generally, if $y = f(x)$ gets closer and closer to a number L, or remains constant at L, as x decreases without bound, then we say that the limit of $f(x)$ is L as x approaches minus infinity. We write

$$\boldsymbol{f(x) \to L \text{ as } x \to -\infty \text{ or } \lim_{x\to-\infty} f(x) = L}$$

The concepts

$$\lim_{x\to a} f(x) = -\infty \qquad \lim_{x\to-\infty} f(x) = \infty$$

$$\lim_{x\to\infty} f(x) = \infty \qquad \lim_{x\to-\infty} f(x) = -\infty$$

$$\lim_{x\to\infty} f(x) = -\infty$$

are defined in a way that is suggested by the previous definitions.

EXAMPLE 1

Determine the behavior of

$$f(x)=4-\frac{3}{x} \text{ as } x \to \infty$$

From Table 3.7 it is clear that as x takes on values that increase without bound, $3/x$ approaches 0, and $f(x)=4-(3/x)$ thus approaches 4. Therefore, we have

$$\operatorname*{limit}_{x\to\infty}\left(4-\frac{3}{x}\right)=4$$

Table 3.7

x	10	100	1000	10,000
$f(x)$	$4-\frac{3}{10}$	$4-\frac{3}{100}$	$4-\frac{3}{1000}$	$4-\frac{3}{10{,}000}$

EXAMPLE 2

Determine the behavior of

$$f(x)=1-\frac{1}{x-3} \text{ as } x \to 3$$

From Table 3.8 it is clear that $f(x)$ is unstable as $x \to 3$. As x takes on values greater than 3 that are approaching 3, $f(x)$ decreases without bound [$f(x)=-9,-99,-999$, and so on]; as x takes on values less than 3 that are approaching 3, $f(x)$ increases without bound [$f(x)$ = 11, 101, 1001, and so on].

Table 3.8

x	$3+\frac{1}{10}$	$3-\frac{1}{10}$	$3+\frac{1}{100}$	$3-\frac{1}{100}$	$3+\frac{1}{1000}$	$3-\frac{1}{1000}$
$f(x)$	−9	11	−99	101	−999	1001

Thus

$$\lim_{x\to 3}\left(1-\frac{1}{x-3}\right) \text{ does not exist.}$$

For a geometric view, see Figure 3.13.

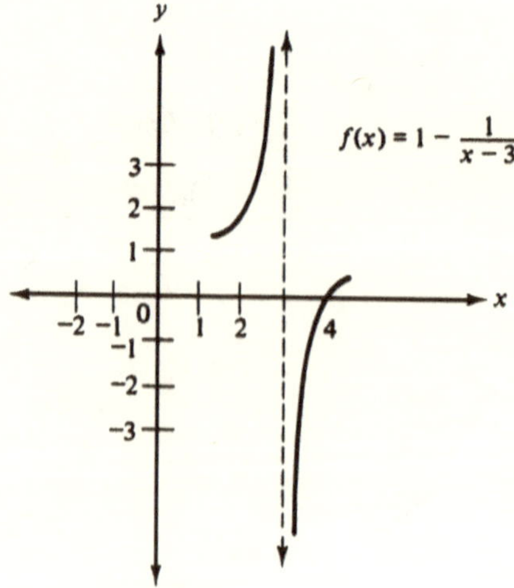

Figure 3.13

EXAMPLE 3

Examine the behavior of

$$f(n)=\left(1+\frac{1}{n}\right)^n \quad \text{as } n\to\infty$$

From Table 3.9 it would seem that $f(n) = [1 + (1/n)]^n$ approaches some value

Table 3.9

n	1	4	10	100	1000	10,000
$\left(1+\frac{1}{n}\right)^n$	2	2.441	2.594	2.705	2.717	2.718

in the neighborhood of 2.718 as $n \to \infty$. This is indeed the case. The limit of $[1 + (1/n)]^n$ exists as $n \to \infty$, and is an irrational number that is denoted by the letter e. Correct to five decimal places, $e \simeq 2.71828$. Thus we have

$$\operatorname*{limit}_{n\to\infty}\left(1+\frac{1}{n}\right)^n = e$$

Exponential functions based on e are discussed in the next section.

It is useful to note a corollary of what we have observed: if

$$g(n) = \left(1+\frac{r}{n}\right)^n$$

where r is any real number, then

$$\operatorname*{limit}_{n\to\infty}\left(1+\frac{r}{n}\right)^n = e^r$$

Thus, for example,

$$\operatorname*{limit}_{n\to\infty}\left(1+\frac{(-2)}{n}\right)^n = e^{-2}$$

$$\operatorname*{limit}_{n\to\infty}\left(1+\frac{0.05}{n}\right)^n = e^{0.05}$$

EXAMPLE 4

Determine the limit behavior of

$$f(x) = \frac{3x^2 + 4x + 1}{4x^2 + x + 3} \text{ as } x \to \infty$$

Both the numerator and denominator components of $f(x)$ are increasing without bound as $x \to \infty$, and this obscures the behavior of their ratio. This difficulty can be overcome by dividing both numerator and denominator by x^2, the term of common highest power. We obtain

$$\underset{x\to\infty}{\text{limit}} \frac{3x^2 + 4x + 1}{4x^2 + x + 3} = \underset{x\to\infty}{\text{limit}} \frac{3 + \frac{4}{x} + \frac{1}{x^2}}{4 + \frac{1}{x} + \frac{3}{x^2}} = \frac{3}{4}$$

since $4/x \to 0$, $1/x^2 \to 0$, $1/x \to 0$, and $3/x^2 \to 0$ as $x \to \infty$.

EXERCISES

Determine the limit behavior of the following functions.

1. $f(x) = 5 + \frac{2}{x}$ as $x \to \infty$
2. $f(x) = -2 - \frac{4}{x}$ as $x \to -\infty$
3. $f(x) = 2 + \frac{1}{x^2}$ as $x \to 0$
4. $f(x) = \frac{1}{x}$ as $x \to 0$
5. $f(x) = \frac{1}{(x-2)^2}$ as $x \to 2$
6. $f(x) = \frac{3}{x-1}$ as $x \to 1$
7. $f(x) = \frac{4x^2 + 1}{x^2 + 2x}$ as $x \to \infty$
8. $f(x) = 2 - \frac{1}{x^2}$ as $x \to 0$

9. $f(x)=\left(1+\frac{4}{x}\right)^{x}$ as $x\to\infty$

10. $f(x)=6+\frac{2}{(x-4)^2}$ as $x\to 4$

11. $f(x)=1-\frac{1}{x^3}$ as $x\to 0$

12. $f(x)=\frac{x^2+3x-2}{3x^2+2x-1}$ as $x\to\infty$

13. $f(x)=5-\frac{2}{x+2}$ as $x\to -2$

14. $f(x)=4+\frac{3}{(x+1)^2}$ as $x\to -1$

15. $f(x)=\frac{2x^3+4x-1}{5x^3-3x^2+2}$ as $x\to -\infty$

16. $f(x)=\frac{x}{x-5}$ as $x\to 5$

17. $f(x)=\frac{1}{x}+2\left(1+\frac{1}{x}\right)^{x}$ as $x\to\infty$

18. $f(x)=5+\frac{2}{x}$ as $x\to -\infty$

19. $f(x)=\frac{3}{x}+\frac{4}{(x-5)^2}$ as $x\to 5$

20. $f(x)=3\left(1-\frac{5}{x}\right)^{x}$ as $x\to\infty$

21. Determine $\underset{L\to\infty}{\text{limit}}\left[1-\left(1-\frac{2}{L}\right)^{KL}\right]$, were K is constant. This problem arises in connection with Robert Feinschreiber's discussion "Accelerated Depreciation: A Proposed New Method," *Journal of Accounting Research* (Spring 1969), pp. 17-21.

CHAPTER 4

An Application of Limits to Finance*

In June 1972 a number of banks proudly announced that as of July 1 interest on savings accounts would be compounded continuously. The competitive advantage that they hoped to achieve was soon blunted by reality. Continuous compounding of interest sounded good, but bank customers didn't understand it and bank officers didn't have a clue either. Soon thereafter it was abandoned for time savings accounts, but it remains an important concept in economics and finance.

4.1. THE MATHEMATICS OF FINANCE

Money Growth Basics

In the world of finance **interest** is simply a fee charged for the use of borrowed money. It is an amount which is stated in terms of some monetary unit (dollars, cents, pounds, francs etc.) Thus, if Mr. Debtor borrows \$100 from Mr. Creditor with the understanding that \$110 is to be repaid one month later, the interest charged Mr. Debtor is \$10. The time period at the end of which the amount borrowed, called the **principal,** and the interest owed are to be paid is called the **interest period**. In the Debtor-

* Continuous compounding of interest is a prerequisite for Sections 7.3 and 12.4 - 12.7.

Creditor transaction the principal is $100 and the interest period is one month. The sum of the principal and interest to be paid at the end of the interest period is called the **amount**. In the Debtor-Creditor transaction the amount is $110.

The ratio of the interest charged during the interest period to the amount of money owed at the beginning of the interest period is called the **interest rate**. The interest rate in the Debtor-Creditor transaction is:

$$\frac{10}{100} = 0.10 \textit{ or } 10\% \textit{ per month}$$

Although it is incorrect to do so, in colloquial language the terms interest and interest rate are used interchangeably.

More generally, let us observe that if P is the principal borrowed and I is the interest to be paid for the underlying interest period, the interest rate i is defined by:

$$i = \frac{I}{P}$$

The interest I for the interest period is given by:

$$I = Pi$$

(Interest) = (Principal) X (rate for the period)

If Mr. Debtor sought to extend his loan at the end of the month when it came due, Mr. Creditor might well argue that in extending the loan another month it would be appropriate to charge interest on the $110 now owed. Ten percent interest on $110 is $11 so that the total interest for the two months would be $10 + $11 = $21. The total amount would be $100 + $21 = $121.

In general, if the interest due is added to the principal at stated intervals of time and itself earns interest thereafter, the sum by which the original principal has been increased at the end of any time is called compound interest. The time interval between successive additions of interest to principal is called the interest period or conversion period. At the end of each conversion period the new principal, consisting of the original principal plus the compound interest, is called the compound amount. In the Debtor-Creditor situation, considered over two months, the compound interest is \$21, the interest or conversion period is a month, and the compound amount is \$121.

To obtain an equation for the compound amount let us suppose that an initial principal P is invested at compound interest at a rate i per interest period. Then at the end of the first interest period the compound amount is:

$$A_1 = P + Pi = P(1 + i)$$

At the end of the second interest period the new amount A_2 is A_1 plus the interest $A_1 i$ obtained from A_1.

$$A_2 = A_1 + A_1 i = A_1(1 + i) = P(1 + i)(1 + i)$$
$$A_2 = P(1 + i)^2$$

At the end of the third interest period the new amount A_3 is A_2 plus the interest $A_2 i$ obtained from A_2.

$$A_3 = A_2 + A_2 i = A_2(1 + i) = P(1 + i)^2(1 + i)$$
$$A_3 = P(1 + i)^3$$

More generally, at the end of the nth interest period the new amount A_n is given by

$$A_n = P(1 + i)^n, \tag{4.1}$$

where P is the principal initially invested, i is the interest rate for the interest period, and n is the number of interest periods.

EXAMPLE 1

\$1000 is to be invested 6 months from now at the rate of 3% per a 6 month period. What is the mount on deposit 24 months from now?

The interest period is 6 months, P = \$1000, i = 0.03, and n = 3. The problem is to find A_3. We have:

$$A_3 = 1000(1.03)^3 = 1000(1.09273)$$

$$A_3 = \$1092.73$$

Values of $(1 + i)^n$ are given in Appendix Table 1, from which we obtain $(1.03)^3 = 1.09273$.

In the world of finance an interest rate cited as 6 percent per annum compounded twice a year (semi-annually) means that 3 percent interest is added every 6 months to the amount accumulated. More generally, the rate r **per annum compounded *m* times a year** envisions the year divided into m interest periods of equal length, with interest at the rate $i = r/m$ being added to the amount accumulated at the end of each period.

Replacing i by r/m in (4.1) yields

$$A = P\left(1 + \frac{r}{m}\right)^n$$

as the compound amount after n interest periods. To express n in terms of years let us first note that the number of interest periods in 1 year is m; thus the number of interest periods n in x years, let us say, is $n = mx$. Thus

$$A = P\left(1 + \frac{r}{m}\right)^{mx} \qquad (4.2)$$

expresses the **compound amount after *x* years** when principal P is invested at the rate r per annum compounded m times a year. x can take on nonnegative integer values (x = 0,1,2, etc.) and fractional values consistent with m ($x = 1/m$, $2m$, etc.)

EXAMPLE 2

Arthur Bryan plans to invest \$1000 at 8% per annum compounded quarterly. What will he have on deposit in 5 years?

The problem, of course, is to recognize what we want and what we have. We want A. We have: $P = \$1000$, $r = 0.08$, $m = 4$, so that $r/m = 0.02$; also $x = 5$, so that $mx = 20$. From (4.2) we obtain:

$$A = 1000(1.02)^{20} = 1000(1.48595)$$
$$A = \$1485.95$$

The situation considered so far involves investing a certain amount P, NOW, and determining what it's worth, A, LATER. Situations also arise where we envision the need for a certain amount of money A, LATER, and wish to find the principal P that we should invest NOW to obtain A later. For instance:

EXAMPLE 3

Piedmont Car Service envisions a need for \$12,000 in 4 years to replace one of its cars. What principal should be invested at 9% per annum compounded 3 times a year if \$12,000 is to available in 3 years.

The problem is to find P. We have $A = \$12{,}000$; $r = 0.09$, $m = 3$, so that $r/m = 0.03$. We have $x = 3$, so that $mx = 9$. Thus:

$$12{,}000 = P(1.03)^9$$

Dividing both sides by $(1.03)^9$ gives us:

$$P = \frac{12{,}000}{(1.03)^9}$$

Let us recall that negative exponents are defined by $a^{-n} = 1/a^n$. Thus, for example, $2^{-3} = 1/2^3 = 1/8$. In our situation we have:

$$\frac{1}{(1.03)^9} = (1.03)^{-9}$$

Thus, we may express P in the form:

$$P = 12{,}000(1.03)^{-9}$$

From Appendix Table 2 we have:

$$(1.03)^{-9} = 0.76642$$

Thus:

$$P = 12{,}000(0.76642)$$
$$P = \$9197.04$$

Piedmont Car Service should invest \$9197.04 at 9% per annum compounded 3 times a year to obtain \$12,000 for a car replacement in 3 years. \$9197.04 is called the present value of \$12,000 with respect to the cited conditions. It expresses the present worth of \$12,000 to be realized 3 years from now.

More generally, solving

$$A = P\left(1 + \frac{r}{m}\right)^{mx} \tag{4.2}$$

by dividing both sides of (4.2) by $(1 + r/m)^{mx}$ yields:

$$P = \frac{A}{\left(1 + \frac{r}{m}\right)^{mx}} = A\left(1 + \frac{r}{m}\right)^{-mx} \tag{4.3}$$

In this setting P is called the **present value** of A. \$$P$ is the sum of money that must be invested now at the rate r per annum compounded m times a year if $\$A$ is to realized x years from now. $\$P$ is what $\$A$, x years in the future, is worth today under the cited interest conditions.

EXERCISES

1. Helen Needy borrowed \$500 from Paula Plenty. The loan was for one year and the interest was \$50.

   (a) What is the interest rate?
   (b) When the loan came due Helen asked Paula to extend the loan for a year at the same rate. State (i) the interest period, (ii) compound interest, (iii) compound amount.

2. James Voss is planning to invest \$2000. How much will he have on deposit at the end of 3 years if this sum of money is invested at 12% per annum compounded (a) 3 times a year? (b) 4 times a year? (c) 6 times a year?
3. Amy Allen has succeeded in saving \$3000 which she now plans to reinvest. How much will she have on deposit at the end of 4 years if she invests this amount at 8% per annum compounded (a) semi-annually? (b) quarterly?
4. The Andrius Company expects that \$5000 will be needed in 5 years to meet the cost of equipment replacement. How much should be initially invested at 12% per annum compounded quarterly to meet this expense?
5. The Lee family wants to have \$8000 available in 6 years for the education of their daughter Michelle. How much should be initially invested at 8% per annum compounded quarterly to meet this expense?

Nominal and Effective Interest Rates

For a given nominal rate r per annum compounded m times a year, the corresponding effective rate v is the rate which compounded annually yields the same interest at the end of the year. The amount yielded by effective rate v at the end of the year is $P[1 + (r/m)]^m$. Since the amounts attained at the end of the year are the same under both arrangements, we have

$$P(1+v) = P\left(1+\frac{r}{m}\right)^m$$

Dividing both sides by P yields

$$1+v = \left(1+\frac{r}{m}\right)^m$$

Thus

$$v = \left(1+\frac{r}{m}\right)^m - 1$$

EXAMPLE 4

Find the effective rate corresponding to a nominal rate of 12 percent per annum compounded six times a year.

$r = 0.12$ and $m = 6$. Thus

$$v = (1.02)^6 - 1$$

From Appendix Table 1, $(1.02)^6 = 1.12616 \simeq 1.126$. Thus

$$v = 1.126 - 1 = 0.126$$

Therefore, 12.6 percent compounded annually yields the same interest at the end of the year as 12 percent per annum compounded six times a year. \$1000 invested at 12 percent per annum compounded six times a year yields a return of $(1000)(0.126) = \$126$ at the end of the year.

EXERCISES

6. Determine the effective rate corresponding to a nominal rate of 8% per annum compounded 8 times a year.
7. Determine the effective rate of 12% per annum compounded 4 times a year.
8. Richard Lawson was offered two investment opportunities: 12% per annum compounded 3 times a year and 11% per annum compounded 11 times a year. Which one, Richard wondered, yields the higher return and by how much.

4.2. CONTINUOUS COMPOUNDING OF INTEREST

The point of view underlying interest relationships considered in the preceding section is that of interest accumulation taking place at the end of a time period. The time period might be a year, a month, a day, a second or even something smaller. Whether small or large, they are all time periods.

But financial institutions may be likened to biological processes and viewed as having a continuous unbroken existence. From such a point of view, it is more natural to regard interest accumulation as a process that is taking place at every moment of time, and not just at the end of certain agreed-to time periods. The problem is to express this point of view in appropriate quantitative terms. To do so, let us note that as m increases without bound ($m \to \infty$) the length of each of the m time periods into which the year is divided approaches zero. This suggests the idea of studying the limit behavior of various interest relationships as $m \to \infty$ as an approach to expressing in precise terms the idea of interest accumulation occurring at each instant of time. Interest accumulation that takes place at each instant of time is called **continuous compounding of interest**.

To begin, let us recall that for the nominal rate r per annum compounded m times a year the corresponding **effective rate v** is that rate which compounded annually yields the same interest at the end of the year. In terms of r and m, v is expressed by

$$v = \left(1 + \frac{r}{m}\right)^m - 1$$

From Section 3.4, Example 3 (p. 112), we have that as $m \to \infty$

$$\left(1 + \frac{r}{m}\right)^m \to e^r.$$

Thus as $m \to \infty$.

$$v = \left(1 + \frac{r}{m}\right)^m - 1 \to e^r - 1$$

> The value
>
> $$e^r - 1$$
>
> is defined as the **effective interest rate corresponding to the nominal rate *r* per annum compounded continuously**. By definition, the rate r per annum compounded continuously will yield the same interest at the end of the year as the effective interest rate $e^r - 1$ compounded annually.

Thus, for example, the rate 7 percent per annum compounded continuously will yield the same interest at the end of the year as the effective rate $e^{0.07} - 1 = 1.07250 - 1 = 7.25$ percent compounded annually. $1000 invested at 7 percent per annum compounded continuously will bring a return of ($1000)(0.7250) = $72.50 at the end of the year.

Asset-Value Functions

From the earlier discussion in this section we have that $1000 invested at 8 percent per annum compounded m times a year accumulates to the asset value

$$1000\left(1+\frac{0.08}{m}\right)^{mx}=1000\left[\left(1+\frac{0.08}{m}\right)^{m}\right]^{x}$$

dollars in x years. As $m\to\infty$,

$$\left(1+\frac{0.08}{m}\right)^{m}\to e^{0.08}$$

and

$$1000\left[\left(1+\frac{0.08}{m}\right)^{m}\right]^{x}\to 1000[e^{0.08}]^{x}=1000e^{0.08x}$$

Thus the function

$$a(x)=1000e^{0.08x}$$

expresses the asset value that \$1000 will grow to be in x years if invested at 8 percent per annum compounded continuously. In 2 years \$1000 would grow to

$$a(2) = 1000e^{0.16} = 1000(1.17351) = \$1173.51$$

More generally, an amount of money A invested at the nominal rate r per annum compounded m times a year accumulates to the asset value

$$A\left(1+\frac{r}{m}\right)^{mx}=A\left[\left(1+\frac{r}{m}\right)^{m}\right]^{x}$$

after x years. As $m\to\infty$,

$$\left(1+\frac{r}{m}\right)^m \to e^r$$

and

$$A\left[\left(1+\frac{r}{m}\right)^m\right]^x \to A(e^r)^x = Ae^{rx}$$

The function

$$a(x) = Ae^{rx}$$

expresses the **asset value** that an amount A will grow to be in x years if invested at r per annum compounded continuously.

The graph of such a function is shown in Figure 4.1.

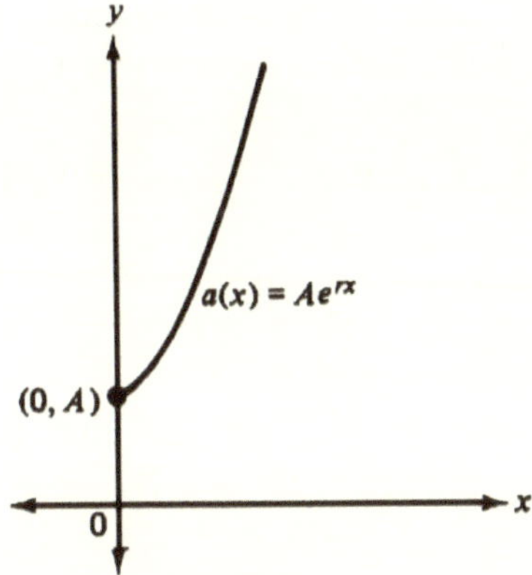

Figure 4.1

Present-Value Functions

From the earlier discussion in this section we have that the principal that must be invested at 7 percent per annum compounded m times a year in order to grow to \$5000 in x years is

$$5000\left(1+\frac{0.07}{m}\right)^{-mx} = 5000\left[\left(1+\frac{0.07}{m}\right)^{m}\right]^{-x}$$

dollars. As $m \to \infty$, $\left(1+\frac{0.07}{m}\right)^{m} \to e^{0.07}$, and

$$5000\left[\left(1+\frac{0.07}{m}\right)^{m}\right]^{-x} \to 5000(e^{0.07})^{-x} = 5000e^{-0.07x}$$

Thus the function

$$b(x) = 5000e^{-0.07x}$$

expresses the amount that must be initially invested at 7 percent per annum compounded continuously if \$5000 is to be available in x years. For \$5000 to be available at the end of 2 years,

$$b(2) = 5000e^{-0.14} = 5000(0.86936) = \$436.80$$

must be initially invested.

More generally, the amount that must be invested at r per annum compounded m times a year in order to grow to an amount A in x years is

$$A\left(1+\frac{r}{m}\right)^{-mx} = A\left[\left(1+\frac{r}{m}\right)^{m}\right]^{-x}$$

As $m \to \infty$, $\left(1+\frac{r}{m}\right)^{m} \to e^{r}$, and

$$A\left[\left(1+\frac{r}{m}\right)^{m}\right]^{-x} \to A(e^{r})^{-x} = Ae^{-rx}$$

Thus the function

$$b(x) = Ae^{-rx}$$

expresses the amount $b(x)$ that must be initially invested at r per annum compounded continuously if amount A is to be available at the end of x years.

The graph of $b(x)$ is shown in Figure 4.2.

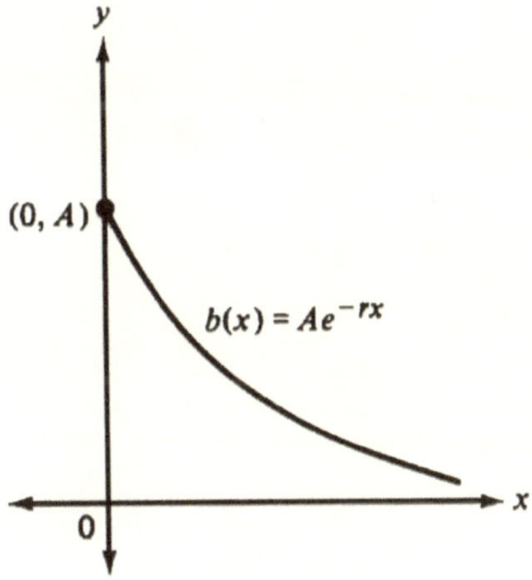

Figure 4.2

EXERCISES

1. Determine the effective rate under conditions of continuous compounding of interest for the following nominal rate r: (a) r = 6%, (b) r = 8%, (c) r = 10%, (d) r = 12%, (Values of e^{x} and e^{-x} are given in Appendix Table 3.)
2. \$8000 is invested at a rate of 8 percent per annum compounded continuously. Find the function that expresses

the amount this sum will grow to in x years. Determine the asset value of $8000 in 2 years, 2.5 years, and 5 years.

3. Find the function that describes the amount that must be initially invested at a rate of 10 percent per annum compounded continuously if $6000 is to be available at the end of x years. Find the amount that must be invested if $6000 is to be available at the end of 2 years, 3 years, 5 years, and 7 years.

SELF-TESTS FOR CHAPTERS 1-4

Allow 75 or so minutes for each self-test. Go over each test before proceeding to the next.

Self-Test 1

1. "Did you know that 0/0 = 5?" John asked his friends Jim and Burt. "It follows from the following mathematical analysis," he exclaimed.

   "Consider the ratio:

$$y = \frac{x^2 + x - 6}{x^2 - 3x + 2}$$

For $x = 2$, calculation yields $y = 0/0$. But the numerator and denominator of the preceding ratio can be simplified by factoring. This gives us:

$$y = \frac{(x-2)(x+3)}{(x-2)(x-1)}$$

Cancellation of $(x-2)$ from the numerator and denominator of the preceding allows us to write y in the following simpler form.

$$y = \frac{x+3}{x-1}$$

For $x = 2$, calculation yields $y = 5$.

Since $y = 0/0$ and 5 for $x = 2$, it follows that $0/0 = 5$.

(a) Do you agree with John's analysis? Explain.

(b) Do you agree with Jim's conclusion that 0/0 =1 and his reasoning? Explain.

(c) Do you agree with Burt's conclusion that 0/0 =0 and his reasoning? Explain?

(d) Do you agree with John's conclusion that "what this means is 5, 0, and 1 are equal"?

2. For $f(x)=\begin{cases} 3x+5, & \text{for } x<4 \\ 12, \text{ for } x=4 & \\ 5x-3, & \text{for } x>4 \end{cases}$ find, if it exists. $\lim\limits_{x\to 4} f(x)$. If this limit does not exist, so state.

3. Explain the basis for your answer to question 2.

4. Is the function $f(x)$ defined in question 2 continuous at 4? Explain the basis of your answer in terms of the definition of continuity.

5. Define e.

6. For $h(x)=\dfrac{3x^2-7x-6}{x-3}$ find, if it exists.
   $\lim\limits_{x\to 3} h(x)$. If this limit does not exist, so state.
7. Explain the basis for your answer to question 6.
8. For $h(x)$ defined in question 6 find, if it exists, $\lim\limits_{x\to 5} h(x)$. If this limit does not exist, so state.
9. Find an equation of the line with slope 3 which passes through the point P(2,5).
10. Sketch the graph of $f(x)=\begin{cases} -2x+1, & \text{for } x\le 1 \\ 3x-1, & \text{for } x>1 \end{cases}$
11. For $f(x)=\dfrac{7x^3+2x^2-4}{3x^3-x+5}$ find, if it exists,
    $\lim\limits_{x\to\infty} f(x)$. If this limit does not exist, so state.
12. Determine and simplify $g(h)=\dfrac{f(3+h)-f(3)}{h}$
    where $f(x) = 2x^2 + x$.

13. For $g(h)$ defined in question (12) find, if it exists, $\lim\limits_{h\to 0} g(h)$. If this limit does not exist, so state.
14. For $f(x)=\dfrac{4}{(x-3)^2}$ find, if it exists, $\lim\limits_{x\to 3} f(x)$ If this limit does not exist or is infinity or minus infinity, so state.
15. Explain the basis for your answer to question 14.
16. State the domain of definition of

$$f(x)=\frac{x-4}{(x-1)\cdot(x+2)}$$

17. Find $(3x+4)(5x-1)$.
18. State the effective rate that corresponds to the rate 15% per annum compounded continuously.

Self-Test 2

In questions 1 through 5 find, if it exists, the limit of the function stated. If the limit does not exist or is ∞ or $-\infty$, so state. Explain the basis for your conclusion in each case.

1. $\lim\limits_{x\to 5} f(x)$, where $f(x)=\begin{cases} 4x+2, & \text{for } x<5 \\ 8, & \text{for } x=5 \\ 3x+3, & \text{for } x>5 \end{cases}$

2. $\lim\limits_{x\to -1} f(x)$, where $f(x)$ is defined above.

3. $\lim\limits_{x\to 6} \dfrac{x^2-36}{x-6}$

4. $\lim\limits_{x\to 2} \dfrac{3}{(x-2)^2}$

5. $\lim\limits_{x\to \infty} \dfrac{4x^3+2x-1}{9x^3-2x^2+2}$

6. State the effective rate corresponding to the rate 8% per annum compounded continuously.
7. The income tax paid in a certain region is \$0.00 if taxable income does not exceed \$10,000, and \$200 plus 15% of the excess over \$10,000 if taxable income exceeds \$10,000. Express tax to be paid as a function of taxable income.
8. Find and simplify $g(h) = \frac{f(3+h) - f(3)}{h}$, where $f(x) = 2x^2 + 1$.
9. For $g(h)$ defined in question 8 find, if it exists, $\underset{h \to 0}{\text{limit}}\, g(h)$. If this limit does not exist, so state.
10. For what values is $f(x) = \frac{x-1}{(x+4)(2x-1)}$ not defined? Explain.
11. Define e.
12. For $h(x) = \frac{4}{x-3}$ find, if it exists, $\underset{x \to 3}{\text{limit}}\, h(x)$. If this limit is infinity, minus infinity, or does not exist, so state. Explain the basis for your conclusion.
13. State the amount that must initially be invested if \$7000 is to be available in five years where the rate is 9% per annum compounded continuously.
14. If it exists, find $\underset{x \to 3}{\text{limit}} \frac{2x^3 + 1}{3x - 2}$. If this limit does not exist, so state.
15. If it exists, find $\underset{x \to 4}{\text{limit}} \frac{3}{(x-4)^2}$. If this limit is infinity, minus infinity, or does not exist, so state. Explain the basis for your conclusion.
16. What is the point of view that underlies the concept of continuous compounding of interest? How is this point of view captured in quantitative terms?
17. Sketch the graph of $f(x) = \begin{cases} 2x+1, & \text{for } x < 1, \\ 5 & \text{for } x = 1 \\ -x+2, & \text{for } x > 1 \end{cases}$

18. Is the function $f(x)$ defined in question 17 continuous at 1? Explain the basis for your conclusion in terms of the definition of continuity.

Self-Test 3

1. Evaluate: $\dfrac{2^{-3}}{3^{-2}}$
2. Find, if it exists, the slope of the line passing through $P(2,1)$ and $Q(2,5)$. If it does not exist, so state. Explain the basis for your conclusion.
3. Simplify $\sqrt{16a^2 - 4b^2}$ to the extent possible.

Consider $f(x) = \begin{cases} 1/x, & \text{for } x \neq 0 \\ 5, & \text{for } x = 0 \end{cases}$. Questions 4 through 8 pertain to this function.

4. Find, if it exists, $\operatorname*{limit}_{x \to 0} f(x)$. If this limit does not exist or is ∞ or $-\infty$, so state.
5. State the basis for your answer to 4.
6. Is $f(x)$ continuous at 0? Explain the basis for your answer in terms of the definition of continuity.
7. Find, if it exists, $\operatorname*{limit}_{x \to \infty} f(x)$. If this limit does not exist or is ∞ or $-\infty$, so state.
8. Explain the basis for your answer to question 7.
9. Find $\operatorname*{limit}_{x \to \infty} \left(1 - \dfrac{3}{x}\right)^x$.
10. $5000 is invested at the rate of 9 percent per annum compounded continuously. State the asset value of this amount after 5 years.

In questions 11 through 18 find, if it exists, the limit of the function stated. If this limit does not exist or is ∞ or $-\infty$, so state. Explain the basis for your conclusion in each case.

11. $\operatorname*{limit}_{x \to 1} \dfrac{x^2 + 2x - 3}{x - 1}$
12. $\operatorname*{limit}_{x \to 6} \dfrac{5}{x - 6}$

13. $\lim_{x\to\infty} \frac{5x^3+2x^2-1}{3x^3-x+2}$

14. $\lim_{x\to 1} \frac{5x-5}{x-1}$

15. $\lim_{x\to\infty} \left(3+\frac{4}{x}-\frac{8}{x^2}\right)$

16. $\lim_{x\to 2} \frac{2x^3+4x-1}{5x^2+1}$

17. $\lim_{x\to 3} f(x)$, where $f(x)=\begin{cases} 2x+6, & \text{for } x<3 \\ 10, & \text{for } x=3 \\ 5x-3, & \text{for } x>3 \end{cases}$

18. $\lim_{x\to 0} \frac{1}{x^2}$

CHAPTER 5

Derivatives of Functions

5.1. DERIVATIVES AND THEIR INTERPRETATION

In our discussion of limits (Section 3.1), limit problems with the following structure emerged in several examples and exercises. Starting with a given function $y = f(x)$, a function of h defined by

$$g(h) = \frac{f(x+h) - f(x)}{h}$$

is set up, and then $\underset{h\to 0}{\text{limit}}\, g(h)$ is determined. Rather than being the weird whim of a mad mathematics teacher, such limit problems lie at the heart of differential calculus. In this section we shall initiate a systematic study of this limit structure. One dimension of its importance will be seen through the exploration of a number of fundamental concepts in such diverse fields as geometry, physics, and economics, which exhibit this structure. We begin with a basic definition.

Derivative of a Function. If $y = f(x)$ is a function, then the function $f'(x)$ defined by

$$f'(x) = \underset{h\to 0}{\text{limit}} \frac{f(x+h) - f(x)}{h}$$

is called the **derivative of $y = f(x)$**. This derivative function $f'(x)$ is defined for all x in the domain of definition of $y = f(x)$ for which the above limit exists.

The derivative at 1, for example, is denoted by $f'(x)$ and defined by

$$f'(1) = \operatorname*{limit}_{h \to 0} \frac{f(1+h) - f(1)}{h}$$

provided that this limit exists. If a function $y = f(x)$ has a derivative at x, then $y = f(x)$ is said to be differentiable at x. The process of finding the derivative of a function is called differentiation.

The $f'(x)$ expression is one of a number of notations for the derivative of $y = f(x)$. It is especially useful for denoting the derivative of $y = f(x)$ at a particular value. The derivative at 1 is denoted by $f'(1)$, the derivative at 0 by $f'(0)$, and so on. Other notations for the derivative of $y = f(x)$ include

$$D_x f(x), \quad \frac{df(x)}{dx}, \quad D_x y, \quad \frac{dy}{dx}, \quad y', \quad y'(x), \quad \text{and} \quad \dot{y}.$$

EXAMPLE 1

For $f(x) = 2x^2 + x + 1$, determine $f'(2)$, provided that it exists.

By definition

$$f'(2) = \operatorname*{limit}_{h \to 0} g(h)$$

where $$g(h) = \frac{f(2+h) - f(2)}{h}$$

Part 1. Phase 1 of our analysis is algebraic. Construct $g(h)$ and simplify:

$$\begin{aligned} f(2+h) &= 2(2+h)^2 + (2+h) + 1 \\ &= 2(4+4h+h^2) + h + 3 \\ &= (8+8h+2h^2) + h + 3 \\ &= 2h^2 + 9h + 11 \\ f(2) &= 2(2)^2 + (2) + 1 = 11 \end{aligned}$$

Subtracting $f(2) = 11$ from $f(2+h) = 2h^2 + 9h + 11$ yields

$$f(2+h) - f(2) = 2h^2 + 9h$$

Dividing by h and simplifying yields $g(h)$:

$$\begin{aligned} g(h) = \frac{f(2+h) - f(2)}{h} &= \frac{2h^2 + 9h}{h} \\ &= \frac{h(2h+9)}{h} \\ &= 2h + 9 \end{aligned}$$

Part 2. Find $\underset{h \to 0}{\text{limit}}\, g(h)$.

$$\begin{aligned} f'(2) &= \underset{h \to 0}{\text{limit}}\, g(h) \\ &= \underset{h \to 0}{\text{limit}}(2h+9) \\ &= 9 \end{aligned}$$

EXAMPLE 2

For

$$f(x) = \begin{cases} x+1, & \text{for } x \geq 0 \\ -x+1, & \text{for } x < 0 \end{cases}$$

determine $f'(0)$, provided that it exists.

By definition

$$f'(0) = \lim_{h \to 0} g(h)$$

where $$g(h) = \frac{f(0+h) - f(0)}{h}$$

Part 1. Construct $g(h)$ and simplify.

$$f(0+h) = f(h) = \begin{cases} h+1, & \text{for } h \geq 0 \\ -h+1, & \text{for } h < 0 \end{cases}$$

$$f(0) = 1$$

Subtracting $f(0) = 1$ from $f(0 + h)$ yields

$$f(0+h) - f(0) = \begin{cases} h, & \text{for } h > 0 \\ -h, & \text{for } h < 0 \end{cases}$$

Dividing by h and simplifying yields

$$g(h) = \frac{f(0+h) - f(0)}{h} = \begin{cases} \frac{h}{h} = 1, & \text{for } h > 0 \\ \frac{-h}{h} = -1, & \text{for } h < 0 \end{cases}$$

$$= \begin{cases} 1, & \text{for } h > 0 \\ -1, & \text{for } h < 0 \end{cases}$$

Part 2. Find $\lim_{h \to 0} g(h)$. Values of $g(h)$ for values of h that are getting closer and closer to 0 are shown in Table 5.1, from which we see that as h takes on values closer

Table 5.1

h	0.1	–0.1	0.01	–0.01	0.001	–0.001
$g(h)$	1	–1	1	–1	1	–1

and closer to 0, $g(h)$ is unstable. $g(h)$ clusters around two values, –1 and 1, rather than approaching one value. Thus

$$f'(0) = \lim_{h \to 0} g(h) \text{ does not exist;}$$

$$f(x) = \begin{cases} x+1, & \text{for } x \geq 0 \\ -x+1, & \text{for } x < 0 \end{cases} \text{ does not have a derivative}$$

at 0.

EXAMPLE 3

For $f(x) = 1$, determine the derivative function $f'(x)$ and the values $f'(0)$, $f'(2)$, and $f'(10)$.

By definition

$$f'(x) = \lim_{h \to 0} g(h)$$

where
$$g(h) = \frac{f(x+h) - f(x)}{h}$$

Part 1. Construct $g(h)$ and simplify.

$$f(x + h) = 1$$
$$f(x) = 1$$

Thus

$$f(x+h) - f(x) = 1 - 1 = 0$$

$$g(h) = \frac{f(x+h) - f(x)}{h} = \frac{0}{h} = 0$$

where $h \neq 0$.

Part 2. Find $\underset{h \to 0}{\text{limit}}\, g(h)$.

$$f'(x) = \underset{h \to 0}{\text{limit}}\, g(h) = \underset{h \to 0}{\text{limit}}\, 0 = 0$$

$f'(x)$ is 0 for all x. Thus $f'(0) = 0$, $f'(2) = 0$, and $f'(10) = 0$.

EXAMPLE 4

For $f(x) = x$, determine the derivative function $f'(x)$ and the values $f'(0)$, $f'(2)$, and $f'(10)$.

By definition

$$f'(x) = \underset{h \to 0}{\text{limit}}\, g(h)$$

where
$$g(h) = \frac{f(x+h) - f(x)}{h}$$

Part 1. Construct $g(h)$ and simplify.

$$f(x + h) = x + h$$
$$f(x) = x$$

Thus

$$f(x+h) - f(x) = x + h - x = h$$
$$g(h) = \frac{f(x+h) - f(x)}{h} = \frac{h}{h} = 1$$

where $h \neq 0$.

Part 2. Find $\underset{h \to 0}{\text{limit}}\, g(h)$.

$$f'(x) = \underset{h \to 0}{\text{limit}}\, g(h) = \underset{h \to 0}{\text{limit}}\, 1 = 1$$

$f'(x)$ is 1 for all x. Thus $f'(0) = 1$, $f'(2) = 1$, and $f'(10) = 1$.

EXAMPLE 5

For $f(x) = x^2$, determine the derivative function $f'(x)$ and the values $f'(-3)$, $f'(0)$, and $f'(4)$.

By definition

$$f'(x) = \underset{h \to 0}{\text{limit}}\, g(h)$$

where
$$g(h) = \frac{f(x+h) - f(x)}{h}$$

Part 1. Construct $g(h)$ and simplify.

$$f(x+h) - (x+h)^2 = x^2 + 2hx + h^2$$
$$f(x) = x^2$$

Thus

$$\begin{aligned} f(x+h) - f(x) &= x^2 + 2hx + h^2 - x^2 \\ &= 2hx + h^2 \\ g(h) &= \frac{f(x+h) - f(x)}{h} = \frac{2xh + h^2}{h} \\ &= \frac{h(2x+h)}{h} \\ &= 2x + h \end{aligned}$$

where $h \neq 0$.

Part 2. Find $\underset{h\to 0}{\text{limit}}\, g(h)$.

$$f'(x) = \underset{h\to 0}{\text{limit}}\, g(h) = \underset{h\to 0}{\text{limit}}(2x + h)$$
$$= 2x$$

Since $f'(x) = 2x$, $f'(-3) = -6$, $f'(0) = 0$, and $f'(4) = 8$.

EXAMPLE 6

For $f(x) = x^3$, determine the derivative function $f'(x)$ and the values $f'(-3)$, $f'(0)$, and $f'(1)$.

By definition

$$f'(x) = \underset{h\to 0}{\text{limit}}\, g(h)$$

where $$g(h) = \frac{f(x+h) - f(x)}{h}$$

Part 1. Construct $g(h)$ and simplify.

$$f(x+h) = (x+h)^3$$

Multiplying $(x+h)$ by itself three times yields

$$(x+h)^3 = x^3 + 3hx^2 + 3h^2x + h^3$$

Thus

$$f(x+h) = x^3 + 3hx^2 + 3h^2x + h^3$$
$$f(x) = x^3$$

$$\begin{aligned} f(x+h)-f(x) &= x^3+3hx^2+3h^2x+h^3-x^3 \\ &= 3hx^2+3h^2x+h^3 \end{aligned}$$

Part 2. Find $\operatorname*{limit}_{h\to 0} g(h)$.

$$\begin{aligned} f'(x) &= \operatorname*{limit}_{h\to 0} g(h) = \operatorname*{limit}_{h\to 0}(3x^2+3hx+h^2) \\ &= 3x^2 \end{aligned}$$

Since $f'(x)=3x^2$, $f'(-3)=27$, $f'(0)=0$, and $f'(1)=3$.

EXAMPLE 7

For $f(x) = x^4$, determine the derivative function $f'(x)$ and the values $f'(-1)$ and $f'(2)$.

By definition

$$f'(x) = \operatorname*{limit}_{h\to 0} g(h)$$

where $$g(h) = \frac{f(x+h)-f(x)}{h}$$

Part 1. Construct $g(h)$ and simplify.

$$\begin{aligned} f(x+h) = (x+h)^4 &= x^4+4hx^3+6h^2x^2+4h^3x+h^4 \\ f(x) &= x^4 \\ f(x+h)-f(x) &= 4hx^3+6h^2x^2+4h^3x+h^4 \end{aligned}$$

Dividing by h to obtain $g(h)$ and simplifying yields

$$g(h) = \frac{f(x+h)-f(x)}{h} = \frac{4hx^3+6h^2x^2+4h^3x+h^4}{h}$$

$$= \frac{h(4hx^3 6hx^2 + 4h^2x + h^3)}{h}$$
$$= 4x^3 + 6hx^2 + 4h^2x + h^3$$

where $h \neq 0$.

Part 2. Find $\lim_{h \to 0} g(h)$.

$$f'(x) = \lim_{h \to 0} g(h) = \lim_{h \to 0}(4x^3 + 6hx^2 + 4h^2x + h^3)$$
$$= 4x^3$$

Since $f'(x) = 4x^3$, $f'(-1) = -4$, and $f'(2) = 32$.

EXAMPLE 8

For $f(x) = x^2 + x$, determine the derivative function $f'(x)$ and the values $f(-1)$, $f'(0)$, and $f'(2)$.

By definition

$$f'(x) = \lim_{h \to 0} g(h)$$

where $$g(h) = \frac{f(x+h) - f(x)}{h}$$

Part 1. Construct $g(h)$ and simplify.

$$f(x+h) = (x+h)^2 + (x+h)$$
$$= x^2 + 2hx + h^2 + x + h$$
$$f(x) = x^2 + x$$
$$f(x+h) - f(x) = x^2 + 2hx + h^2 + x + h - x^2 - x$$
$$= 2hx + h^2 + h$$

Dividing by h to obtain $g(h)$ and simplifying yields

$$g(h)=\frac{f(x+h)-f(x)}{h}=\frac{2hx+h^2+h}{h}$$
$$=\frac{h(2x+h+1)}{h}$$
$$=2x+h+1$$

where $h \neq 0$.

Part 2. Find $\underset{h\to 0}{\text{limit}}\, g(h)$.

$$f'(x)=\underset{h\to 0}{\text{limit}}\, g(h)=\underset{h\to 0}{\text{limit}}(2x+h+1)$$
$$=2x+1$$

Since $f'(x)=2x+1$, $f'(-1)=-1$, $f'(0)=1$, and $f'(2)=5$.

A Problem in the World of Geometry

Classical geometry provides us with tangent-line concepts for such curves as the circle and parabola. An important concern of geometricians at one time was the problem of defining a tangency concept for a greater variety of curves than could be accomplished by using the methods provided by classical geometry. As we shall see, this tangent-line definition problem is one that is accessible to the limit ideas that we have discussed.

We begin our analysis by considering the special case of a curve C that is the graph of $f(x) = x^2$ (Figure 5.1), and the problem of defining the tangent line to C at the origin (0, 0). The x-axis, I think we would agree, is the most natural candidate for the office of tangent line to C at (0, 0). And so we could handle this situation by agreeing to define the x-axis ($y = 0$) as the tangent line to C at (0, 0). This approach, however, is of limited usefulness; it does not provide us with a method for defining the tangent line to C at other points. However, we at least have a feeling for what to aim at in this situation. Any general approach that we develop should lead us to what we expect, or, if it leads elsewhere, should provide us with some insight into why what we expect is unreasonable.

Consider a point Q on C that is near the origin O, and let OQ denote the line determined by O and Q (see Figure 5.2). Let h denote the difference between the x-coordinate of Q and the x-coordinate of the origin. In terms of h, Q has coordinates $Q(0 + h, f(0 + h) = Q(h, h^2)$. As h approaches zero, $Q(h, h^2)$ moves toward the origin

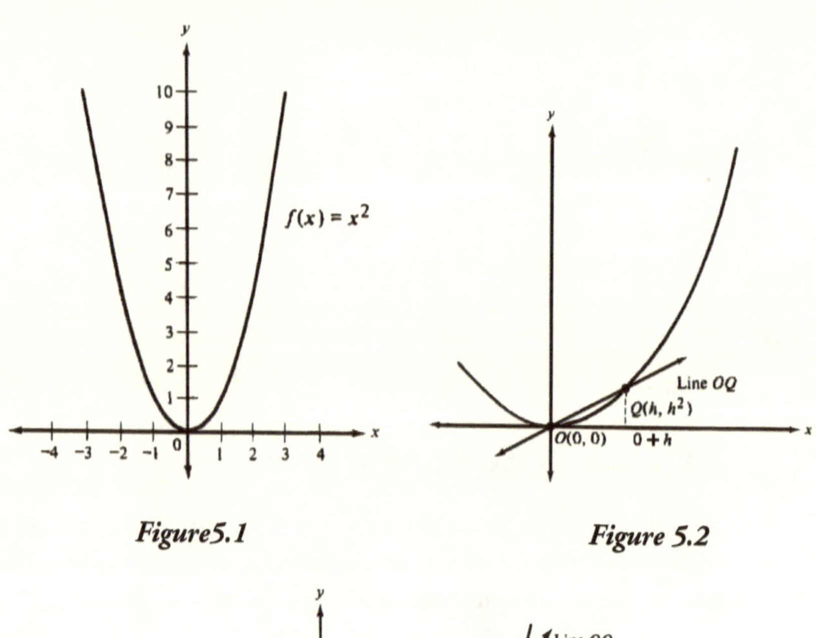

Figure5.1

Figure 5.2

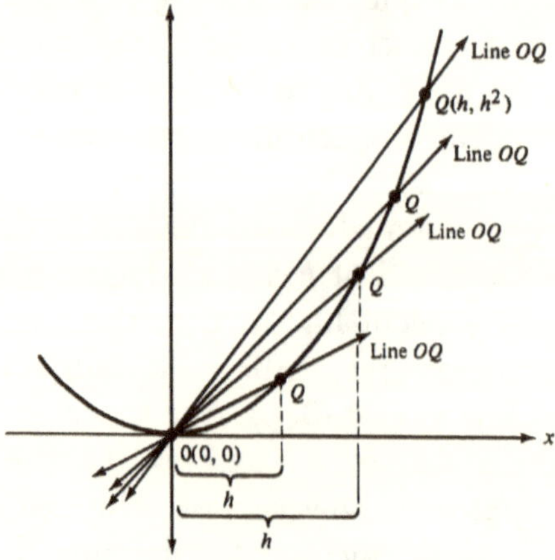

Figure 5.3

O on C, and the line OQ swings around and approaches the x-axis, our candidate for tangent line to C at O (see Figure 5.3). We can describe this geometric behavior in more precise analytic terms by working in terms of the slope of OQ. The slope function

$$m(h) = \frac{h^2 - 0}{h - 0} = h$$

where $h \neq 0$, describes the slope of OQ. As h approaches 0, $m(h)$ = h approaches O. That is,

$$\operatorname*{limit}_{h \to 0} m(h) = 0$$

But the x-axis, our candidate for the tangent line to C at the origin, has slope 0. Thus if we define the tangent line to C at the origin to be the line passing through the origin with slope

$$\operatorname*{limit}_{h \to 0} m(h) = 0$$

we obtain the x-axis. This approach yields what we agreed is most natural in this situation and is applicable to other situations as well.

To illustrate the general applicability of our approach, consider the point $P(1, f(1)) = P(1, 1)$ on C. Let $Q(1 + h, f(1 + h))$ denote a neighboring point on C (see Figure 5.4). h is the difference between the x-coordinate of Q and the x-coordinate of $P(1, 1)$, so that, in terms of h, Q has coordinates $(1 + h, f(1 + h))$. The slope function

Dividing by h to obtain $g(h)$ and simplifying yields

$$g(h) = \frac{f(x+h) - f(x)}{h} = \frac{3hx^2 + 3h^2x + h^3}{h}$$

$$= \frac{h(3x^2 + 3hx + h^2)}{h}$$

$$= 3x^2 + 3hx + h^2$$

where $h \neq 0$.

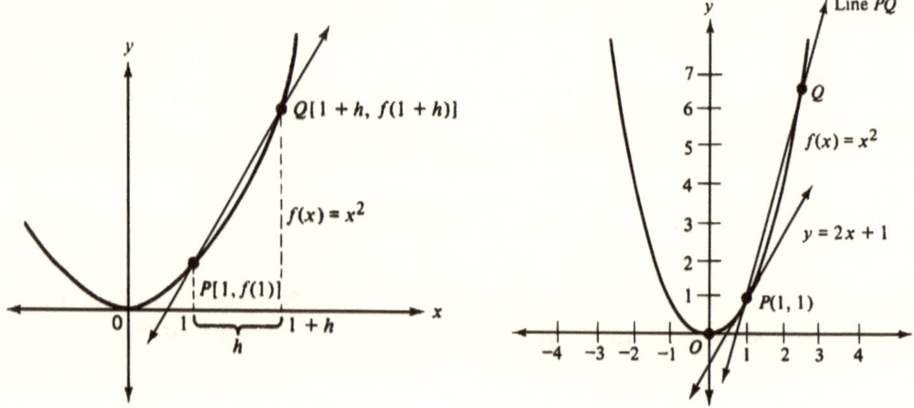

Figure 5.4 *Figure 5.5*

$$m(h) = \frac{f(1+h) - f(1)}{h}$$

describes the slope of line *PQ*. Now

$$f(1+h) = (1+h)^2 = 1 + 2h + h^2$$
$$f(1) = 1$$

so that

$$f(1+h) - f(1) = 1 + 2h + h^2 - 1$$
$$= 2h + h^2$$

Dividing by *h* yields *m*(*h*):

$$m(h) = \frac{2h+h^2}{h} = \frac{h(2+h)}{h} = 2+h$$

where $h \neq 0$.
As $h \to 0$, $m(h) = 2+h \to 2$. That is,

$$\operatorname*{limit}_{h\to 0} m(h) = 2$$

We define the tangent line to *C* at *P*(1, 1) to be the line that passes through *P*(1, 1) and has slope 2 (see Figure 5.5). An equation of this tangent line is

$$y = 2x - 1$$

In geometric terms, as $h \to 0$, line *PQ* swings around and approaches the tangent line $y = 2x - 1$.

At the same time, we should not fail to observe that

$$\operatorname*{limit}_{h\to 0} m(h) = \operatorname*{limit}_{h\to 0} \frac{f(1+h) - f(1)}{h} = 2$$

is, by definition, $f'(1)$, the derivative of $f(x) = x^2$ at 1. Thus the tangent line to the graph of $f(x) = x^2$ at *P*(1, 1) has slope $f'(1) = 2$.

More generally, consider a curve *C* that is the graph of a function $y = f(x)$ and a point $P(x, f(x))$ on *C*. Let $Q(x + h, f(x + h))$ denote a point on *C* is near *P* (see Figure 5.6). *h* is the difference between the *x*-coordinate of *Q* and the *x*-coordinate of

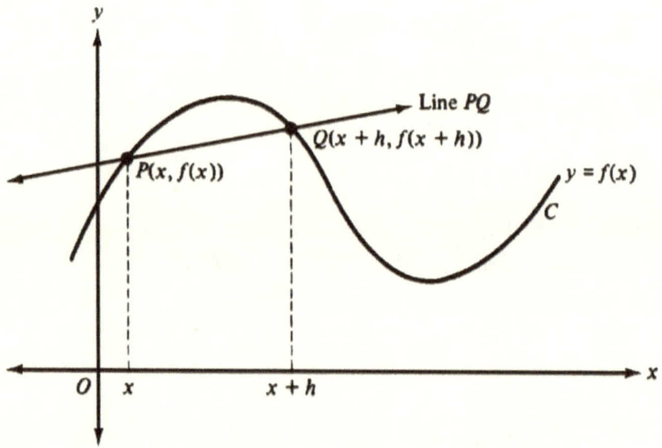

Figure 5.6

$P(x, f(x))$, so that, in terms of h, Q has coordinates $(x + h, f(x + h))$. The slope of line PQ is given by the slope function

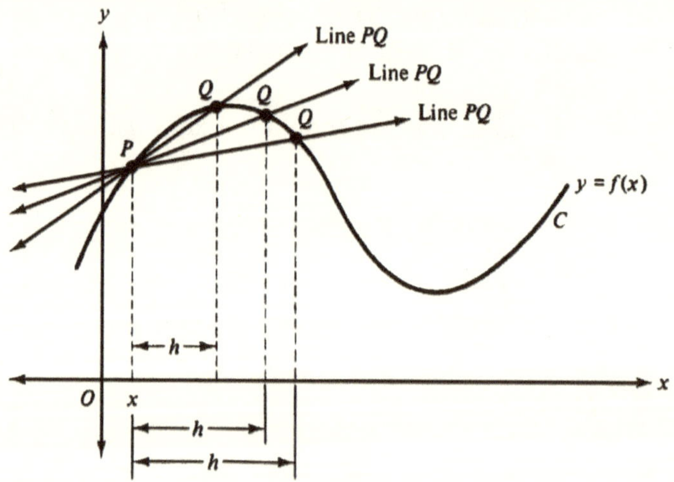

Figure 5.7

$$m(h) = \frac{f(x+h) - f(x)}{h}$$

As h approaches zero, Q approaches P on C (see Figure 5.7).

If, as h approaches zero, the slope function $m(h)$ approaches a number L; that is,

$$\lim_{h\to 0} m(h) = \lim_{h\to 0}\frac{f(x+h)-f(x)}{h} = L$$

then the **tangent line to C at $P(x, f(x))$** is defined as the line that passes through $P(x, f(x))$ and has slope L.

We immediately recognize that L is the derivative of the function $y = f(x)$, which has curve C as its graph. Thus the determination of the tangent line to the graph of a function $y = f(x)$ at a point reduces to determining the derivative of $y = f(x)$ at the x-coordinate of the point.

A Problem in the World of Physics

If an object, such as a stone, is dropped from a building to the street below, then

$$y = 16t^2$$

is the time-distance function that describes the motion of the object. Time t is measured in seconds, and distance y is measured in feet.

The familiar concept of average velocity, we recall, is defined as the ratio

$$\frac{\text{distance moved by the object}}{\text{time taken to move the distance}}$$

Thus, for example, if we consider the time interval from 2 seconds to $\frac{5}{2}$ seconds, the average velocity of the object for this time interval is

$$\frac{16\left(\frac{5}{2}\right)^2 - 16(2)^2}{\frac{5}{2} - 2} = 72 \text{ ft/sec}$$

For the time interval from 2 seconds to $\frac{9}{4}$ seconds, the average velocity of the object is

$$\frac{16\left(\frac{9}{4}\right)^2 - 16(2)^2}{\frac{9}{4} - 2} = 68 \text{ ft/sec}$$

For the time interval from 2 seconds to 2 + h seconds, the average velocity of the object is

$$\begin{aligned} v(h) &= \frac{16(2+h)^2 - 16(2)^2}{h} \\ &= \frac{16(4+4h+h^2) - 64}{h} \\ &= \frac{64+64h+16h^2 - 64}{h} \\ &= \frac{h(64+16h)}{h} \\ &= 64+16h \end{aligned}$$

where $h \neq 0$.

The average-velocity concept gives us a motion characteristic of an object with respect to a time interval. The value obtained varies with the time interval, and this feature is bothersome in many situations of importance. One way of obtaining a unique value is to examine the behavior of the average velocity function $v(h)$ as the time interval gets smaller and smaller, that is, as h approaches zero. For the situation in question, we obtain

$$\operatorname*{limit}_{h\to 0} v(h) = \operatorname*{limit}_{h\to 0}(64+16h) = 64$$

The value 64 is a motion characteristic of the object with respect to the time value (as opposed to interval) of 2 seconds; 64 feet per second is said to be the instantaneous velocity of the object for time 2 seconds.

Let us observe that

$$\operatorname*{limit}_{h\to 0} v(h) = \operatorname*{limit}_{h\to 0}\frac{16(2+h)^2 - 16(2)^2}{h} = 64$$

is, by definition, the derivative of the time-distance function $y = 16t^2$ at 2.

More generally, consider the time interval from t seconds to $t + h$ seconds. The average velocity function

$$\begin{aligned} v(h) &= \frac{16(t+h)^2 - 16t^2}{h} \\ &= \frac{16(t^2+2th+h^2) - 16t^2}{h} \\ &= \frac{16t^2 + 32th + 16h^2 - 16t^2}{h} \\ &= \frac{h(32t+16h)}{h} \\ &= 32t + 16h \end{aligned}$$

where $h \neq 0$, describes the average velocity of the object in this time interval. By taking the limit of $v(h)$ as h approaches zero, we obtain

$$\operatorname*{limit}_{h\to 0} v(h) = \operatorname*{limit}_{h\to 0}(32t+16h) = 32t$$

the derivative of the time-distance function $y = 16t^2$, as the instantaneous velocity of our object at time t seconds. At time $t =$

1, the instantaneous velocity of the object is 32 feet per second. At time $t = 3$, the instantaneous velocity of the object is 96 feet per second.

Still more generally, consider an object whose motion along a line is described by a time-distance function $y = f(t)$. The function

$$v(h) = \frac{f(t+h) - f(t)}{h}$$

specifies the average velocity of this object in the time interval from time t to time $t + h$.

If, as h approaches zero, the average-velocity function $v(h)$ approaches a number L, that is,

$$\lim_{h \to 0} v(h) = \lim_{h \to 0} \frac{f(t+h) - f(t)}{h} = L$$

then L is called the **instantaneous velocity of the object at time t.** Let us note that L is, by definition, the derivative $f'(t)$ of the time-distance function $y = f(t)$, which describes the motion of the object in question.

In connection with the significance of the concept of instantaneous velocity, we should note that its usefulness is established within the world of physics, not within the world of mathematics. The role of mathematics here is to provide a framework for the suitable expression of the instantaneous-velocity concept, but its significance is to be shown in the world of physics. We can only observe at this point that this concept and its cousin, instantaneous acceleration, are central to the statement of Newton's laws of motion, which lie at the heart of the study of bodies in motion.

Two Basic Concepts from the World of Economics

Consider a firm that produces a certain commodity, and let

$$R = f(x)$$

denote the total-revenue function of the firm, where x is output per unit time and R is the total revenue derived from output x. If output is increased from x to $x + h$ units, the change in total revenue is given by the difference

$$f(x+h) - f(x)$$

The average change in total revenue is expressed by

$$\frac{f(x+h) - f(x)}{h}$$

If

$$f'(x) = \operatorname*{limit}_{h \to 0} \frac{f(x+h) - f(x)}{h}$$

exists, then $f'(x)$ is called the **marginal revenue for an output of *x* units.**

At the same time if

$$c = f(x)$$

denotes the total cost function of the firm, where x is output per unit time and c is the total cost for output x, and output is increased from x to $x + h$ units, the change in total cost is given by the difference

$$f(x+h) - f(x)$$

The average change in total cost is expressed by

$$\frac{f(x+h)-f(x)}{h}$$

If

$$f'(x) = \lim_{h \to 0} \frac{f(x+h)-f(x)}{h}$$

exist, then $f'(x)$ is called the **marginal cost for an output of x units**.

Marginal revenue and marginal cost are profit characteristics of a firm with respect to an output level x that tell us something about by how much total revenue and total cost are gaining or losing as we increase x. As is the case with any concept that belongs to another world of application but is defined in mathematical terms, the significance of the marginal-revenue and marginal-cost concepts is to be established in the world of economics itself. We can only note at this point that the marginal-revenue and marginal-cost concepts are basic to the study of the profit-maximization behavior of the firm. A central proposition of economics is that profit is maximized at that output for which marginal revenue equals marginal cost. This proposition and some of its applications are further discussed in Section 7.2.

In summary then, if $R = R(x)$ and $c = c(x)$ are the total-revenue and total-cost functions of a firm, then the derivatives of these functions

$$R'(x) \text{ and } c'(x)$$

are called **marginal revenue and marginal cost for output x.**[*]

[*] These definitions of marginal revenue and marginal cost presuppose that the total-revenue and total-cost functions have derivatives, which in turn

Tangent line to a curve at a point, instantaneous velocity, marginal revenue, and marginal cost are four of many basic concepts in diverse fields of study whose point of contact is the derivative structure. Figure 5.8 shows the situation at a glance. The central core is the abstract concept of derivative of a function $y = f(x)$. As $y = f(x)$ is given different interpretations, $f'(x)$ in turn acquires different interpretations. The determination of a tangent line to a curve, instantaneous velocity, marginal revenue, marginal cost, and a variety of other quantities all reduce to the determination of the derivative of a function.

In applications involving the derivative of the function $y = f(x)$, the derivative is sometimes called the **rate of change of the function** or the **instantaneous rate of change of *y* with respect to *x***. In terms of this language, marginal revenue, for example, would also be termed the instantaneous rate of change of total revenue with respect to output.

After a look at the **relationship between differentiability** (the property of having a derivative) **and continuity** in the next section, we turn our attention in Sections 6.1-6.7 to developing tools (theorems) that would enable us to determine derivatives as quickly and painlessly as possible. Properly equipped, we return to applied concepts with a derivative structure and show how they can be put to work in Section 6.8.

Other dimensions of derivatives put to work are explored in Chapter 7, 8, and 11.

---

presupposes that the output is divisible—that h can be as small as desired. For some commodities (airplanes, for example), output must be restricted to integral values and the smallest value that h can be given is 1. For such situations and in basic economics courses that do not presuppose calculus, marginal revenue is defined as the increment of total revenue that comes when we increase output x by an increment of one unit. Marginal cost is defined analogously.

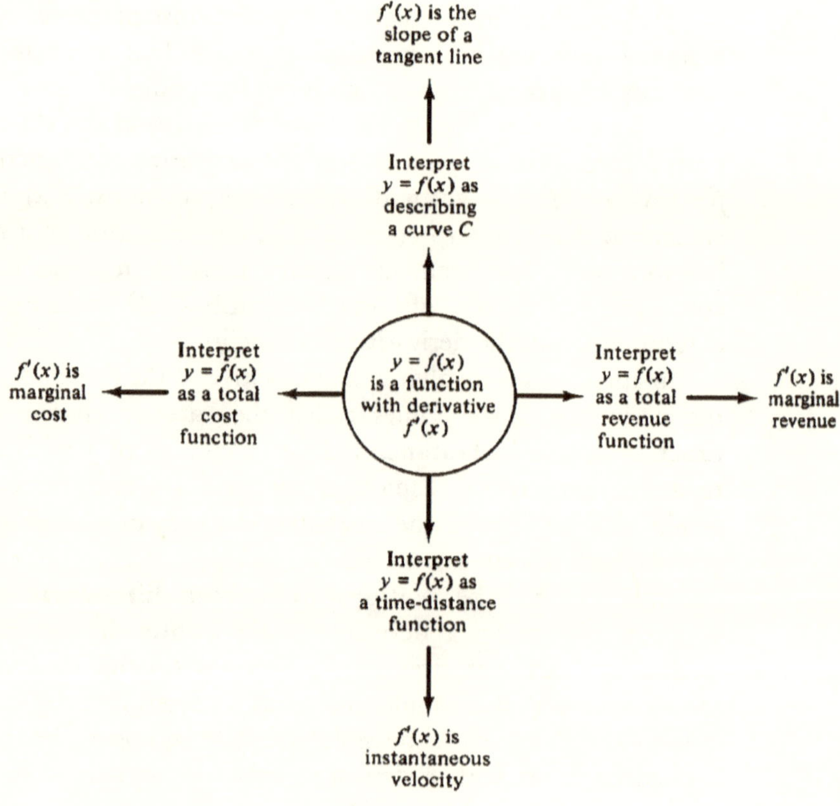

Figure 5.8

EXERCISES

Find the following derivatives.

1. $f'(1)$ where $f(x) = 3x + 2$
2. $f'(3)$ where $f(x) = x^2 + 50$
3. $f'(2)$ where $f(x) = x^2 + 3x + 100$
4. $f'(2)$ where $f(x) = 3x^2 - 2x + 1$
5. $f'(x)$ where $f(x) = 4x^2 - 2$
6. $f'(x)$ where $f(x) = 2x + 5$
7. $f'(x)$ where $f(x) = 10$
8. $f'(x)$ where $f(x) = x^2 - 2x - 1$
9. From Exercise 2, $f'(3) = 6$. State four interpretations of this result.

10. From Exercise 3, $f'(2) = 7$. State four interpretations of this result.

5.2. DIFFERENTIABILITY AND CONTINUITY

Are continuity and differentiability equivalent, related in some way, or unrelated? The answer to this question is most easily seen in geometric terms. To begin, we return to the function

$$f(x) = \begin{cases} x+1, & \text{for } x \geq 0 \\ -x+1, & \text{for } x < 0 \end{cases}$$

whose graph is reproduced in Figure 5.9. In Section 3.3, Example 4 (p. 83), we showed that $f(x)$ is continuous at 0, and in Section 5.1, Example 2, we showed that $f(x)$ is not differentiable (does not have a derivative) at 0. This example shows that a function that is continuous at a value need not necessarily have a derivative at the value. Continuity does not imply differentiability. Looking at the situation geometrically, we see from Figure 5.9 that, although the graph of $f(x)$ does not have any gaps, jumps, or

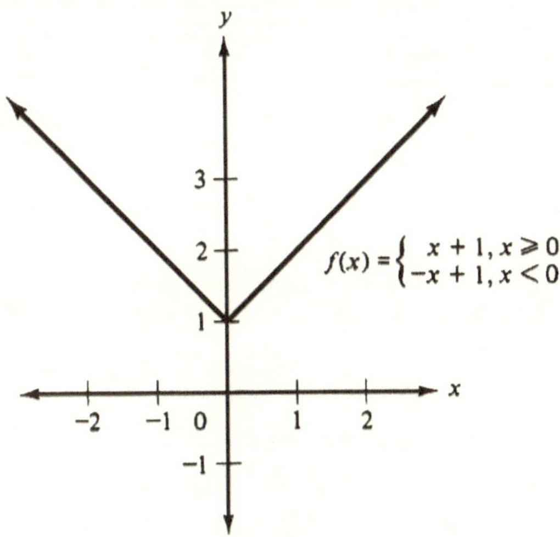

Figure 5.9

breaks at 0, the corner at (0, 0) is too sharp to support a tangent line. Continuity implies a graph that is in one piece, but says nothing about the smoothness of the graph. The graph may have sharp, jagged corners that cannot support a tangent line. Differentiability, on the other hand, implies a graph that is in one piece and at the same time is also smooth enough to support a tangent line. Thus differentiability implies continuity; a function that has a derivative at a value must be continuous at the value.

EXERCISES

1. The graph of a function $y = f(x)$ defined on an interval from a to b is shown in Figure 5.10.

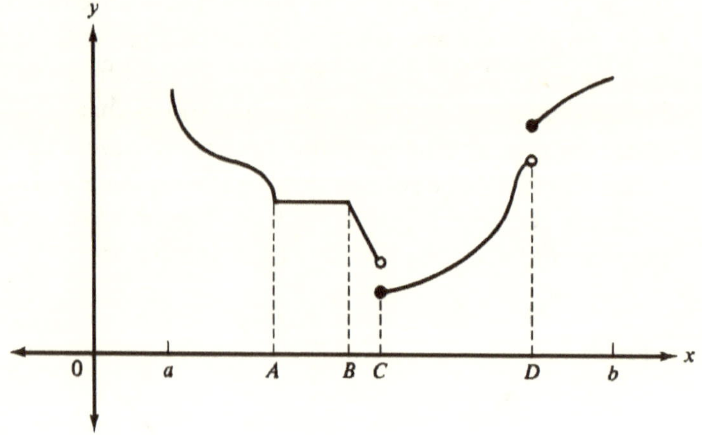

Figure 5.10

(a) Is $y = f(x)$ differentiable at all values between a and b? Explain.
(b) Is $y = f(x)$ continuous at all values between a and b? Explain.
(c) Are there any values at which $y = f(x)$ is continuous but not differentiable? Explain.
(d) Are there any values at which $y = f(x)$ is differentiable but not continuous? Explain.

2. Is $f(x)=\begin{cases} x, \text{ for } x \geq 0 \\ -x, \text{ for } x < 0 \end{cases}$ differentiable at 0? Explain.

   Is $f(x)$ continuous at 0? Explain.

3. Is $f(x)=\begin{cases} x, \text{ for } x \geq 0 \\ x^2, \text{ for } x < 0 \end{cases}$ differentiable at 0? Explain.

   Is $f(x)$ continuous at 0? Explain.

4. Is $f(x) = 2$ differentiable at 0? Explain. Is $f(x) = 2$ continuous at 2? Explain.

CHAPTER 6

Tools for Computing Derivatives

6.1. CONSTANT AND POWER FUNCTIONS

The calculation of derivatives by use of the definition of derivative is feasible only in the simplest of situations, and thus we are led to develop tools to facilitate the calculation of derivatives. To begin, we return to several situations considered in Section 5.1. In Example 3 (p. 141) we showed that, if $f(x) = 1$, then $f'(x) = 0$. Such is the case when constant functions in general.

Theorem. If $f(x) = c$, where c is a constant, then $f'(x) = 0$.

Thus for example,

If $f(x) = 2$, $f'(x) = 0$.
If $f(x) = -5$, $f'(x) = 0$.
If $f(x) = \frac{1}{7}$, $f'(x) = 0$.

Another result that should be committed to memory is established in Example 4 (p. 142).

Theorem. If $f(x) = x$, then $f'(x) = 1$.

In Examples 5 through 7 (pp. 143-145), the following results are established.

If $f(x) = x^2$, then $f'(x) = 2x$.
If $f(x) = x^3$, then $f'(x) = 3x^2$.
If $f(x) = x^4$, then $f'(x) = 4x^3$.

The pattern exhibited by these functions and their derivatives is this. If $f(x) = x^n$, then $f'(x) = nx^{n-1}$. We have shown that this result holds for $n = 2$, 3, and 4. That this result holds for all rational numbers can be established on the basis of the definition of the derivative. We thus have the following theorem:

Theorem. If n is a rational number and $f(x) = x^n$, then $f'(x) = nx^{n-1}$ for all x at which $f'(x)$ is defined.

Thus, for example,

For $f(x) = x^5$, $f'(x) = 5x^4$.
For $f(x) = x^{42}$, $f'(x) = 42x^{41}$.

These are obvious examples. But this theorem for derivatives of what are termed **power functions** ($f(x) = x^n$; independent variable x raised to a fixed power) are also applicable to functions such as $f(x) = \sqrt{x}$ and $f(x) = 1/x$ since they can be stated in the form of x raised to a power (see Section 1.5). Thus:

$$\text{For } f(x) = \sqrt{x} = x^{1/2}, f'(x) = \frac{1}{2}x^{-1/2} = \frac{1}{2\sqrt{x}}$$

$$\text{For } f(x) = \frac{1}{x} = x^{-1}, f'(x) = -x^{-2} = -\frac{1}{x^2}$$

EXERCISES

Find the derivatives of the following functions and their values at the stated numbers.

1. $f(x)=4, x=2$
2. $f(x)=x^7, x=2$
3. $f(x)=x^{12}, x=-1$
4. $f(x)=1/x^2, x=1$
5. $f(x)=\sqrt[3]{x}, x=8$
6. $f(x)=e, x=\pi$
7. $f(x)=1/x^3, x=-2$
8. $f(x)=\sqrt[4]{x}, x=16$
9. $f(x)=x^8, x=2$
10. $f(x)=x^{2/3}, x=-1$
11. $f(x)=\sqrt{8}, x=3$
12. $f(x)=1/\sqrt{x}, x=4$

6.2. SUMS, DIFFERENCES, PRODUCTS, AND QUOTIENTS

In developing tools for computing limits, we found it fruitful to view a function as being made up of component parts held together by the operations of addition, subtraction, multiplication, and division, and to examine the limit behavior of the component parts. The analogous approach that is suggested here is to view a function as being made up of components, determine the derivatives of the components, and put these derivatives together in a suitable way to obtain the derivative of the original function. Such an approach is fruitful and gives rise to a number of theorems, which we now state and illustrate. Some of these theorems are structurally analogous to the general limit theorems developed in Section 3.2; others are similar in spirit but different in structure.

In much of the following discussion it will be convenient to use the $D_x f(x)$ notation for the derivative of $y=f(x)$. In terms of this notation, the derivative of $f(x)=x^2$, for example, is denoted by $D_x(x^2)$.

Derivative of a Sum or Difference of Functions

In Section 5.1, Examples 4 and 5 (pp. 142-143), we saw that

$$D_x(x^2) = 2x \text{ and } D_x(x) = 1$$

and in Example 8 (p. 146), we saw that

$$D_x(x^2 + x) = 2x + 1$$

Thus, in this case, we have that $D_x(x^2 + x) = D_x(x^2) + D_x(x)$. $= 2x + 1$ That is, the derivative of the sum of the two functions is the sum of the derivatives of the component parts. That such is the case in general is the content of the following theorem:

> **Theorem**: If $h(x) = f(x) + g(x)$ and $D_x f(x)$ and $D_x g(x)$ exist, then
>
> $$D_x h(x) = D_x f(x) + D_x g(x)$$
>
> If $h(x) = f(x) - g(x)$, then
>
> $$D_x h(x) = D_x f(x) - D_x g(x)$$

The derivative of a sum or difference of functions is the sum or difference of the derivatives of the component parts, provided that the component parts have derivatives. This theorem holds for the sum or differences of any number of component parts.

EXAMPLE 1

Determine the derivative of $h(x) = x^4 + x^2 - 3$.

$$\begin{aligned} D_x(x^4 + x^2 - 3) &= D_x(x^4) + D_x(x^2) - D_x(3) \\ &= 4x^3 + 2x - 0 \\ &= 4x^3 + 2x \end{aligned}$$

EXAMPLE 2

Determine the derivative of $h(x) = \sqrt{x} + \frac{1}{x}$.

In terms of components, we have

$$h(x) = \sqrt{x} + \frac{1}{x} = x^{1/2} + x^{-1}$$

$$\begin{aligned} D_x(x^{1/2} + x^{-1}) &= D_x(x^{1/2}) + D_x(x^{-1}) \\ &= \frac{1}{2}x^{1/2-1} + (-1)x^{-1-1} \\ &= \frac{1}{2}x^{-1/2} - x^{-2} \\ &= \frac{1}{2\sqrt{x}} - \frac{1}{x^2} \end{aligned}$$

Derivative of a Product of Functions

Whereas the limit of a product of functions is the product of the limits of the component parts, the analogous result for derivatives (the derivative or a product of functions is the product of the derivatives of the component parts) is incorrect. To see this, consider the function $h(x) = x^2 = x \cdot x$. Taking the derivatives of the component parts yields $D_x(x) = 1$, $D_x(x) = 1$. Multiplying these component derivatives yields $1 \cdot 1 = 1$. At the same time, the derivative of $h(x) = x^2$ is $2x$, and $2x \neq 1$ for all x other than $\frac{1}{2}$. There is, however, a derivative theorem for products that is analogous in spirit to the limit theorem for products, but different in structure.

Theorem. If $h(x) = f(x) \cdot g(x)s$ and $D_x f(x)$ and $D_x g(x)$ exist, then

$$D_x h(x) = f(x) D_x g(x) + g(x) D_x f(x)$$

If $f(x)$ and $g(x)$ are called the first and second members of the product, respective, the product theorem can be stated in colloquial language as follows:

$$D_x h(x) = \begin{pmatrix}\text{first}\\ \text{function}\end{pmatrix}\begin{pmatrix}\text{derivative of the}\\ \text{second function}\end{pmatrix} + \begin{pmatrix}\text{second}\\ \text{function}\end{pmatrix}\begin{pmatrix}\text{derivative of the}\\ \text{first function}\end{pmatrix}$$

For $h(x) = x^2 = x \cdot x$, we have

$$\begin{aligned} D_x(x^2) &= xD_x(x) + xD_x(x) \\ &= x \cdot 1 + x \cdot 1 = 2x \end{aligned}$$

EXAMPLE 3

Determine the derivative of $h(x) = (x^3 + x^2)(x^4 - x + 1)$.

By the product theorem

$$D_x h(x) = (x^3 + x^2) D_x(x^4 - x + 1) + (x^4 - x + 1) D_x(x^3 + x^2)$$

Now

$$\begin{aligned} D_x(x^4 - x + 1) = D_x(x^4) - D_x(x) + D_x(1) &= 4x^3 - 1 + 0 \\ &= 4x^3 - 1 + 0 \\ D_x(x^3 + x^2) = D_x(x^3) + D_x(x^2) &= 3x^2 + 2x \end{aligned}$$

Thus we have

$$D_x h(x) = (x^3 + x^2)(4x^3 - 1) + (x^4 - x + 1)(3x^2 + 2x)$$

Derivative of a Constant Function Times a Function

If $f(x) = c$ is a constant function, and $h(x) = c \cdot g(x)$, then a special case worthy of particular importance arises. Applying the product theorem yields

$$D_x h(x) = cD_x g(x) + g(x) D_x(c)$$

Since $D_x(c) = 0$, we obtain

$$\boxed{D_x h(x) = c \cdot D_x g(x)}$$

In colloquial language, the derivative of a constant (function) times a function equals the constant times the derivative of the function. For example:

If $h(x) = 3x,\;\; D_x h(x) = 3D_x(x) = 3(1) = 3$.

If $h(x) = 4x^2,\;\; D_x h(x) = 4D_x(x^2) = 4(2x) = 8x$.

If $h(x) = 8\sqrt{x},\;\; D_x h(x) = 8D_x(x^{1/2}) = 8(\frac{1}{2}x^{-1/2}) = 4x^{-1/2} = \frac{4}{\sqrt{x}}$.

If $h(x) = \frac{1}{2}x^8,\;\; D_x h(x) = \frac{1}{2}D_x(x^8) = \frac{1}{2}(8x^7) = 4x^7$.

Derivative of a Quotient of Functions

The derivative theorem for a quotient of functions is also one that is analogous in spirit to the limit theorem for quotients, but different in structure.

Theorem. If $h(x) = f(x)/g(x)$ and $D_x f(x)$ and $D_x g(x)$ exist, the

$$D_x h(x) = \frac{g(x) \cdot D_x f(x) - f(x) \cdot D_x g(x)}{[g(x)]^2}$$

In colloquial language,

$$D_x h(x) = \frac{(\text{denominator})\begin{pmatrix}\text{der. of the}\\ \text{numerator}\end{pmatrix} - (\text{numerator})\begin{pmatrix}\text{der. of the}\\ \text{denominator}\end{pmatrix}}{(\text{denominator})^2}$$

EXAMPLE 4

Find the derivative of

$$h(x) = \frac{x^3 + 3x - 4}{4x^2 + 2}$$

The quotient theorem yields

$$D_x h(x) = \frac{(4x^2 + 2)D_x(x^3 + 3x - 4) - (x^3 + 3x - 4)D_x(4x^2 + 2)}{(4x^2 + 2)^2}$$

For the component parts, we have

$$D_x(x^3 + 3x - 4) = D_x(x^3) + 3D_x(x) - D_x(4) = 3x^2 + 3 - 0 = 3x^2 + 3$$
$$D_x(4x^2 + 2) = 4D_x(x^2) + D_x(2) = 4(2x) + 0 = 8x$$

Thus we obtain

$$D_x h(x) = \frac{(4x^2 + 2)\,(3x^2 + 3) - (x^3 + 3x - 4)(8x)}{(4x^2 + 2)^2}$$

EXERCISES

Differentiate the following functions.

1. $f(x) = 5x^7$
2. $f(x) = 2x^{10}$
3. $f(x) = -3x^{15}$
4. $f(x) = 3x^6$
5. $f(x) = 8x^{1/4}$
6. $f(x) = -3x^{1/3}$

7. $f(x)=\dfrac{3}{x^3}$ 8. $f(x)=4x^{-1/3}$ 9. $f(x)=\dfrac{5}{x^4}$

10. $f(x)=\dfrac{3}{x^{2/3}}$ 11. $f(x)=3x^2+2x-1$

12. $f(x)=4x^3-5x^2+2$ 13. $f(x)=4x^5-3x+\dfrac{1}{x}$

14. $f(x)=3x^6+5x-\dfrac{1}{x^2}$ 15. $f(x)=\dfrac{1}{10}x^2-5x+4$

16. $f(x)=\dfrac{1}{4}x^3-\dfrac{1}{2}x^2-11$ 17. $f(x)=\dfrac{3}{4}x^2-10x-\dfrac{4}{x}$

18. $f(x)=\dfrac{1}{x^3}-\dfrac{1}{x}-\sqrt{x}$ 19. $f(x)=(4x^2+2)(3x^4+2x)$

20. $f(x)=(x^2-1)(x^2+2)$ 21. $f(x)=(x^3-3x)(x^2-4x)$

22. $f(x)=x^3(1-2x)$ 23. $f(x)=5x^2(2-4x+\dfrac{1}{3}x^6)$

24. $f(x)=(3x-2)(4x^3+5)$ 25. $f(x)=(5x^2+3)x^4$

25. $f(x)=(3x^4-2x)(4x^7+13)$ 27. $f(x)=\dfrac{x+1}{1-x}$

28. $f(x)=\dfrac{3x^2+2}{1-x^2}$ 29. $f(x)=\dfrac{300x^2}{500x^2+10}$

30. $f(x)=\dfrac{1+x^3}{3x+2}$ 31. $f(x)=\dfrac{5-3x}{1-x^2}$

32. $f(x)=\dfrac{900x^3}{200+x^4}$ 33. $f(x)=\dfrac{x^6+\sqrt{x}}{5-7x}$

34. $f(x)=\dfrac{4x-500}{500+2x^4}$

6.3. THE CHAIN RULE

Consider the problem of finding the derivative of

$$y = (x^2 + 1)^2$$

One way of handling the problem is to multiply $(x^2 + 1)$ by itself and take the derivative of each term. We obtain

$$y = x^4 + 2x^2 + 1$$
$$D_x y = 4x^3 + 4x$$

The chain rule, as it is sometimes called, provides us with another approach to this problem, and to problems that are similar but cannot be handled by the device of multiplying out and taking the derivative of each term. Consider again

$$y = (x^2 + 1)^2$$

The chain rule is based on the idea of introducing two appropriate "helping functions" whose derivatives are related to the derivative we are seeking. We take as our first helping function $u = x^2 + 1$. Then from

$$y = \underbrace{(x^2 + 1)^2}_{u}$$

we obtain the second helping function, $y = u^2$. We now have three functions on the scene, the original function

$$y = (x^2 + 1)^2$$

and the two helping functions

$$u = x^2 + 1 \qquad \text{and} \qquad y = u^2$$

The chain rule expresses a remarkably simple relationship among the derivatives of these functions: the derivative of the original function, $D_x y$, is the product of the derivatives of the helping functions, $D_u y$ and $D_x u$.

$$\underbrace{D_x y}_{\substack{\text{derivative}\\ \text{of the original}\\ \text{function}}} \quad = \quad \underbrace{D_u y \cdot D_x u}_{\substack{\text{product of the}\\ \text{derivatives of}\\ \text{the helping functions}}}$$

$$y = u^2, \qquad D_u y = 2u$$
$$u = x^2 + 1, \quad D_x u = 2x$$

Thus from the chain rule we obtain

$$D_x y = 2u(2x)$$

Replacing u by $x^2 + 1$ yields

$$D_x y = 2(x^2 + 1)(2x)$$

By multiplying, we obtain

$$D_x y = 4x^3 + 4x$$

which agrees with the result obtained earlier by multiplying (x^2+1) by itself and taking the derivative of each term.

EXAMPLE 1

Find the derivative of $y = (x^4 - 2x + 1)^5$.

A clue to the appropriate introduction of a helping function is provided by the observation that if we had u^5 to contend with as opposed to $(x^4 - 2x + 1)^5$, our differentiation problem would be enormously simplified. This observation suggests taking $u = x^4 - 2x + 1^5$ as our first helping function. Then from

$$y = \underbrace{(x^4 - 2x + 1)^5}_{u}$$

the second helping function $y = u^5$ is immediately suggested.

Our helping functions, then, are

$$y = u^5, \qquad u = x^4 - 2x + 1$$

and their derivatives are expressed by

$$D_u y = 5u^4, \qquad D_x u = 4x^3 - 2$$

From the chain rule we have

$$\begin{aligned} D_x y &= D_u y \cdot D_x u \\ &= 5u^4(4x^3 - 2) \end{aligned}$$

Replacing u by $x^4 - 2x + 1$ yields

$$D_x y = 5(x^4 - 2x + 1)^4(4x^3 - 2)$$

More generally, the functions considered in the preceding examples have the structure $y = [f(x))]^n$. To differentiate functions with the

$$y = \underbrace{[f(x)]^n}_{u}$$

structure, we use the chain rule. Take $u = f(x)$ as the first helping function, and $y = u$" as the second helping function. The derivatives of

$$y = u^n \quad \text{and} \quad u = f(x)$$

are

$$D_u u^n \quad \text{and} \quad D_x f(x)$$

The chain rule yields

$$\begin{aligned} D_x y &= D_u u^n \cdot D_x f(x) \\ &= n u^{n-1} D_x f(x) \end{aligned}$$

Replacing u by $f(x)$ yields

$$D_x y = n[f(x)]^{n-1} D_x f(x)$$

In colloquial language, if

$$y = (\text{function of } x)^n$$

then

$$D_x y = n(\text{function of } x)^{n-1} \cdot \begin{pmatrix} \text{derivative of the} \\ \text{function of } x \end{pmatrix}$$

EXAMPLE 2

Find the derivative of $y = (3x^2 - 3x + 5)^4$.

By the chain rule

$$\begin{aligned} D_x y &= 4(3x^2 - 3x + 5)^3 D_x(3x^2 - 3x + 5) \\ &= 4(3x^2 - 3x + 5)^3 (6x - 3) \end{aligned}$$

EXAMPLE 3

Find the derivative of $y = \sqrt[3]{5x^4 + 2x - 3}$.

$$
\begin{aligned}
y &= (5x^4 + 2x - 3)^{1/3} \\
D_x y &= \frac{1}{3}(5x^4 + 2x - 3)^{-2/3} D_x(5x^4 + 2x - 3) \\
&= \frac{1}{3}(5x^4 + 2x - 3)^{-2/3}(20x + 2) \\
&= \frac{20x + 2}{3\sqrt[3]{(5x^4 + 2x - 3)^2}}
\end{aligned}
$$

EXAMPLE 4

Find the derivative of $y = (x^3 + 2)\sqrt{x^2 + 1}$.

The question is, which theorem do we apply first, the chain rule or the product theorem? The product theorem takes precedence; we first apply the product theorem to break the components apart. This makes it possible to apply the chain rule to obtain the derivative of the second component of the product.

Applying the product theorem yields:

$$D_x y = (x^3 + 2) D_x (x^2 + 1)^{1/2} + (x^2 + 1)^{1/2} D_x (x^3 + 2) \qquad (6.1)$$

$$D_x (x^3 + 2) = 3x^2$$

From the chain rule, we have

$$
\begin{aligned}
D_x (x^2 + 1)^{1/2} &= \frac{1}{2}(x^2 + 1)^{-1/2} D_x (x^2 + 1) \\
&= \frac{1}{2}(x^2 + 1)^{-1/2}(2x) \\
&= \frac{x}{\sqrt{x^2 + 1}}
\end{aligned}
$$

Substituting these results into (6.1) yields

$$D_x y = \frac{(x^3+2)x}{\sqrt{x^2+1}} + 3x^2\sqrt{x^2+1}$$

EXERCISES

Differentiate the following functions.

1. $y=(x^2+2)^7$
2. $y=(2x^4+x)^6$
3. $y=(4x^5-3x+1)^5$
4. $y=(2x^3-5x^2+1)^3$
5. $y=\sqrt{4x^3+2x-10}$
6. $y=\sqrt{5x^4-7x^2+13}$
7. $y=(4x^6-7x^4+10)^{1/4}$
8. $y=(3x^5-4x^4+2)^4$
9. $y=x^2(8x^6+4x^2-1)^3$
10. $y=\dfrac{5x+1}{(3x^2-4x)^4}$
11. $y=\dfrac{(4x^3-2x)^3}{4x^2}$
12. $y=(4x^2+1)\sqrt{x^3+6}$
13. $y=(5x^3-3)^4(3x^2-1)$
14. $y=x^2(3x+7)^6$
15. $y=\dfrac{4x+3}{\sqrt{x^2+1}}$
16. $y=\dfrac{\sqrt{4x+7}}{x^4}$

6.4. IMPLICIT DIFFERENTIATION*

A function $y = f(x)$ is said to be defined **explicitly** if its dependent variable y is expressed directly in terms of x. All the functions that we have had occasion to consider were defined explicitly. As illustrations of explicitly defined functions, we have

* This topic may be omitted without loss of continuity.

$$y = x^2 + 2x - 3$$

and
$$y = \begin{cases} x+1, & \text{for } x \geq 0 \\ -x+1, & \text{for } x < 0 \end{cases}$$

Not all functions are defined explicitly. For example, an equation in x and y such as

$$\frac{y}{x^2} - 1 = 0 \qquad (6.2)$$

is said to define a function $y = f(x)$ if

$$\frac{f(x)}{x^2} - 1 = 0$$

holds for all x in the domain of definition of $y = f(x)$. Such a function $y = f(x)$ is said to be defined **implicitly** by the given equation. Sometimes an explicit expression for the implicitly defined function can be obtained. Solving (6.2) for y yields

$$\frac{y}{x^2} - 1 = 0, \quad \frac{y}{x^2} = 1, \quad y = x^2$$

so that $y = x^2$, $x \neq 0$, is an explicit expression for the function $y = f(x)$ defined implicitly by (6.2). Sometimes it is either very difficult or impossible to obtain an explicit expression for an implicitly defined function. The derivative of such a function can be determined by differentiating each term in the given equation in x and y, keeping in mind that y represents an implicitly defined function $f(x)$. The result obtained can often be solved for the derivative of the implicitly defined function. This process is called **implicit differentiation**. For example, for the function $y = x^2$, $x \neq 0$, defined implicit by

$$\frac{y}{x^2}-1=0$$

implicit differentiation yields the following:

$$D_x\left(\frac{y}{x^2}\right)-D_x(1)=D_x(0)$$

Since $D_x(1)=0$ and $D_x(0)=0$, we have

$$D_x\left(\frac{y}{x^2}\right)=0$$

The quotient theorem yields

$$\frac{x^2D_xy-yD_x(x^2)}{(x^2)^2}=0$$

$$\frac{x^2D_xy-2xy}{x^4}=0$$

Solving for D_xy yields

$$\begin{aligned} x^2D_xy-2xy&=0\\ x^2D_xy&=2xy\\ D_xy&=\frac{2xy}{x^2}\\ &=\frac{2y}{x} \end{aligned} \tag{6.3}$$

Usually, this would be as far as we would go, since $y = f(x)$ is not explicitly known. Here $y = x^2$, $x \neq 0$, and substitution of this result into (6.3) yields the expected result:

$$D_x y = \frac{2(x^2)}{x} = 2x\,, \qquad x \neq 0$$

EXAMPLE 1

The equation $x^2 + y^2 = 5\,x^2{+}y^2{=}5$ defines two differentiable functions, $y = \sqrt{5 - x^2}$ and $y = -\sqrt{5 - x^2}$. Determine the derivatives of these functions by implicit differentiation.

Term-by-term differentiation of $x^2 + y^2 = 5$, thinking of y as a function of x, yields

$$\begin{aligned} D_x(x^2 + y^2) &= D_x(5) \\ 2x + 2yD_x y &= 0 \\ 2yD_x y &= -2x \\ D_x y &= -\frac{x}{y} \end{aligned}$$

Thus

$$D_x\sqrt{5 - x^2} = -\frac{x}{\sqrt{5 - x^2}} \text{ and } D_x(-\sqrt{5 - x^2}) = \frac{x}{\sqrt{5 - x^2}}$$

EXERCISES

Determine $D_x y$ for the functions implicitly defined by the following equations.

1. $x^2 y - 10 = 0$
2. $xy - 1 = 0$
3. $x^2 y - y - 4 = 0$
4. $x^3 y + 12 = 0$
5. $x^2 y - x + 8 = 0$
6. $x^2 + y^2 = 1$

7. $x^2 + y^2 = 16$

8. $4x^2 + 9y^2 = 25$

9. $xy + x^2 = 20$

10. $xy = 4$

11. $y^2 - 6x = 0$

12. $x^3 - x = y^3 - y^2 + 10$

6.5. INVERSE FUNCTIONS AND THEIR DERIVATIVES

Consider a market within which two variables, x and y, have been singled out for study. Variable y represents the quantity of coal in thousands of tons per month purchased in the region, and x represents the price of coal in dollars per ton. In this, and in many other situations as well, it is useful to express y as a function of x, and also to express x as a function of y.

> If $y = f(x)$ defines y as a function of x and x can be expressed as a function of y, $x = g(y)$, then $x = g(y)$ is said to be the **inverse** of $y = f(x)$, and vice versa.

If, for example, the function

$$y = -2x + 150$$

describes the quantity of coal purchased as a function of price, then solving for x in terms of y yields its inverse

$$x = -\frac{1}{2}y + 75$$

which expresses the market price of coal as a function of the quantity purchased by the market.

Let us observe that both $y = -2x + 150$ and its inverse, $x = -\frac{1}{2}y + 75$, have derivatives, given by

$$D_x y = -2 \qquad \text{and} \qquad D_y x = -\frac{1}{2}$$

and that these derivatives are related by

$$D_x y = \frac{1}{D_y x}$$

Such a relationship between the derivatives of functions that are inverses of each other holds in general.

Derivative of the Inverse of a Function. If $y = f(x)$ and $x = g(y)$ are inverses of each other, and $D_x y$ and $D_y x$ exist, then

$$D_y x = \frac{1}{D_x y}$$

in any interval in which $D_x y \neq 0$.

EXAMPLE 1

Find the derivative of the inverse of $y = x^3 + 4$.

$$D_y x = \frac{1}{D_x y} = \frac{1}{D_x(x^3 + 4)} = \frac{1}{3x^2}$$

EXERCISES

Find the inverse function of each of the following functions.

1. $y = x + 10$
2. $y = 5x + 120$
3. $x = 27y^3$
4. $y = \frac{x+1}{x}$
5. $y = \log_{10} x$
6. $y = \sqrt{x+4}$

7. If $x = -5y + 30$ describes the quantity of beef purchased by a market in millions of pounds per month as a function of the price of beef in dollars per pound, then find the inverse function that describes the price of beef per pound as a function of the quantity of beef in millions of pounds purchased per month by the market.

Find the derivative of the inverse of each of the following functions by using the theorem on the derivative of the inverse of a function.

8. $y = 3x + 2$
9. $y = 4x + 12$
10. $y = (x+1)^3$
11. $y = x^3 + 27$
12. $x = 27y^3$
13. $x = (y+4)^3$
14. $y = \dfrac{x+1}{x}$

6.6. EXPONENTIAL FUNCTIONS AND THEIR DERIVATIVES

The functions

$$y = 2^x, \quad y = 10^x. \quad y = (\frac{1}{2})^x, \quad \text{and} \quad y = e^x$$

Where $e \simeq 2.71828$ and is defined by $\lim_{n\to\infty} [1 + (1/n)]^n$ (see Section 3.4, Example 3, p. 112), illustrate functions with the general structure

$$y = b^x$$

where b is a positive constant other than 1. Functions of this type, with a fixed base and variable exponent, are called **exponential functions.** To obtain a geometric view of exponential functions, consider $y = 2^x$ and $y = (\frac{1}{2})^x$. Values of y for given values of x for these functions are given in Tables 6.1 and 6.2, from which we obtain the graphs shown in Figures 6.1 and 6.2.

Table 6.1

x	−4	−3	−2	−1	0	1	2	3	4
2^x	$\frac{1}{16}$	$\frac{1}{8}$	$\frac{1}{4}$	$\frac{1}{2}$	1	2	4	8	16

Table 6.2

x	−4	−3	−2	−1	0	1	2	3	4
$(\frac{1}{2})^x$	16	8	4	2	1	$\frac{1}{2}$	$\frac{1}{4}$	$\frac{1}{8}$	$\frac{1}{16}$

More generally, the graph of $y = b^x$, where $b > 1$, has the form shown in Figure 6.1, and the graph of $y = b^x$, where b is between 0 and 1 ($0 < b < 1$), has the form shown in Figure 6.2.

Exponential functions, functions of exponential type (built up from exponential functions), and their derivatives appear in a number of situations in the worlds of biology, business, economics, finance, physics, and psychology. The psychologist C. L. Hull* was led to the function

$$h(x) = 100(1 - e^{-ax})$$

where a is a positive constant, to describe habit strength in terms of repetitions. The graph of $h(x)$ is shown in Figure 6.3. In his now classic work on population dynamics, G. F. Gause† obtained the function

---

* C. L. Hull, *Principles of Behavior* (New York: Appleton-Century-Crofts, 1945).

† G. F. Gause, *The Struggle for Existence* (Baltimore, Md.: Williams & Wilkins Co., 1934; reprinted by Dover Publications, Inc., New York, 1971), p. 55.

$$y = a(1 - e^{-kx})$$

where a and k are positive constants, to describe the relative increase of predators as a function of the concentration of prey.

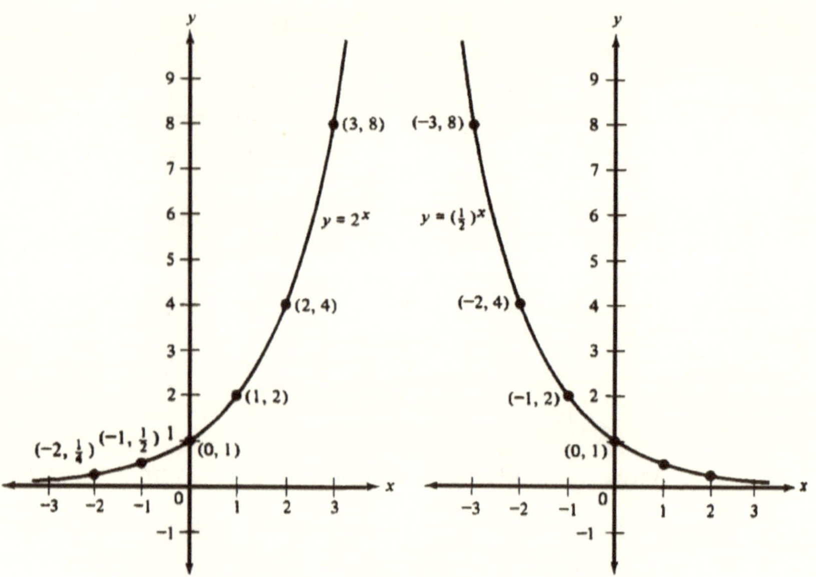

Figure 6.1 *Figure 6.2*

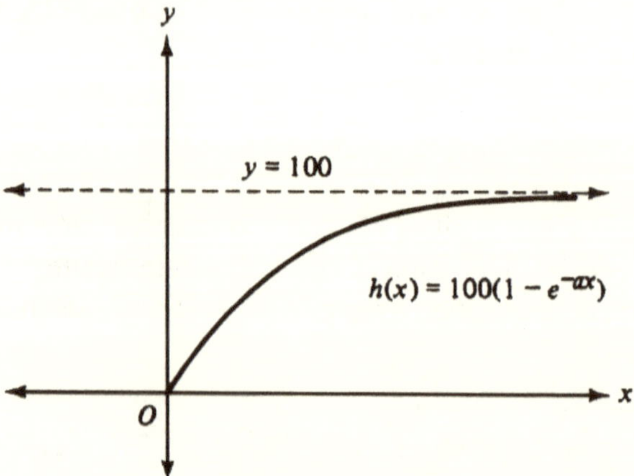

Figure 6.3

In the world of finance and economics, continuous compounding of interest problems give rise to exponential-type functions of the form $a(x) = Ae^{rx}$ and $b(x) = Ae^{-rx}$, which are discussed in Section 4.2

Derivatives of Exponential Functions

We focus our attention on the exponential function

$$f(x) = e^x$$

whose central importance is made clear by the preceding examples. The derivative function is, by definition,

$$f'(x) = \underset{h \to 0}{\text{limit}} \frac{f(x+h) - f(x)}{h}$$

provided that this limit exists.

$$f(x+h) = e^{x+h}, \qquad f(x) = e^x$$

Thus

$$\begin{aligned} f(x+h) - f(x) &= e^{x+h} - e^x \\ &= e^x e^h - e^x \\ &= e^x(e^h - 1) \end{aligned}$$

Dividing by h yields

$$\frac{f(x+h) - f(x)}{h} = \frac{e^x(e^h - 1)}{h}$$

Therefore,

$$f'(x) = \lim_{h\to 0} \frac{e^x(e^h - 1)}{h}$$
$$= e^x \lim_{h\to 0} \frac{e^h - 1}{h}$$

Since e^x is constant as far as h is concerned, the problem of determining $f'(x)$ is thus reduced to finding $\lim_{h\to 0} (e^h - 1)/h$. Values of e^h and $(e^h - 1)/h$ for values of h that are getting closer and closer to 0 are shown in Table 6.3, from which it would seem that $(e^h - 1)/h$ approaches 1, or some value near 1, as $h \to 0$. This is indeed the case;

Table 6.3

h	0.10	-0.10	0.05	-0.05	0.01	-0.01	0.001	-0.001
e^h	1.10517	0.90484	1.05127	0.95123	1.01005	0.99005	1.0010005	0.9990005
$\frac{e^h - 1}{h}$	1.052	0.952	1.025	0.975	1.005	0.995	1.0005	0.9950

It can be shown that

$$\lim_{h\to 0} \frac{e^h - 1}{h} = 1$$

so that

$$f'(x) = e^x \lim_{h\to 0} \frac{e^h - 1}{h}$$
$$= e^x(1)$$
$$= e^x$$

Thus we have the following remarkable result:

For $f(x) = e^x$, $\qquad f'(x) = e^x$

EXAMPLE 1

Find the derivative of $y = \frac{e^x}{x^2+4}$.

By the quotient theorem, we have

$$D_x y = \frac{(x^2+4)D_x(e^x) - e^x D_x(x^2+4)}{(x^2+4)^2}$$
$$= \frac{(x^2+4)e^x - e^x(2x)}{(x^2+4)^2}$$

EXAMPLE 2

Find the derivative of $y = e^{4x+3}$.

The exponent is a function of x, a structure that calls for use of the chain rule. We take $u = 4x+3$ as our first helping function; then $y = e^u$ is immediately obtained as the second helping function. In summary, our helping functions and their derivatives are given by

$$y = e^u, \qquad u = 4x+3$$
$$D_u y = e^u, \qquad D_x u = 4$$

From the chain rule,

$$D_x y = D_u y \quad D_x u$$

we obtain

$$D_x y = e^u(4) = 4e^u$$

Replacing u by $4x + 3$ yields

$$D_x y = 4e4^{x+3}$$

More generally, if

$$y = e^{f(x)}$$

where $f(x)$ is a function of x, let $u = f(x)$. Then $y = e^u$ is immediately obtained. We have then the helping functions

$$y = e^u, \quad u = f(x)$$

with derivatives

$$D_u y = e^u, \quad D_x f(x)$$

The chain rule yields

$$\begin{aligned} D_x y &= D_u y \;\; D_x f(x) \\ &= e^u D_x f(x) \end{aligned}$$

By replacing u by $f(x)$, we obtain

$$D_x y = e^{f(x)} \quad D_x f(x)$$

In colloquial language, if

$$y = e^{(\text{function of } x)}$$

then

$$D_x y = e^{(\text{function of } x)} \; D_x \, (\text{function of } x)$$

Thus, for example:

If $y = e^{x^2-1}, D_x y = e^{x^2-1} D_x(x^2 - 1) = e^{x^2-1}(2x)$.

If $y = e^{-x}, D_x y = e^{-x} D_x(-x) = e^{-x}(-1) = -e^{-x})$.

EXAMPLE 3

Find the derivative of $y = x^4 e^{3x-2}$.

First and foremost, we have a product of functions. Thus the product theorem is first in order; it yields

$$D_x y = x^4 D_x(e^{3x-2}) + e^{3x-2} D_x(x^4) \tag{6.4}$$

By the chain rule

$$D_x(e^{3x-2}) = e^{3x-2} D_x(3x-2) = 3e^{3x-2}$$

Also

$$D_x(x^4) = 4x^3$$

Substituting these results into (6.4) yields

$$D_x y = 3x^4 e^{3x-2} + 4x^3 e^{3x-2}$$

EXERCISES

1. Sketch the graph of $f(x) = 3^x$.

2. Sketch the graph of $f(x) = (\frac{1}{3})^x$.

Differentiate the following functions.

3. $y = x^2 + e^x$
4. $y = 4e^x$
5. $y = x^2 e^x$
6. $y = \frac{e^x}{x^2}$
7. $y = 3x^4 - 5e^x$
8. $y = 2x^3 e^x$
9. $y = 4xe^x$
10. $y = e^{-2x+1}$
11. $y = 3x^2 + 2x + xe^x$
12. $y = 2x^3 e^x$
13. $y = 4xe^x + x^4$
14. $y = 3e^x + 10$
15. $y = \frac{3e^x}{x+4}$
16. $y = \frac{x^2+1}{e^x}$
17. $y = \frac{2x+5}{xe^x}$
18. $y = e^{3x+8}$
19. $y = 4e^{6x}$
20. $y = x^2 e^{5x}$
21. $y = \frac{2e^x+1}{2x-3}$
22. $y = 7e^{-x}$
23. $y = x^2 e^{-x}$
24. $y = \frac{(4x+3)}{3e^x}$
25. $y = e^{-x^2}$
26. $y = 3x^2 e^{-x^2}$

6.7. LOGARITHMIC FUNCTIONS AND THEIR DERIVATIVES

Logarithms

Logarithms were originally developed in the early seventeenth century in response to a need for simplifying extremely tedious computations that arose in astronomy. Although sophisticated calculators and computers have now made logarithms obsolete as a calculation tool, logarithmic functions arise in higher mathematics and its applications—calculus, in particular—and familiarity with the logarithm concept is essential.

> If b is a positive number, $b \neq 1$, and $b^y = x$, then the exponent y is called the **logarithm of x to the base b**. In symbols we write
>
> $$y = \log_b\ x$$

Thus the logarithm of x to the base b is the exponent y to which b must be raised to obtain x.

For example,

$$\begin{aligned} \log_{10}\ 100 &= 2, \quad \text{since } 10^2 = 100 \\ \log_3\ 81 &= 4, \quad \text{since } 3^4 = 81 \\ \log_{10}\ 1 &= 0, \quad \text{since } 10^0 = 1 \\ \log_4\ 4 &= 1, \quad \text{since } 4^1 = 4 \end{aligned}$$

EXERCISES

Express the following exponential statements in logarithmic notation.

1. $6^2 = 36$ 2. $10^3 = 1000$ 3. $25^{1/2} = 5$

4. $2^3 = 8$ 5. $27^{1/3} = 3$ 6. $10^{-2} = \frac{1}{100}$

7. $5^{-3} = \frac{1}{125}$ 8. $32^{-1/5} = \frac{1}{2}$ 9. $9^{3/2} = 27$

10. $5^0 = 1$ 11. $81^{-3/4} = \frac{1}{27}$ 12. $10^{-3} = \frac{1}{1000}$

Express the following logarithmic statements in exponential form.

13. $\log_2 16 = 4$ 14. $\log_3 1 = 0$ 15. $\log_5 125 = 3$

13. $\log_{12} 144 = 2$ 17. $\log_{10} 0.1 = -1$ 18. $\log_8 4 = \frac{2}{3}$

19. $\log_7 343 = 3$ 20. $\log_{10} 10,000 = 4$ 21. $\log_{10} 0.01 = -2$

Find the value of each of the following.

22. $\log_2 8$ 23. $\log_{10} 1000$ 24. $\log_{10} 0.001$

25. $\log_3 3$ 26. $\log_9 27$ 27. $\log_8 \frac{1}{2}$

Logarithms with respect to base e, called **natural logarithms**, are of particular importance in mathematics and its applications. They are the focus of our attention here. The $\log_e x$ is denoted by $\ln x$.

Logarithmic Functions

A function of the form

$$y = \log_b x$$

where base b is a positive constant and $b \neq 1$, is called a **logarithmic function**. Logarithmic functions are defined only for positive real numbers. To obtain a geometric view of logarithmic functions, consider $y = \log_2 x$ and $y = \log_{1/2} x$. Values of y for selected values of x for these functions are given in Tables 6.4 and 6.5, from which we obtain the graphs shown in Figure 6.4 and 6.5. More generally, the graph of $y = \log_b x$, where $b > 1$ (shown in Figure 6.6) has the same form as $y = \log_2 x$, and the graph of $y = \log_b x$, where $0 < b < 1$ (shown in Figure 6.7), has the same form as $y = \log_{1/2} x$.

Table 6.4

x	$2^{-2} = \frac{1}{4}$	$2^{-1} = \frac{1}{2}$	$2^0 = 1$	$2^1 = 2$	$2^2 = 4$	$2^3 = 8$
$\log_2 x$	-2	-1	0	1	2	3

Table 6.5

x	$(\frac{1}{2})^3 = \frac{1}{8}$	$(\frac{1}{2})^2 = \frac{1}{4}$	$(\frac{1}{2})^1 = \frac{1}{2}$	$(\frac{1}{2})^0 = \frac{1}{1}$	$(\frac{1}{2})^{-1} = 2$	$(\frac{1}{2})^{-2} = 4$	$(\frac{1}{2})^{-3} = 8$
$\log_{1/2} x$	3	3	1	0	–1	–2	–3

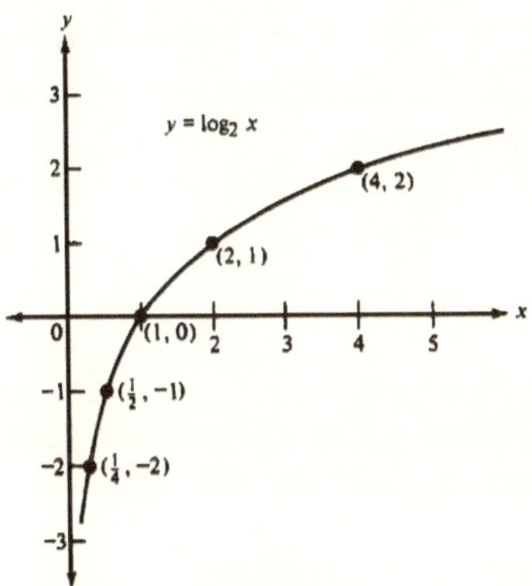

Figure 6.4

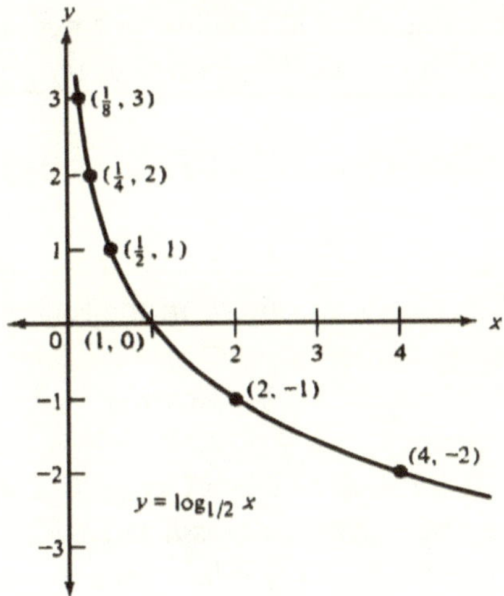

Figure 6.5

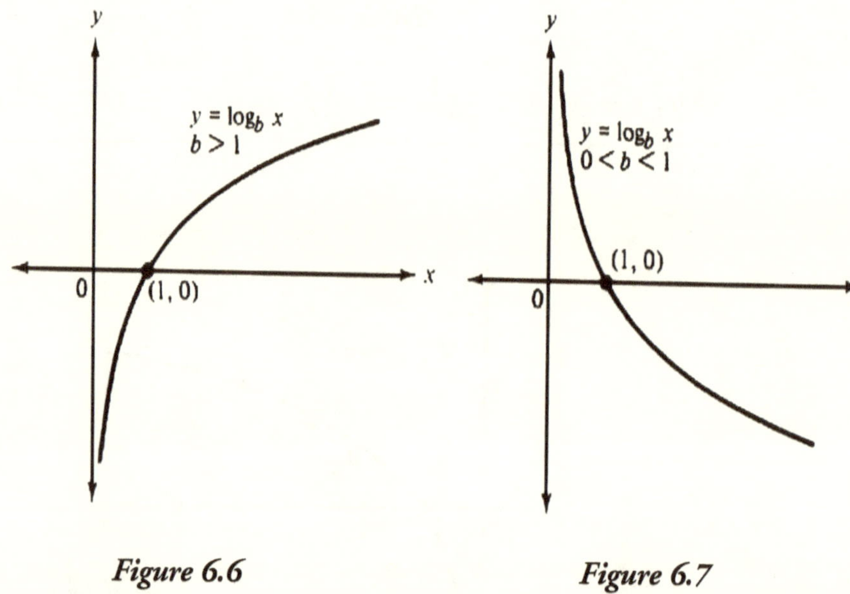

Figure 6.6 *Figure 6.7*

An interesting situation that gives rise to a logarithmic-type function involves potassium-argon dating. The potassium-argon dating technique is based on the fact that potassium 40, in the course of undergoing radioactive decay, is transformed into argon 40. This technique can be used to date potassium-bearing rocks and minerals that were as recently formed as 10,000 years ago and as old as 4 billion years. The potassium-argon age function is

$$t = (1.885 \cdot 10^{9}) \ln (9.068x + 1)$$

where t is age in years and x is the ratio of the amount of argon 40 to the amount of potassium 40 in the rock today.

A major archeological event occurred in 1959 when Dr. and Mrs. Louis Leakey discovered a fossil hominid skull, which they termed *Zinjanthropus boisei*, at Olduvai Gorge in Tanzania. Potassium-argon dating of the volcanic material associated with the hominid remains of Olduvai Gorge played a key role in establishing that the *Zinjanthropus* remains were about 1.75 million years old.

EXERCISES

28. Sketch the graph of $y = \log_3 x$.

29. Sketch the graph of $y = \log_{1/3} x$.

Derivatives of Logarithmic Functions

The logarithmic function $y = \log_e x$, expressed by the notation

$$y = \ln x,$$

is the inverse of $x = e^y$. Whenever one appears the other cannot be far behind. Both of these functions have derivatives, and to establish the following theorem on the derivative of $y = \ln x$ we employ the theorem on the derivative of the inverse of a function.

Theorem. If $y = \ln x$, then $D_x y = \frac{1}{x}$

Analysis. $y = \log_e x$ is, by definition of logarithm, the inverse of $x = e^y$. Both functions have derivatives. Thus:

$$D_x y = \frac{1}{D_y x} = \frac{1}{e^y} = \frac{1}{x}$$

EXAMPLE 1

Find the derivative of $y = \dfrac{x^3 + 1}{\ln x}$.

By the quotient theorem, we have

$$D_x y = \frac{(\ln x) D_x (x^3 + 1) - (x^3 + 1) D_x (\ln x)}{(\ln x)^2}$$

$$= \frac{(\ln x)(3x^2) - (x^3 + 1) \cdot \frac{1}{x}}{(\ln x)^2}$$

Multiplying numerator and denominator by x yields

$$D_x y = \frac{3x^3 \ln x - (x^3 + 1)}{x(\ln x)^2}$$

EXAMPLE 2

Find the derivative of $y = \ln(2x^2 - 4)$.

A situation such as this, in which the logarithm of a function of x is before us, calls for use of the chain rule. Let u be the function of x; that is, $u = 2x^2 - 4$. Then $y = \ln u$ is immediately obtained.

We have then the helping functions

$$y = \ln u, \quad u = 2x^2 - 4$$

with derivatives

$$D_u y = \frac{1}{u}, \quad D_x u = 4x$$

From the chain rule

$$D_x y = D_u y \cdot D_x u$$

we have

$$D_x y = \frac{1}{u} \cdot (4x)$$

Replacing u by $2x^2 - 4$ yields

$$D_x y = \frac{4x}{2x^2 - 4}$$

More generally, if

$$y = \ln f(x)$$

let $u = f(x)$ be the first helping function. Then $y = \ln u$ is obtained as the second helping function. We have the helping functions

$$y = \ln u, \qquad u = f(x)$$

with derivatives

$$D_u y = \frac{1}{u}, \qquad D_x f(x)$$

The chain rule yields

$$\begin{aligned} D_x y &= D_u y \cdot D_x u \\ &= \frac{1}{u} D_x f(x) \end{aligned}$$

Replacing u by $f(x)$ yields

$$D_x y = \frac{D_x f(x)}{f(x)}$$

In colloquial language, if

$$y = \ln \text{(function of } x\text{)}$$

then

$$D_x y = \frac{D_x \text{(function of } x\text{)}}{\text{function of } x}$$

Thus, for example:

$$\text{If } y = \ln(3x-1),\quad D_x y = \frac{D_x(3x-1)}{3x-1} = \frac{3}{3x-1}.$$

$$\text{If } y = \ln(x^3+4),\quad D_x y = \frac{D_x(x^3+4)}{x^3+4} = \frac{3x^2}{x^3+4}.$$

EXERCISES

Differentiate the following functions.

30. $y = x^2 \ln x$
31. $y = \dfrac{\ln x}{3x+1}$
32. $y = \dfrac{4x+3}{\ln x}$
33. $y = 3e^x \ln x$
34. $y = 4x^2 + e^{x^2} + \ln x$
35. $y = \ln(3x+10)$
36. $y = \ln(4x^2+5)$
37. $y = \sqrt{\ln x}$
38. $y = \ln \dfrac{50}{x}$
39. $y = \dfrac{x + \ln x}{2x+1}$
40. $y = 3x \ln(x^2+1)$
41. $y = (\ln x)^2$
42. $y = e^{x^2} \ln x$
43. $y = \dfrac{3e^x}{(\ln x)^2}$
44. $y = \dfrac{3x^2 + e^{-x}}{\ln x}$

6.8. RETURN TO DERIVATIVE STRUCTURES IN APPLIED SETTINGS

Now that we have developed tools to help us determine derivatives, let us return to some of the concepts which exhibit the derivative structure and apply the results to their study. Concepts which exhibit the derivative structure include the following.

Instantaneous Velocity. If $d = f(t)$ is a time-distance function which describes the motion of some object, then $f'(t)$ is defined as the **instantaneous velocity of the object at time *t*.**

For $f(t) = 3t^3 + 4$, the instantaneous velocity at time t is $f'(t) = 9t^2$. The **instantaneous velocity at time** $t = 2$ is $f'(2) = 36$.

Instantaneous Acceleration. If $v = v(t)$ describes the instantaneous velocity at time *t* of an object in motion, then $v'(t)$ is defined as the instantaneous acceleration of the object at time *t*.

For $v(t) = 9t^2$, the instantaneous acceleration at time t is $v'(t) = 18t$. The instantaneous acceleration at time $t = 2$ is $v'(2) = 36$.

Tangent Line to a Curve. The tangent line to the graph of a function *f* at point *P*(*x*, *f*(*x*)) is defined as the **line through** *P* **with slope** $f'(x)$.

EXAMPLE 1

Find the tangent line to the graph of $y = 1/x^2$ at *P*(1, 1).

The slope of this tangent line is the derivative of $f(x) = 1/x^2 = x^{-2}$ at *x* = 1. We begin by determining $f'(x)$ and then obtaining $f'(1)$.

$$f(x) = x^{-2}, \quad f'(x) = -2x^{-2-1} = -2x^{-3} = \frac{-2}{x^3}$$

Thus $f'(1) = -2$. The tangent line passes through *P*(1, 1) and has slope -2; it is given by the equation $y = -2 + 3$ and shown in Figure 6.8.

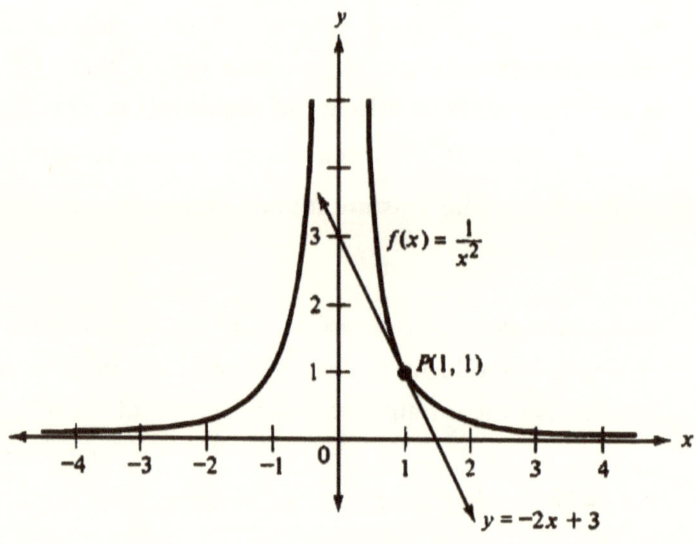

Figure 6.8

Marginal Revenue. If $R = R(x)$ is the total revenue function of a producer supplying x units of a certain commodity to a market per unit time, then $R'(x)$ is defined as the **marginal revenue for an output of x units.**

To illustrate, for the total revenue function $R(x) = 800x - 3x^2$ (in dollars), the marginal revenue for an output of x tons is $R'(x) = 800 - 6x$. The marginal revenue for an output of 100 tons is $R'(100) = 200$. We say that revenue is increasing at the instantaneous rate of $200 per ton of the commodity when 100 tons are produced.

EXAMPLE 2

The total revenue function of a tea producer is

$$R(x) = \frac{1000x}{x+20} - 5x$$

where x is output in hundreds of pounds per day and $R(x)$ is revenue in dollars per 100 pounds. Find the marginal revenue function and the marginal revenue for an output of 3000 pounds.

The marginal revenue function is the derivative of the total revenue function. Use of the differences and quotient theorems for derivatives yields

$$D_x R(x) = \frac{(x+20)D_x(1000x) - (1000x)D_x(x+20)}{(x+20)^2} - D_x(5x)$$

Since $D_x(1000x) = 1000$, $D_x(x+20) = 1$, and $D_x(5x) = 5$, we obtain

$$\begin{aligned} D_x R(x) &= \frac{(x+20)(1000) - 1000x}{(x+20)^2} - 5 \\ &= \frac{20,000}{(x+20)^2} - 5 \end{aligned}$$

The marginal revenue for an output of 3000 pounds per day is

$$R'(30) = \frac{20,000}{(50)^2} - 5 = 3$$

EXAMPLE 3

The total revenue function for a fruit-juice producer is given by

$$R(x) = 50x \cdot \ln \frac{60,000}{x}$$

where x is output in gallons per day. Find the marginal-revenue function.

First and foremost, $R(x)$ is a product of functions, so that to obtain $D_x R(x)$ the product theorem is the first tool to be employed. Thus

$$D_x R(x) = 50x D_x \ln\frac{60,000}{x} + \left(\ln\frac{60,000}{x}\right) D_x(50x)$$

From the chain rule, we have

$$D_x \ln\frac{60,000}{x} = \frac{D_x\left(\frac{60,000}{x}\right)}{\frac{60,000}{x}} = \frac{\frac{-60,000}{x^2}}{\frac{60,000}{x}} = \frac{-1}{x}$$

Also

$$D_x(50x) = 50$$

Substituting these results into (6.5) yields

$$D_x R(x) = 50x\left(\frac{-1}{x}\right) + \left(\ln\frac{60,000}{x}\right)50$$
$$= -50 + 50\ln\frac{60,000}{x}$$

Marginal Cost If $c = c(x)$ is the total cost function of a firm producing x units of a certain commodity per unit time, then $c'(x)$ is defined as the marginal cost for an output of x units.

For the total cost function $c(x) = \frac{1}{5}x^2 + 10x + 200$ (in dollars) the marginal cost for an output of x tons is $c'(x) = \frac{2}{5}x + 10$. The marginal cost for an output of 50 tons is $c'(50) = 30$. We say that cost is increasing at the instantaneous rate of $30 per ton of the commodity when 50 tons are produced.

EXAMPLE 4

The cost function for a steel plant is given by

$$c(x) = \sqrt{1000x + 100,000}$$

where x is output in tons per day and $c(x)$ is cost in hundreds of dollars. Find the marginal-cost function and the marginal cost for an output of 100 tons per day.

To differentiate $c(x)$ we first write it in exponent terms and then follow up with the chain rule.

$$c(x) = (1000x + 100,000)^{1/2}$$

By applying the chain rule we obtain:

$$c'(x) = \frac{1}{2}(1000x + 10,000)^{-1/2}(1000)$$
$$= \frac{500}{\sqrt{1000x + 100,000}}$$

The marginal cost for an output of 100 tons per day is

$$c'(100) = \frac{500}{\sqrt{200,000}} = \frac{500}{447.21} = 1.12\,.$$

Price Elasticity of Demand Let $x = f(p)$ denote the demand function for a commodity in a certain market which expresses the quantity purchased per unit time, x, as a function of the unit price p. If the price changes by an amount h from p to $p + h$, then the change in the quantity purchased is $f(p+h) - f(p)$. The percentage changes in quantity and price are given by

$$\frac{f(p+h)-f(p)}{f(p)} \text{ x } 100 \text{ percent and } \frac{h}{p} \text{ x } 100 \text{ percent}$$

The ratio

$$\frac{\dfrac{f(p+h)-f(p)}{f(p)}(100)}{\dfrac{h}{p}(100)} = \frac{[f(p+h)-f(p)]100}{f(p)} \cdot \frac{p}{100h}$$

$$= \frac{p}{f(p)} \cdot \frac{f(p+h)-f(p)}{h}$$

expresses the ratio of the percentage change in quantity demanded to the given percentage change in price. If $f'(p)$ exists, then

$$\operatorname*{limit}_{h \to 0} \frac{p}{f(p)} \cdot \frac{f(p+h)-f(p)}{h} = \frac{p}{f(p)} f'(p) = \frac{p}{x} f'(p)$$

If $x = f(p)$ is the demand function of a commodity, then **the point elasticity of demand for price** p, denoted by the Greek letter η (eta), is defined by

$$\eta = -\frac{p}{x} f'(p)$$

The minus sign is introduced to make η a positive quantity. Its use is a matter of preference; some economists prefer not to introduce it.

The price elasticity of demand is a quantitative measure of the responsiveness of demand to changes in price, and it gives the approximate percent change in demand that corresponds to a change of 1 percent in the price when the price is p.

EXAMPLE 5

The demand function for sugar in Bell City is

$$f(p) = 120 - \frac{1}{5}p,$$

where $f(p)$ is demand in hundreds of pounds (per day) and p is the price of sugar in dollars per hundred pounds. Find the price elasticity of demand for a price of p_0 dollars per hundred pounds.

$f'(p_0) = -\frac{1}{5}$. Thus the point elasticity of demand for price p_0 is

$$\eta = -\frac{p_0}{f(p_0)} f'(p_0) = -\frac{p_0}{120 - \frac{1}{5}p_0}\left(-\frac{1}{5}\right) = \frac{p_0}{600 - p_0}.$$

The elasticity of demand for a market price of \$50 per hundred pounds is

$$\eta = \frac{50}{600 - 50} = 0.091.$$

This means that for a market price of \$50 per hundred pounds of sugar a small percent change in price will bring an approximate .091 percent change in demand.

There are three elasticity situations to be distinguished.

1. If $\eta > 1$, demand is said to be **elastic**. A small percentage change in price brings with it a greater percentage change in quantity demanded.
2. If $\eta = 1$, demand is said to have **unit elasticity**. A small percentage change in price brings with it an equal percentage change in quantity demanded.
3. If $\eta < 1$, demand is said to be **inelastic**. A small percentage change in price brings with it a smaller percentage change in demand.

To illustrate, let us return to the demand function $f(p) = 120 - \frac{1}{5}p$. $0 < p < 600$ (see Figure 6.9). As we have noted, point elasticity of demand for price p is

$$\eta = \frac{p}{600 - p}.$$

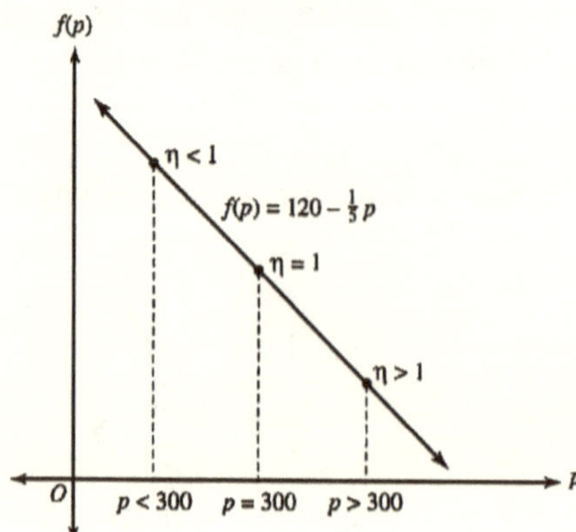

Figure 6.9

For $\eta = 1$ we have

$$\frac{p}{600 - p} = 1, \quad p = 600 - p, \quad 2p = 600, \quad p = 300.$$

Thus $\eta = 1$ when $p = 300$. Also, $\eta > 1$ when $p > 300$, and $\eta < 1$ when $p < 300$. Demand, in this situation, is inelastic when $p < 300$, of unit elasticity when $p = 300$, and elastic when $p > 300$.*

* For further discussion of price elasticity of demand, see R. G. D. Allen, *Mathematical Analysis for Economists* (New York: Macmillan Company, 1939). Section 10.6; and works on price theory and microeconomic theory such as D. S. Watson, *Price Theory and Its Uses*, 2d ed. (Boston: Houghton Mifflin Co., 1968), especially p. 137: C. E. Furguson, *Microeconomic Theory* (Homewood, Illinois: Richard D. Irwin, Inc., 1969); J. M. Henderson and R. E. Quandt, *Microeconomic Theory: A Mathematical Approach* (New York: McGraw-Hill, 1958).

Marginal Propensity to Consume. In macroeconomic theory, if $C = C(Y)$ expresses consumption C as a function of income Y, then the **marginal propensity to consume for an income** Y is defined as $c = \dfrac{dC}{dY}$

EXERCISES

1. Find the tangent line to the graph of $y = 8/x^3$ at the point $P(2, 1)$.

2. Find the tangent line to the graph of $y = 2x^2 - x + 4$ at the point $P(-1, 7)$.

3. Find the instantaneous velocity at time $t = 1$ for the object whose motion is described by the time-distance function $f(t) = 16t^2 + 100t$.

4. The cost in dollars of producing x tons of sugar per week is given by the cost function $c(x) = \frac{1}{4}x^2 + 2x + 850$. Find the marginal-cost function and the marginal cost for an output of 50 tons per week.

5. $R(x) = 1500x - 50x^2$ is the total revenue function of a coffee producer, where x is output in tons per day and $R(x)$ is revenue in dollars per ton. Find the marginal revenue for an output of 20 tons per day.

6. A tea plantation produces x tons of tea per day at a cost of

$$c(x) = \frac{x^2(x+50)}{50(x+200)} + 75$$

hundred dollars. Find the marginal-cost function and the marginal cost for an output of 10 tons of tea per day.

7. Find the tangent line to the graph of $f(x)=4x/(x+1)$ at the point $P(1, 2)$.

8. $R(x)=\dfrac{2000x}{x+40}-10x$ is the total revenue function of a chocolate producer, where x is output in tons per week and $R(x)$ is revenue in dollars per ton. Find the marginal-revenue function and the marginal revenue for an output of 10 tons per week.

9. Find the tangent line to the graph of $f(x)=10/(x^2+1)$ at the point $P(2, 2)$.

10. $R(x)=20x\ \ln(70{,}000/x)$ is the total revenue function for an orange-juice producer, where x is output in gallons per day. Find the marginal-revenue function.

11. The time-distance function of an object for which the resistance of air is proportional to the velocity of the object is

$$y=\frac{-100}{k}e^{-kt}+\frac{100}{k}$$

where k is a positive constant. Determine the instantaneous-velocity function.

12. Determine the instantaneous acceleration function for the object whose time-distance function is given in 11.

13. For Cameron City it is found that x, the number of gallons of ice tea purchased a day in the summer months, depends on p, the price in cents per gallon, according to

$$x=(90{,}000)e^{-(p/30)}.$$

Determine the elasticity of demand for a price of 30¢ a gallon.

14. The inverse of the ice tea demand function $x=(90{,}000)e^{-(p/30)}$ discussed in the previous exercise is

$$p = 30\ln\frac{90,000}{x}.$$

The total revenue function for the ice tea producers is obtained by multiplying the price of a unit p by output x. Thus we have

$$R = xp = 30x\ln\frac{90,000}{x}.$$

Determine the marginal revenue function for these ice tea producers.

15. The demand for chocolate in Chocolateville is described by the demand functions

$$x = \frac{800}{p+4} - 10 \quad \text{and} \quad p = \frac{800}{x+10} - 4$$

where x represents the market demand for chocolate in thousands of pounds per day and p is the price of chocolate in dollars per thousand-pounds. Determine the elasticity of demand for a price of \$36 per thousand-pounds. Assuming that the chocolate is produced by a monopolist, determine his total revenue and marginal revenue functions. What is the marginal revenue corresponding to a price of \$36 per thousand-pounds?

16. The motion of a certain object is described by the time-distance function $d = 3t^3 + 5t$. Find the instantaneous velocity and instantaneous acceleration of the object at time t.

17. A tea plantation produces x tons of tea a day at a cost of

$$c(x) = \frac{x^2(x+50)}{50(x+200)} + 75$$

hundred dollars. Determine the marginal cost function and the marginal cost for an output of 10 tons of tea a day.

18. The demand for milk in Milktown is described by the demand functions

$$x = \frac{50 - \sqrt{p}}{2} \quad \text{and} \quad p = (50 - 2x)^2$$

where x represents the average market demand for milk per household in gallons per month and p is the price of milk in cents per gallon. Determine the elasticity of demand for a price of 64¢ per gallon. Assuming that the milk supply is controlled by a monopolist, determine his total revenue and the marginal revenue function. What is the marginal revenue corresponding to a price of 64¢ per gallon?

19. Find the tangent line to the graph of $f(x) = 3x/(2x+1)$ at the point $P(1,1)$.

20. Show that $\dfrac{dR}{dx} = p(1 - \dfrac{1}{\eta})$,

where $p = g(x)$ and its inverse $x = f(p)$ are demand functions which relate p, the unit price, and x, the quantity bought by the market per unit time. $R = xp$ is the total revenue function for the firm producing output x, $\dfrac{dR}{dx}$ is marginal revenue, and

$\eta = -\dfrac{p}{x} \cdot \dfrac{dx}{dp}$ is elasticity of demand.

6.9. HIGHER-ORDER DERIVATIVES

If $f'(x)$ is the derivative of the function $y = f(x)$, then the derivative of $f'(x)$, provided that it exists, is called the second derivative of $y = f(x)$, and is denoted by $f''(x)$; the derivative of $f''(x)$, provided that it exists, is called the third derivative of $y = f(x)$ and is denoted by $f'''(x)$; and so on. In terms of other notations for derivatives, such higher-order derivatives at x are denoted as follows:

$$
\begin{array}{llll}
f(x), & f'(x), & f''(x) & f'''(x), \ldots \\
f(x), & D_x f(x), & D_x^2 f(x), & D_x^3 f(x), \ldots \\
y, & \dfrac{dy}{dx}, & \dfrac{d^2y}{dx^2}, & \dfrac{d^3y}{dx^3}, \ldots
\end{array}
$$

EXAMPLE 1

Find the first four derivatives of $f(x) = 3x^2 + 2x - 1$ and their values at $x = 2$.

$$
\begin{aligned}
&f(x) = 3x^2 + 2x - 1, \quad f'(x) = 6x + 2, \quad f''(x) = 6 \\
&f'''(x) = 0, \quad f^{(iv)}(x) = 0 \\
&f'(2) = 14, \quad f''(2) = 6, \quad f'''(2) = 0, \quad f^{(iv)}(2) = 0
\end{aligned}
$$

EXERCISES

Find, if they exist, the first three derivatives of the following functions and their values at $x = -2$.

1. $f(x) = 3x^2 - 2x + 1$
2. $f(x) = 4x^3 - 2x^2 + 4$
3. $f(x) = 5x^4 + 2x^3 - 2x - 100$
4. $f(x) = \frac{1}{10}x^2 - 12x + 400$
5. $f(x) = \ln x$
6. $f(x) = \frac{3}{2}x^2 + 4x - \frac{1}{x}$

CHAPTER 7

Optimization Problems

7.1. EXTREME VALUES OF FUNCTIONS

There are many situations in which one is led to seek the largest or least value of a function. To illustrate, consider a chocolate producer for which

$$c(x) = x^2 + 2x + 100$$

is the total dollar cost of producing x tons of chocolate a day. The average cost function, obtained by dividing cost $c(x)$ by output x, is

$$\overline{c}(x) = x + 2 + \frac{100}{x}.$$

For what output is the average cost function $\overline{c}(x)$ a minimum? If the profit function of this firm is

$$P(x) = -101x^2 + 1198x - 100$$

for what output is the profit function $P(x)$ a maximum?

In this chapter we develop methods for answering questions of this sort. We begin by introducing some basic definitions.

A function f is said to have an **absolute maximum value at c** if $f(c) \geq f(x)$ for all values of x in the domain of definition of f. (See Figure 7.1.) $f(c)$ itself is said to be the **maximum value of f**.

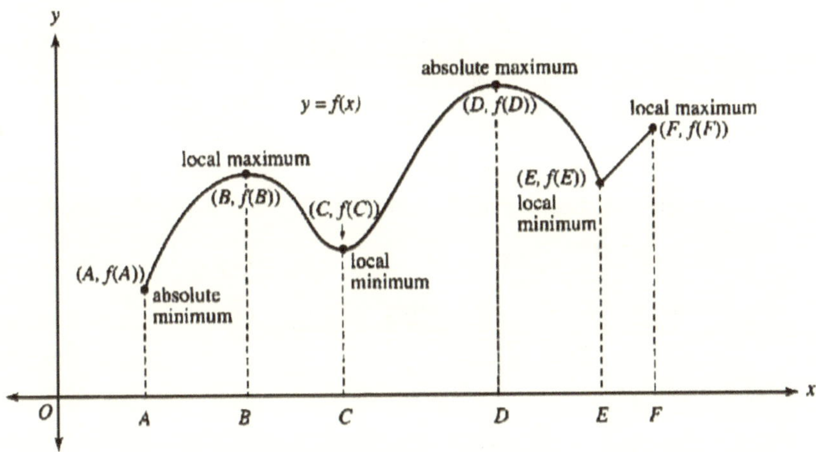

Figure 7.1

Function f is said to have a **local maximum value at c** if there is an interval I containing c such that $f(c) \geq f(x)$ for any value of x in interval I which is also in the domain of definition of f. $f(c)$ is said to be a **local maximum value of f**.

The concepts of **absolute minimum value** and **local minimum value** are defined in an analogous way.

The function f shown in Figure 7.1 has an absolute minimum value at A, an absolute maximum value at D, local minimum values at C and E, and local maximum values at B and F.

The maximum and minimum values of a function are called its extreme values. When does a function have extreme values? One answer can be given in terms of the concept of continuity. It can be shown that if a function is continuous at every point of an interval I, including the end points of I, then the function has an absolute maximum value and an absolute minimum value in I. If a function is not continuous at every point in an interval, then it may or may not have extreme values.

Where in its domain of definition does a function assume its extreme values? To obtain some insight into where extreme values occur, consider the functions whose graphs are shown in Figures 7.2-7.5. The extreme values of $f(x) = x + 1, 0 \leq x \leq 2$ (Figure 7.2) occur at 0 and 2; the extreme value of $f(x) = x^2 + 1$ (Figure 7.3) occurs at 0, as is also the case with $f(x)=-x^2+1$ (Figure 7.4) and $f(x)=\begin{cases} x+1, & \text{for } x\geq 0 \\ -x+1, & \text{for } x<0 \end{cases}$ (Figure 7.5).

What properties do these numbers have? 0 and 2 are end points of the domain of definition of $f(x)=x+1,\ 0\leq x\leq 2$. 0 in the domain of definition of $f(x)=x^2+1$ yields the point (0, 1) at which the tangent line to the graph of $f(x)=x^2+1$

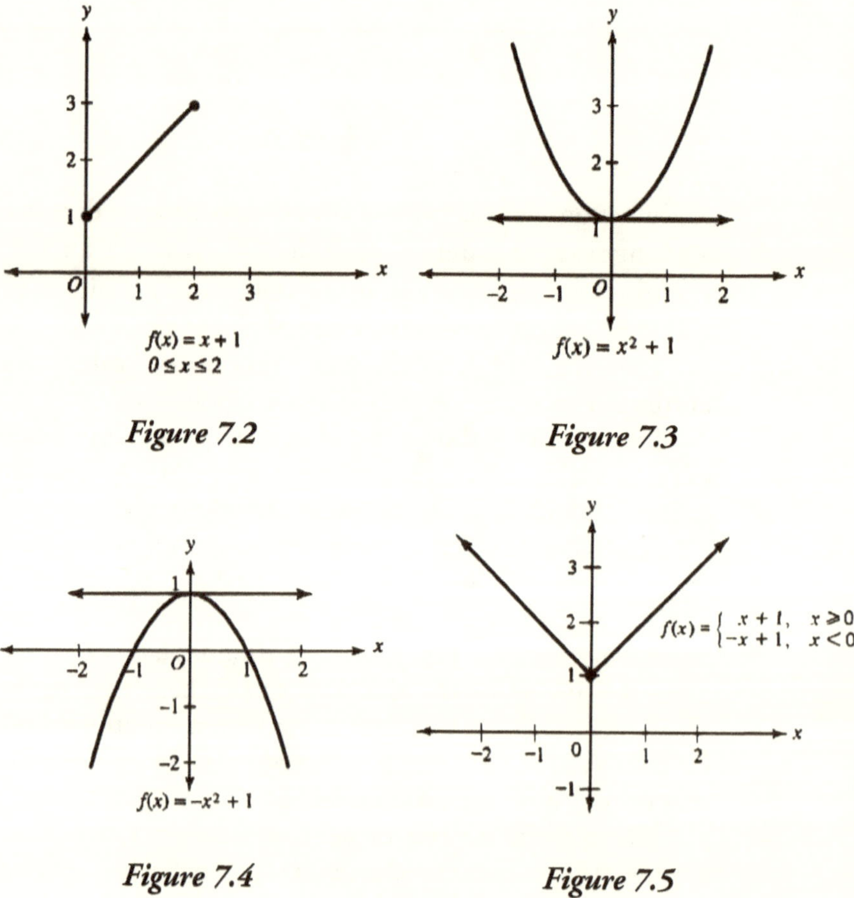

Figure 7.2

Figure 7.3

Figure 7.4

Figure 7.5

is horizontal, which means that the derivative of $f(x)=x^2+1$ is zero at this value. A similar situation is the case for 0 in the domain of definition of $f(x) = -x^2 + 1$. On the other hand, the derivative of $f(x) = \begin{cases} x+1, & \text{for } x \geq 0 \\ -x+1, & \text{for } x < 0 \end{cases}$ is not defined at 0 (see Sec. 5.1, Ex. 2, p. 139).

These examples show that extreme values can occur at numbers of the following types in the domain of *f*.

1. A number *c* such that $f'(c) = 0$
2. A number *c* such that $f'(c)$ does not exist.
3. A number *c* which is an end point of the domain of definition of function *f*.

It can be shown that a function can have an extreme value only at numbers of these three types. We will call such numbers critical values of the function. Extreme values occur only at critical values. Thus the search for extreme values begins with a search for critical values. However, while every extreme value occurs at a critical value, not every critical value need necessarily yield an extreme value.

As an illustration, consider $f(x) = x^3 + 1$, whose graph is shown in Figure 7.6.

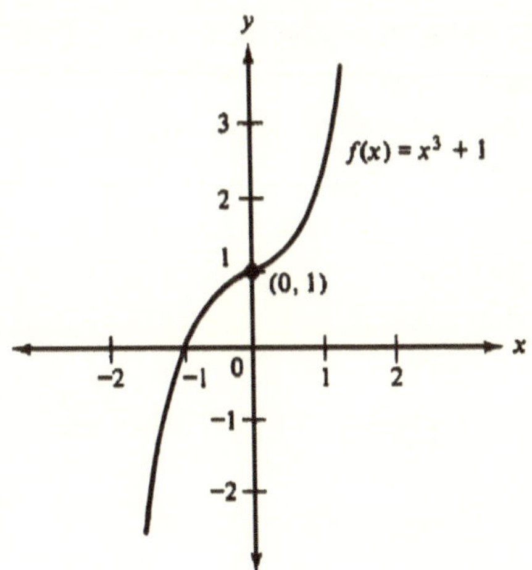

Figure 7.6

0 is a critical value (type 1) since $f'(0)=0$. $f(0)=1$ but 1 is not a maximum value since it is not true that $f(0)\geq f(x)$ for x in some interval containing 0; $f(x)>1$ when x is positive. $f(0)$ is not a maximum value since it is not true that $f(0)\leq f(x)$ for x in some interval containing 0; $f(x)<1$ when x is negative. Thus, while 0 is a critical value for $f(x)=x^3+1$, $f(0)=1$ is not an extreme value.

Therefore to find the extreme values of a function, we must first find its critical values and then distinguish those which yield extreme values from those which do not. The real problem is with critical values of type 1 (c such that $f'(c)=0$).

In fact, a method of distinguishing critical values of type 1 which give rise to extreme values from those which do not is suggested by two situations that we have already examined, $f(x)=x^2+1$ and $f(x)=-x^2+1$ (shown in Figures 7.3 and 7.4). 0 is a critical value for both of these functions. $f(0)$ is a minimum value of $f(x)=x^2+1$ and $f(0)$ is a maximum value of $f(x)=-x^2+1$. In both cases $f'(0)=0$. But how does $f'(x)$ behave near the critical value 0?

Let us observe that in the case of $f(x)=x^2+1$, with a minimum value at 0, the derivative $f'(x)=2x$. Thus the slope of the tangent line is negative for x less than the critical value 0 and positive for x greater than the critical number 0. That is, $f'(x)=2x$ goes from negative to positive as x goes from values less than the critical value 0 to values greater than the critical value 0 (see Figure 7.7).

On the other hand, for $f(x)=-x^2+1$, just the opposite happens. The derivative $f'(x)=-2x$, and thus the slope of the tangent line is positive for x less than the

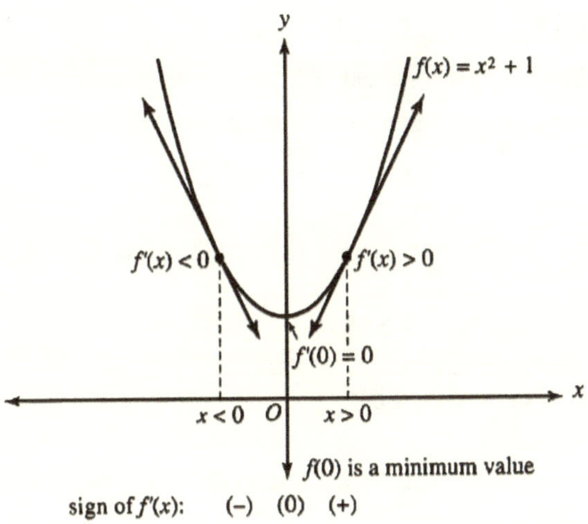

Figure 7.7

critical value 0 and negative for x greater than the critical value 0. That is, $f'(x) = -2x$ goes from positive to negative as x goes from values less than the critical value 0 to values greater than the critical value 0 (see Figure 7.8).

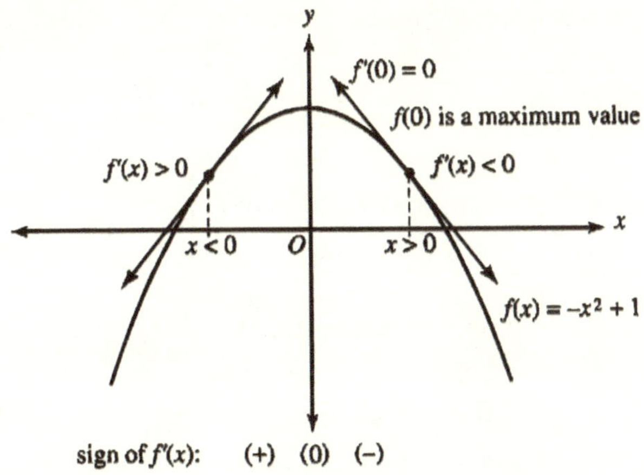

Figure 7.8

These examples suggest the following first derivative test, which is indeed valid.

First Derivative Test. Let c denote a type 1 critical value $(f'(c)=0)$. If there is an interval I containing c such that $f'(x)$ goes from negative to positive as x goes from values less than c in I to values greater than c in I, then $f(c)$ is a local minimum value of f. If $f'(x)$ goes from positive to negative, then $f(c)$ is a local maximum value of f. If $f'(x)$ does not change sign, then $f(c)$ is not an extreme value of f.

In summary we have:

First Derivative Test

$f(c)$ is a local maximum value if $f'(c)=0$ and $f'(x)$ goes from $+$ to $-$;
$f(c)$ is a local maximum value if $f'(c)=0$ and $f'(x)$ goes from $-$ to $+$;
$f(c)$ is not an extreme value when $f'(c)=0$ but $f'(x)$ does not change sign.

EXAMPLE 1. AVERAGE COST FUNCTION MINIMIZATION

The average cost function for a chocolate producer is

$$\overline{c}(x)=x+2+\frac{100}{x}$$

dollars for an output of x tons of chocolate per day. Determine the output for which the average cost function is a minimum.

The domain of definition of $\overline{c}(x)$ is $x>0$, subject to plant limitation. There are no end point critical values and there are no values of $x > 0$ at which $\overline{c}(x)$ does not exist. To determine type 1 critical values (values at which $\overline{c}(x)=0$), we must determine $\overline{c}(x)$, set it equal to zero, and solve for x.

$$\overline{c}(x)=1-\frac{100}{x^2}$$

Setting $\overline{c}(x)$ equal to zero and solving for x yields

$$1-\frac{100}{x^2}=0$$
$$x^2=100$$
$$x=10.$$

Thus $x = 10$ is a type 1 critical value.

To determine if $\overline{c}(10)$ is a minimum value, we must examine the behavior of

$$\overline{c}(x)=1-\frac{100}{x^2}=\frac{x^2-100}{x^2}$$

for $x < 10$ and $x > 10$.

When $x<10$, $x^2<100$ and $\overline{c}(x)<0$.

When $x>10$, $x^2>100$ and $\overline{c}(x)>0$.

Since $\overline{c}(10)=0$ and $\overline{c}(x)$ goes from negative to positive, it follows by the first derivative test that $\overline{c}(10)$ is a minimum value. Thus average cost function is minimized when 10 tons of chocolate are produced daily.

EXAMPLE 2. PROFIT FUNCTION MAXIMIZATION

The chocolate producer's profit function is

$$P(x)=-101x^2+1198x-100$$

dollars for an output of x tons of chocolate a day. Determine the output for which the profit function is maximized.

The domain of definition of $P(x)$ is $x > 0$, subject to plant limitation. There are no end point critical values and there are no values of

$x > 0$ at which $P'(x)$ does not exist. To find type 1 critical values, we must determine $P'(x)$, set it equal to zero, and solve for x.

$$P'(x) = -202x + 1198$$

Setting $P'(x)$ equal to zero and solving for x yields

$$-202x + 1198 = 0$$
$$x = \frac{-1198}{-202} = 5.93.$$

Thus $x = 5.93$ is a type 1 critical value.

To determine if $P(5.93)$ is a maximum value, we must examine the behavior of

$$P'(x) = -202x + 1198 = -202(x - 5.93)$$

for $x < 5.93$ and $x > 5.93$. When $x < 5.93$,

$$P'(x) = \underbrace{-202}_{\text{neg.}}\underbrace{(x - 5.93)}_{\text{neg.}}.$$

$P'(x)$ is the product of two negative numbers and thus is positive. When $x > 5.93$,

$$P'(x) = \underbrace{-202}_{\text{neg.}}\underbrace{(x - 5.93)}_{\text{pos.}}.$$

$P'(x)$ is the product of a negative number and a positive number and thus is negative. Since $P'(5.93) = 0$ and $P'(x)$ goes from positive to negative, it follows by the first derivative test that $P(5.93) = 3453$ is a maximum value.

Does this mean that if the output level is 5.93 tons per day, the profit derived will be $ 3453, or approximately so, and that this is the largest profit that can be had?

Math vs. Reality: The Strings Attached

This question is about the real-world accuracy of the math conclusion obtained. Alas, math analysis does not deal directly with issues of real-world accuracy; it deals directly with issues of validity with respect to a starting point, in this case the function $P(x) = -101x^2 + 1198x - 100$. The math analysis establishes that, mathematically speaking, 5.93 yields the maximum value $P(5.93) = 3453$ for the afore function $P(x)$.

The string attached between the math and real worlds is simple, but vital for us to keep in mind.

A Very Big IF

IF the function $P(x) = -101x^2 + 1198x - 100$ *realistically* describes profit, then the valid conclusion we derived, $P(5.93) = 3453$ is the maximum value of $P(x)$, *realistically* describes maximum profit.

It is also vital that we keep in mind the other color of this string attached.

IF, however, $P(x) = -101x^2 + 1198x - 100$ *does NOT* realistically describe profit, $P(5.93) = 3453$ is still the maximum value of $P(x)$, but don't count on it describing maximum profit.

Our **fundamental assumption** in subsequent discussions of applications is that functions taken as profit functions, cost functions, and the like, **realistically describe profit, cost, and the like.**

As an attempt to keep tangled language to a minimum we will allow ourselves to write such statements as $P(5.93) = \$ 3453$ is maximum profit, as opposed to the correct, but tangled statement, $P(5.93) = 3453$ is the maximum value of $P(x)$ which expresses the maximum profit \$ 3453.

EXAMPLE 3

Determine the extreme values of $f(x) = e^{-x^2}$.

$$f'(x) = e^{-x^2} D_x(-x^2) = e^{-x^2}(-2x)$$

To determine type 1 critical values, we set $f'(x)$ equal to zero and solve for x.

$$e^{-x^2}(-2x) = 0$$
$$-2x = 0$$
$$x = 0$$

Thus 0 is a critical value, and we must investigate the behavior of $f'(x)$ for x immediately before 0 ($x < 0$) and x immediately after 0 ($x > 0$). Note that e^{-x^2} is positive for all x (Why?)

For $x < 0$, $f'(x) = \underbrace{e}_{\text{pos.}}\underbrace{^{-x^2}(-2x)}_{\text{pos.}}$ is positive.

For $x > 0$, $f'(x) = \underbrace{e}_{\text{pos.}}\underbrace{^{-x^2}(-2x)}_{\text{neg.}}$ is negative.

Since $f'(x)$ goes from positive to negative, $f(0) = e^0 = 1$ is a local maximum value.

There are no end-point critical values, since $f(x)$ is defined for all x. Also $f'(x)$ is defined for all x, so that type 2 critical values do not emerge.

EXERCISES

Find the extreme values of the following functions by using the first-derivative test.

1. $f(x) = x^2 + 4x + 5$
2. $f(x) = 3x^2 + 12x + 2$

3. $f(x) = 2x^2 + 4x + 3$

4. $f(x) = -3x^2 - 6x + 8$

5. $f(x) = 3x^4 - 4x^3$

6. $f(x) = 2x^3 + 3x^2$

7. $f(x) = \frac{1}{3}x^3 - \frac{1}{2}x^2 - 2x + 1$

8. $f(x) = \frac{1}{3}x^3 + x^2 - 3x + 4$

9. $f(x) = x^3 + 4x^2 - 3x + 6$

10. $f(x) = x^3 - 2x^2 - 4x + 3$

The Second Derivative Test

In some situations, another test, called the second-derivative test because it is based on the behavior of the second derivative, is easier to apply than the first-derivative test. To establish the background for the second-derivative test, consider the functions shown in Figures 7.9 and 7.10. The graph of $y = f(x)$ shown in Figure 7.9 is said to be **concave upward at the point** $P(c, f(c))$, because the graph is above the tangent line at P for values in some interval I centered at c. The graph of $y = f(x)$ shown in Figure 7.10 is said to be **concave downward at the point** $P(c, f(c))$, because the graph is below the tangent line at P for values in some interval I centered at c. The graph of a function $y = f(x)$ is said to **be concave upward over an interval**

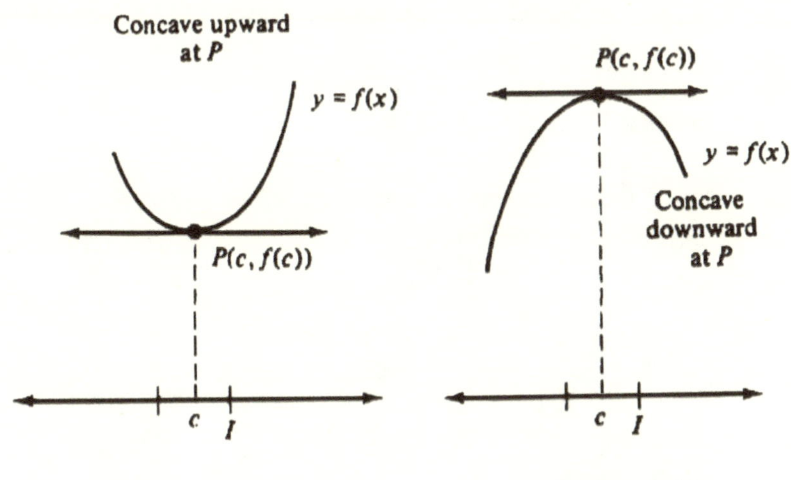

Figure 7.9 *Figure 7.10*

(or concave downward over an interval) if it is concave upward (or concave downward) at points on the graph for all values in the interval.

The concavity behavior of the graph of $y = f(x)$ is indicated by the sign of $f''(x)$, the second derivative. If $f''(x)$ is positive for all values in an interval, then the graph of $y = f(x)$ is concave upward for each value in the interval; if $f''(x)$ is negative for all values in an interval, then the graph of $y = f(x)$ is concave downward for each value in the interval.

Thus if c is a type 1 critical value $[f'(c) = 0]$, and $f''(x) > 0$ for x in some interval centered at c, so that the graph of $y = f(x)$ is concave upward in this interval, then the situation described by Figure 7.9 is the case, and $f(c)$ is a minimum value; if $f''(x) < 0$ for x in some interval centered at c, so that the graph of $y = f(x)$ is concave downward in this interval, then the situation described by Figure 7.10 is the case, and $f(c)$ is a maximum value. If the third derivative $f'''(c)$ exists, then it suffices to consider the sign of the second derivative at c, $f'''(c)$, to determine the concavity behavior of the graph of $y = f(x)$ for some interval centered at c. In summary, then, we have the following second-derivative test.

Second Derivative Test

Let c denote a type 1 critical number $[f'(c) = 0]$. If the third derivative at c, $f'''(c)$, exists, and there is some interval centered at c within which $f''(x)$ exists, then

1. *$f(c)$ is a local maximum value if* $f''(c) < 0$.
2. *$f(c)$ is a local minimum value if* $f''(c) > 0$.
3. N*o conclusion can be drawn if* $f''(c) = 0$ [*$f(c)$ may or may not be an extreme value*].

EXAMPLE 4

Determine the extreme values of $f(x) = 2x^2 + 8x + 10$.

$f'(x) = 4x + 8$. Setting $f'(x)$ equal to zero and solving for x yields

$$4x + 8 = 0$$
$$x = -2$$

Thus -2 is a type 1 critical value. This part, involving the determination of type 1 critical values, is common to both the first-derivative and second-derivative tests. To continue via the second-derivative test, we must determine $f''(-2)$.

$$f'(x) = 4x + 8, \quad f''(x) = 4, \quad f''(-2) = 4$$

Also $f'''(x)$ exists (its value is 0 for all x). Since $f''(-2) = 4$, which is positive, it follows from the second-derivative test that $f(-2)$ is a local minimum value.

Since $f(x) = 2x^2 + 8x + 10$ is a quadratic function and the graphs of quadratic functions are parabolas, we may further conclude that $f(-2) = 2$ is an absolute minimum value.

Let us further observe for the sake of completeness that there are neither end-point critical values nor critical values for which the derivative does not exist to be considered.

EXAMPLE 5. RETURN TO THE THOMAS COMPANY'S PROBLEM

We return to the Thomas fast-food chain situation described in example 3 of Section 2.3 (pg. 79). The problem is to determine the number of new restaurants that should be opened in the New England region so that the daily average income of all restaurants in the chains is maximized.

(a) Can the methods developed in this chapter be used to address this problem? Explain

(b) If there are difficulties, is there any way around them? Explain.

(c) If the answer to (b) is yes, then proceed to address the problem formulated.

(d) Under what conditions would you implement the result obtained in (c). Explain.

(a) The methods developed in this chapter cannot be used to address this problem because the Thomas Company's total daily average function

$$I\ (x) = (20{,}000 - 200x)\ (60 + x)$$

$$= -200x\ + 8000x + 1{,}200{,}000$$

is defined for the non-negative integer values 0, 1, ..., 99 and is therefore not differentiable.

(b) The way around this difficulty is to consider the first cousin of this function, mathematically speaking,

$$I_1\ (x) = -200x^2 + 8000x + 1{,}200{,}000$$

defined for $0 \leq x \leq 99$, which is differentiable, and apply the results obtained to its cousin, assuming of course that they are applicable.

This device for handling such situations is employed without explicit comment when non-differentiable functions arise. Keep it in mind and press the red alert button when such situations arise in further examples and exercises.

(c) For the first cousin differentiable function $I_1\ (x)$

$$I'_1\ (x) = -400x + 8000$$

which yields $x = 20$. $I_1\ (20) = 1{,}280{,}000$. The end point critical values 0 and 99 yield 1,200,000 and 31,800, respectively.

Application of the second derivative test establishes that I_1 (20) is the maximum value of I_1 (x). Since 20 is in the domain of definition of I (x), I (20) = 1,280,000 is the maximum value of I(x).

(d) What Does this Mean to the Thomas Company?

Does it mean that if the Thomas Company were to implement this finding by adding 20 additional restaurants to its chain the total daily average income of all restaurants in its chain will be maximized? We cannot say for sure, Mathematically speaking, we have a valid conclusion with respect to the assumptions made. How close it is to reality's mark is another issue. This takes us back to the question of how realistic are the two starting points of the mathematical analysis, the explicitly stated assumption,

P1: The average amount of business done by each restaurant in the chain will drop by $200 per day for each new restaurant opened,

and the not explicitly stated assumption, but implicit in all such analyses.

P: Other factors that bear on the situation are not being considered. This may be the case because other factors that may bear on the situation are viewed as negligible, or it is not understood how to come to grips with them, or a combination of both.

If the management of the Thomas Company is satisfied that these assumptions are realistic, then it would be reasonable to implement the mathematical conclusion by adding 20 additional restaurants to its chain.

EXAMPLE 6

The director of the studies abroad program at Ecap University is planning to have 5000 copies of a poster describing his forthcoming program in Rome and Athens printed. The type plate containing the printed material is to be 96 square inches, the top and bottom margins are each to be 1 inch, and the side margins are each to be $1\frac{1}{2}$ inches. An all-out effort is being made to advertise the program, and the poster is to be printed on high-quality paper. Of interest is the question of what the dimensions of the type plate should be so as to use the least possible amount of paper.

Let x and y denote the length and width, respectively, of the type plate (see Figure 7.11). The length of the poster is x +2 inches, and its width is y + 3 inches. Thus the area of the poster, which we want to minimize, is given by the product.

$$(x+2)(y+3) = xy+2y+3x+6$$

To express this area in terms of one variable, we need a relationship between x and y. Since the area of the type plate is to be 96 square inches, we have

$$xy = 96, \quad y = \frac{96}{x}$$

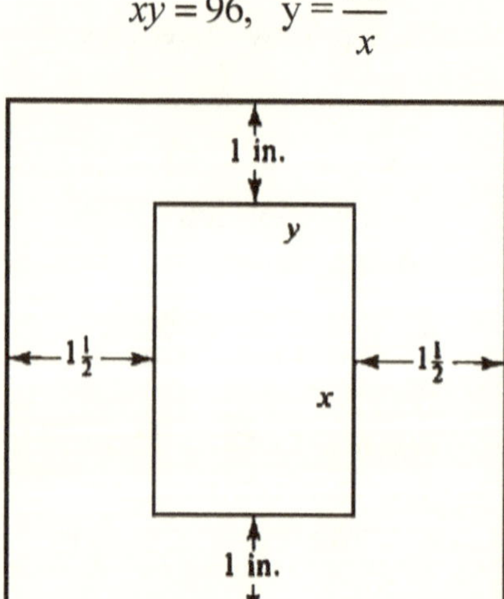

Figure 7.11

Substituting $96/x$ for y in our expression for the area of the poster yields

$$\begin{aligned} A(x) &= x\left(\frac{96}{x}\right)+2\left(\frac{96}{x}\right)+3x+6 \\ &= 96+\frac{192}{x}+3x+6 \\ &= 102+3x+\frac{192}{x}, \quad x>0 \end{aligned}$$

No type 2 or type 3 critical values arise. To determine type 1 critical values we determine $A'(x)$, set it equal to 0, and solve for x.

$$A'(x)=3-\frac{192}{x^2}$$

$$3-\frac{192}{x^2}=0$$

$$x^2=64$$

$$x=8$$

Thus $x = 8$ is a type 1 critical value. To determine if it yields the expected minimum value let us apply the second derivative test.

$$A''(x)=\frac{384}{x^3}$$

$$A''(8)=\frac{384}{8^3}>0$$

Thus $x = 8$ yields a minimum value. The type plate which is 8 inches long and 12 inches wide uses the least amount of paper, with area 96 square inches.

If the second derivative of a function is difficult to determine or evaluate at the critical value in question, such as in Example 3, practical considerations may favor the first-derivative test. As we have noted, an important limitation is that the second derivative at the critical value must be nonzero. If it is zero, then anything can happen, as is illustrated by the functions $f(x) = x^3 + 1$, $f(x) = x^4 + 1$, and $f(x) = -x^4 + 1$, shown in Figures 7.12 through 7.14, respectively.

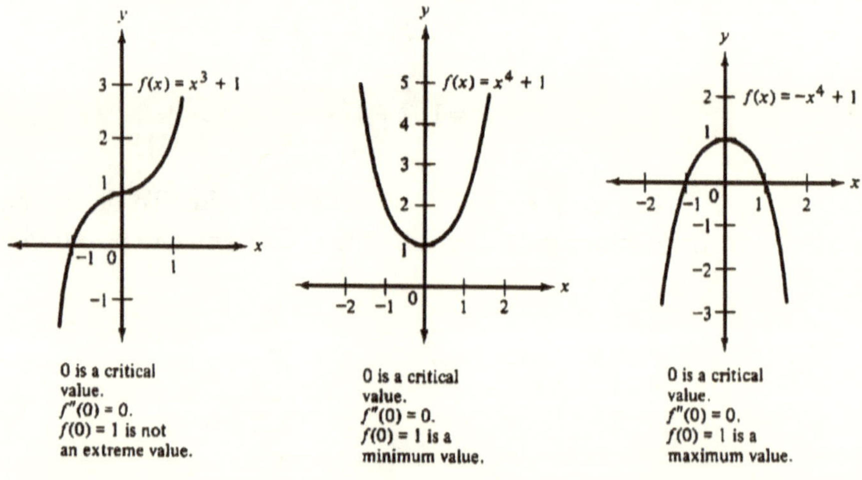

Figure 7.12 ***Figure 7.13*** ***Figure 7.14***

EXERCISES

Find the extreme values of the functions stated in 11-22.

11. $f(x) = 5x^2 + 10x + 3$
12. $f(x) = x^3 - 7x^2 - 24x + 2$
13. $f(x) = 2x^5 - 5x^4$
14. $f(x) = x^4 + \frac{7}{3}x^3 - x^2 + 4$
15. $f(x) = \dfrac{x^2 - 5}{x - 3}$
16. $f(x) = x + 10 + \frac{100}{x}$
17. $f(x) = x^4 + \frac{2}{3}x^3 - x^2 + 1$
18. $f(x) = \frac{1}{4}x + 5 + \frac{1}{x}$
19. $f(x) = xe^x$
20. $f(x) = x(\ln x)$
21. $f(x) = \dfrac{\ln x}{x}$
22. $f(x) = \dfrac{1}{\sqrt{2\pi}}e^{-x^2/2}$

23. A lamp manufacturer has fixed expenses of \$1000 per week. His production cost is \$5 per lamp. It is estimated that, if the selling price of a lamp is x dollars, then $450-10x$ lamps would be sold per week. Determine the selling price that would maximize weekly profit, the maximum weekly profit, and the optimum number of lamps made and sold per week under the given conditions.

24. In connection with the background described in Example 6, what interest rate should be set (a) if the association can earn 18 percent on the money it raises; (b) if the association can earn rate r on the money it raises?

25. A special edition of Lithuanian poetry is to be published. The type plate for each page is to be 24 square inches, the top and bottom margins are each to be $\frac{3}{4}$ inch, and the side margins are each to be 1/2 inch. What should the dimensions of the type plate to be so that the area of the page is minimized?

26. A rectangular bicycle parking lot with a perimeter of 500 feet is to be constructed. What are the dimensions of the lot with the largest area?

27. The design of the Institute for Baltic Studies at Ecap University calls for a rectangular structure with 60,000 square feet. The building material to be used for the front of the building is, per linear foot, twice as expensive as the material to be used for the other three sides. What should the dimensions of the building be for the material cost of the four walls to be a minimum?

28. Forty wells, each producing an average of 500 barrels of oil per hour, are being operated in an oil field. It is estimated that, for each new well that is drilled in the field, the average output of each well will drop by 10 barrels per

hour. Assuming this to be the case, how many new wells should be drilled in the oil field to maximize the total average hourly output?

29. From a rectangular sheet of metal (below), 32 by 20 inches, identical squares are to be cut out from the four corners of the sheet. The sides are then to be bent so as to form an open rectangular box. What should the side length of these squares be so that the volume of the box is maximized?

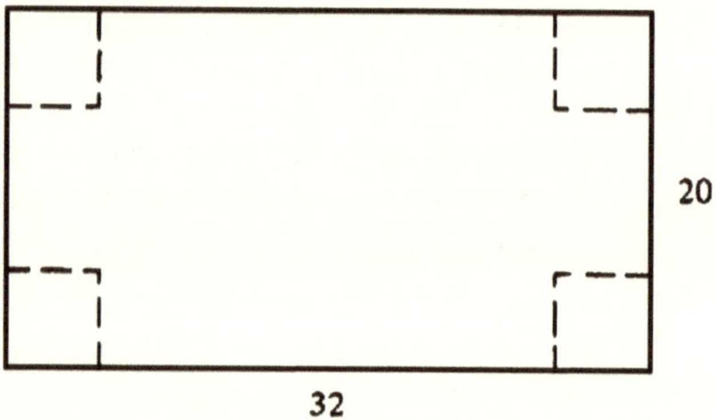

30. What two numbers, whose sum is 12, yield the largest possible product?

7.2. PROFIT MAXIMIZATION BEHAVIOR OF THE FIRM

Let us recall that, if $R = R(x)$ and $c = c(x)$ are the total-revenue and total-cost functions of a firm, then the derivatives of these functions, $R'(x)$ and $c'(x)$, are called marginal revenue and marginal cost for output x. A central proposition of economics, which we establish here under suitable assumptions, is the following:

Profit is maximized at the output level at which marginal revenue equals marginal cost.

Proof: The profit function $P(x)$ of the firm is the difference between the total-revenue and total-cost functions

$$P(x) = R(x) - c(x)$$

We assume that $R'(x)$ and $c'(x)$ are defined for all x under consideration. The derivative of $P(x)$ is given by

$$P'(x) = R'(x) - c'(x)$$

Setting $P'(x)$ equal to zero yields

$$R'(x) - c'(x) = 0$$
$$R'(x) = c'(x)$$

Thus the critical number for which $P'(x)$ equals zero is that value of x for which marginal revenue equals marginal cost.

How can we be certain that this critical number yields a maximum value for $P(x)$, the profit function? The economic argument is that to say $P(x)$ is a profit function means certain conditions must be satisfied, one of which is that $P(x)$ has a unique extreme value, that is, a maximum value. Having described the critical number, we are thus guaranteed by economic assumptions that it yields a maximum value. For those who would like a mathematical condition that guarantees a maximum value, one can easily be given in terms of the second-derivative test. The second derivative of $P(x)$ is given by

$$P''(x) = R''(x) - c''(x)$$

If we denote the critical number by a, so that $P'(a) = 0$, or, equivalently, $R'(a) = c'(a)$, then the condition

$$R''(a) - c''(a) < 0$$

or, equivalently,

$$R''(a) < c''(a)$$

ensures that $P(x)$ has a maximum value at the critical number a.

The condition $R''(a) < c''(a)$ makes sense from an economic point of view in that it says that profit is maximized where marginal revenue equals marginal cost, and also marginal revenue is increasing more slowly than marginal cost.

EXAMPLE 1

A sugar refinery has revenue and cost functions

$$R(x) = 405x - 2x^2, \quad c(x) = \frac{1}{2}x^2 + 5x + 3000$$

where x is output in tons per week, and $R(x)$ and $c(x)$ are revenue and cost in dollars. Find the optimal output level and anticipated profit.

Marginal revenue and marginal cost are given by

$$R'(x) = 405 - 4x, \quad c'(x) = x + 5$$

Setting marginal revenue equal to marginal cost and solving for x yields

$$\begin{aligned} 405 - 4x &= x + 5 \\ -5x &= -400 \\ x &= 80 \end{aligned}$$

Thus the optimal output level is 80 tons per week, $R(80)$ = 19,600 and $c(80)$ = 6600, so the anticipated profit is $\$19,600 - \$6600 = \$13,000$ per week.

Profit Maximization After Taxation

Since we live in an age of taxes, the effects of various tax structures on output and profit are worthy of consideration. One tax, called a lump-sum or licensing tax, consists of a fixed fee that the producer must pay the government, irrespective of output.

EXAMPLE 2

Consider again the sugar refinery of Example 1 with revenue and cost functions

$$R(x) = 405x - 2x^2, \quad c(x) = \frac{1}{2}x^2 + 5x + 3000$$

and suppose that a lump sum tax of $6000 per week is imposed. What is the effect on the optimal output level, which, before the imposition of this tax, is 80 tons per week?

Profit after taxes is

$$P(x) = R(x) - c(x) - 6000$$

The derivative of $P(x)$ is given by

$$P'(x) = R'(x) - c'(x) - 0$$

since the derivative of a constant function is 0. Setting $P'(x)$ equal to zero yields the same condition obtained before:

$$R'(x) = c'(x)$$

The optimal output level is not disturbed by the imposition of a lump-sum tax of $6000 per week or any other amount for that matter. It is still that output for which marginal revenue equals marginal cost.

For the sugar refinery, this means that the optimal output level remains at 80 tons per week. The profit of the firm is, however, reduced by the amount of the tax, so that the anticipated profit after taxes is $\$13,000-\$6000=\$7000$ per week.

Another possible taxation mechanism involves charging the firm a percentage of profit, that is, a percentage of the difference between $R(x)$ and $c(x)$. Let us suppose that a tax of 30 percent of the difference $R(x)-c(x)$ is imposed. Then the profit after taxes is

$$\begin{aligned} P(x) &= R(x)-c(x)-0.30[R(x)-c(x)] \\ &= R(x)-c(x)-0.30R(x)+0.30c(x) \\ &= 0.70R(x)-0.70c(x) \end{aligned}$$

The derivative of $P(x)$ is

$$\begin{aligned} P'(x) &= 0.70R'(x)-0.70c'(x) \\ &= 0.70[R'(x)-c'(x)] \end{aligned}$$

Setting $P'(x)$ equal to zero yields the same condition obtained before:

$$\begin{aligned} 0.70[R'(x)-c'(x)] &= 0 \\ R'(x)-c'(x) &= 0 \\ R'(x) &= c'(x) \end{aligned}$$

The optimal output level is not disturbed by the imposition of a percentage of profit tax. It is still that output for which marginal revenue equals marginal cost.

For the sugar refinery, this means that the optimal output level still remains at 80 tons per week. The anticipated profit before taxes is, as we observed in Example 1, $13,000 per week. With a 30 percent of profit tax, the tax due is (0.30)(13,000) = $3900, and the profit after taxes is $\$13,000-\$39000=\$9100$.

Finally, we turn our attention to the output tax, whereby a certain fixed amount t dollars per unit output is imposed on the producer.

EXAMPLE 3

We return once more to the sugar refinery of Example 1, and assume that an output tax of \$100 per ton has been imposed. Then the tax to be paid on an output of x tons is $100x$ dollars, and the profit function of the firm is

$$P(x) = R(x) - c(x) - 100x$$

The derivative of $P(x)$ *is*

$$P'(x) = R'(x) - c'(x) - 100$$

Setting $P'(x)$ equal to zero yields

$$R'(x) - c'(x) - 100 = 0$$

from which we obtain the condition

$$R'(x) - c'(x) + 100$$

The optimal output level is that output for which marginal revenue equal marginal cost plus the unit output tax value of 100. For the sugar refinery,

$$R(x) = 405x - 2x^2, \quad c(x) = \frac{1}{2}x^2 + 5x + 3000$$

so that

$$R'(x) = 405 - 4x, \quad c'(x) = x + 5$$

The optimal output condition $R'(x) = c'(x) + 100$ yields

$$405-4x=x+5+100$$
$$-5x=-300$$
$$x=60$$

An output tax does affect the optimal output level, and for our sugar refinery the effect is to lower the optimal output level from 80 to 60 tons per week. For an output level of 60 tons, $R(60)$ - 17,100 and $c(60)$ = 5100, so that profit before taxes is $R(60)-c(60)=\$12,000$. At \$100 per ton, the tax on an output of 60 tons is 100(60) = \$6000, and the profit after taxes is $\$12,000-\$6000=\$6000$.

Based on an output tax of \$100 per ton, the tax revenue derived is \$6000 per week. If a \$6000 per week lump-sum tax were imposed on the sugar refinery instead of an output tax of \$100 per ton, the tax revenue would be the same, \$6000, and for this reason we call this lump-sum tax the **lump-sum equivalent of the \$100 per ton output tax**.

If the lump-sum equivalent of the \$100 per ton output tax were imposed in its place, the optimal output level would remain at 80 tons per week, instead of dropping to 60 tons per week, the profit of the firm after taxes (based on the 80 ton per week output level) would be \$7000 per week (see Example 2), as opposed to \$6000 per week with the output tax, and either way the government would obtain the same tax revenue, \$6000. It is not difficult to see why the lump-sum equivalent of an output tax is preferable to the output tax. Output is higher, the profit of the firm is higher, the government obtains the same revenue, and, in general, the market price of the commodity is lower.

We conclude our analysis of the effect of an output tax on the optimal output level of a firm by showing the following:

If an output tax of t dollars per unit output were imposed, the optimal output would be that value of x for which

$$R'(x)=c'(x)+t$$

where $R = R(x)$ and $c = c(x)$ are the revenue and cost functions of the producer.

Proof: For output x the tax to be paid is tx, and thus the profit function of the firm is

$$P(x) = R(x) - c(x) - tx$$

The derivative of $P(x)$ is

$$P'(x) = R'(x) - c'(x) - t$$

Setting $P'(x)$ equal to zero yields the desired relationship:

$$\begin{aligned} R'(x) - c'(x) - t &= 0 \\ &= c'(x) + t \end{aligned}$$

Tax-Revenue Maximization

We have seen that if an output tax of \$100 per ton were imposed on the sugar producer, the tax revenue generated would be \$6000 per week. Meanwhile, back at tax headquarters the question being asked is, how can we squeeze the most sugar out of this firm? That is, what output tax t dollars per ton generates the largest tax revenue?

To answer this question, we must first formulate an appropriate tax-revenue function that expresses tax revenue in terms of output tax t. To begin, let's set down the obvious.

tax revenue = (unit output tax t) × (output)

Thus the question is, what is the output of the sugar producer when output tax t is imposed? The answer is given by the relationship derived earlier. Output x satisfies the condition:

$$R'(x) - c'(x) + t$$

Let us recall that for the sugar producer the revenue and cost functions are

$$R(x) = 405x - 2x^2 \text{ and } c(x) = \frac{1}{2}x^2 + 5x + 3000$$

so that marginal revenue and marginal cost are given by

$$R'(x) = 405 - 4x \text{ and } c'(x) = x + 5$$

Thus we obtain

$$405 - 4x = x + 5 + t$$

Solving for output x yields

$$-5x = -400 + t$$
$$x = 80 - \frac{1}{5}t$$

Substituting $x = 80 - \frac{1}{5}t$ for output in the relationship

tax revenue = (unit output tax t) × (output)

yields the tax-revenue function

$$T(t) = t(80 - \frac{1}{5}t)$$
$$= 80t - \frac{1}{5}t^2$$

where $t > 0$.

To obtain the optimal tax level, we find $T'(t)$, set $T'(t)$ equal to zero, and solve for t.

$$T'(t) = 80 - \frac{2}{5}t$$

Setting $T'(t)$ equal to zero yields

$$80-\frac{2}{5}t=0$$
$$-\frac{2}{5}t=-80$$
$$t=200$$

Thus t = 200 is a critical number. Let us note that $T''(t)=-\frac{2}{5}$, so that $T''(200)=-\frac{2}{5}$, which is negative. By the second-derivative test, t = 200 in the optimal unit output tax level. The tax revenue generated is $T(200)$ = \$8000.

The effect of this output tax is to reduce the optimal output level from x = 80 tons to $x=80-\frac{1}{5}(200)=40$ tons per week. For this output level, the sugar producer's revenue and cost are $R(40)$ = \$13,000 and $c(40)$ = \$4000, and his profit after taxes is $R(40)-c(40)-8000=\$1000$. I a lump-sum tax of \$8000 per week were imposed instead of the \$200 per ton output tax, the tax revenue generated would be the same, but the optimal output level would remain at 80 tons per week. $R(80)$ = 19,600 and $c(80)$ = 6600, so that the producer's profit after taxes would be $R(80)-c(80)-8000=\$5000$, as opposed to \$1000 when the tax is imposed as an output tax of \$200 per ton.

EXERCISES

1. The revenue and cost functions of a coffee producer are $R(x)=204x-x^2$ and $c(x)=x^2+4x+1000$, where x is output in tons per day, and $R(x)$ and $c(x)$ are revenue and cost in dollars.
   (a) What are the coffee producer's optimal output level and profit before tax consideration?
   (b) If a lump-sum tax of \$1500 per day were imposed, what would the optimal output level and profit after taxes be?

(c) If a tax of 20 percent of profit were imposed, what would the optimal output level and profit after taxes be?

(d) If an output tax of \$40 per ton were imposed, what would the optimal output level and profit after taxes be?

(e) If the lump-sum tax equivalent of the \$40 per ton output tax were imposed, what would the optimal output level and profit after taxes be?

(f) What output tax should be imposed to maximize tax revenue?

(g) If this output tax were imposed, what would the optimal output level and profit after taxes be?

(h) If the lump-sum tax equivalent of this output tax were imposed, what would the optimal output level and profit after taxes be

2. The revenue and cost functions of a tea producer are $R(x) = 310x - 2x^2$ and $c(x) = \frac{1}{2}x^2 + 10x + 1500$, where x is output in tons per day, and $R(x)$ and $c(x)$ are revenue and cost in dollars.

(a) What are the tea producer's optimal output level and profit before tax considerations?

(b) If a lump-sum tax of \$1200 per day were imposed, what would the optimal output level and profit after taxes be?

(c) If a tax of 40 percent of profit were imposed, what would the optimal output level and profit after taxes be?

(d) If an output tax of \$60 per ton were imposed, what would the optimal output level and profit after taxes be?

(e) If the lump-sum tax equivalent of \$60 per ton output tax were imposed, what would the optimal output level and profit after taxes be?

(f) What output tax should be imposed to maximize tax revenue?

(g) If this output tax were imposed, what would the optimal output level and profit after taxes be?

(h) If the lump-sum tax equivalent of this output tax were imposed, what would the optimal output level and profit after taxes be?

3. The revenue and cost functions of a chocolate producer are $R(x) = 614x - 3x^2$ and $c(x) = x^2 + 6x + 1200$, where x is output in tons per day, and $R(x)$ and $c(x)$ are revenue and cost in dollars.

(a) What are the chocolate producer's optimal output level and profit before tax considerations?

(b) If a tax of 30 percent of profit were imposed, what would the optimal output level and profit after tax consideration be?

(c) What output tax should be imposed to maximize tax revenue?

(d) If this output tax were imposed, what would the optimal output level and profit after taxes be?

(e) If the lump-sum tax equivalent of this output tax were imposed, what would the optimal output level and profit after taxes be?

7.3. AN OPTIMAL STORAGE TIME PROBLEM

At a specified time, which becomes our time point of reference, a certain commodity is obtained. It is stored and it increases in value as time passes. The selling price of the commodity x years after its initial introduction is a known function,

$$S = f(x)$$

where x is the time elapsed and S is the selling price per unit at time x. Wine and timber are classic examples of commodities whose values increase with time. New wine is bought by a dealer to be sold at a later time. An investor acquires land on which timber has been planted and which is to be sold at a later time. The problem is to determine the optimal time at which to sell the commodity.

Specifically, suppose timber-beating land has been obtained, and

$$S = 1000e^{\sqrt{x}}$$

expresses S, the dollar selling price of the land, as a function of x, the number of years after its acquisition. Before we can formulate our optimization problem, a question of interest must be considered. The problem is that the sale value at one time is not directly comparable to the sale value at another time. Because of the time lag, growth due to interest accumulation must be taken into account.

Let us assume that interest accumulates at a rate r compounded continuously. To be very specific, suppose the problem is to compare a sale value of $2000 two years after acquisition of a commodity with a sale value of $3000 three years after the acquisition of the commodity, where interest is accumulated at a rate of 5% compounded continuously. Then the two sums can be compared by reference to the amounts that must initially be invested at a rate of 5% compounded continuously if at the end of two years $2000 is to be obtained and if at the end of three years $3000 is to be attained.

In Section 4.2 we saw that

$$A\,e^{-rx}$$

expresses the amount that must initially be invested at rate r compounded continuously if at the end of x years sum A is to be attained. Thus

$$2000\,e^{-0.05(2)} = 2000(0.9048) = \$1809.60$$

grows to a value of $2000 in two years, and

$$3000\,e^{-0.05(3)} = 3000(0.8607) = \$2582.10$$

grows to a value of $3000 in three years. Therefore, if the stated interest conditions are taken into consideration, a comparison of a sale value of \$2000 after two years with a sale value of \$3000 after three years is equivalent to comparing \$1809.60 with \$2582.10.

Thus, returning to our timber-land situation, if

$$S = 1000e^{\sqrt{x}}$$

expresses the dollar sale value of the land as a function of the number of years after its acquisition and interest is accumulated at a rate of r compounded continuously, then

$$y = Se^{-rx} = 1000e^{\sqrt{x}}e^{-rx} = 1000e^{\sqrt{x}-rx}$$

expresses the value of the timber land x years after its acquisition when the given interest conditions have been taken into consideration. Our optimization problem is to determine the value of x for which y is a maximum. To do this we must determine $\frac{dy}{dx}$, set it equal to zero, and solve for x. By using the chain rule we obtain

$$\frac{dy}{dx} = 1000e^{\sqrt{x}-rx}\left(\frac{1}{2\sqrt{x}} - r\right).$$

Setting $\frac{dy}{dx}$ equal to zero yields

$$1000e^{\sqrt{x}-rx}\left(\frac{1}{2\sqrt{x}} - r\right) = 0.$$

Since $1000e^{-rx} \neq 0$, this condition can only be satisfied if

$$\frac{1}{2\sqrt{x}} - r = 0.$$

Solving for x yields

$$x = \frac{1}{4r^2}.$$

To see if $x = 1/4r^2$ yields a maximum value, we use the first derivative test. If $x < 1/4r^2$.

$$\frac{dy}{dx} = \underbrace{1000e^{\sqrt{x}-rx}}_{\text{positive}}\underbrace{\left(\frac{1}{2\sqrt{x}} - r\right)}_{\text{negative}}$$

$\frac{dy}{dx}$ is the product of two positive factors and is thus positive. If $x > 1/4r^2$,

$$\frac{dy}{dx} = \underbrace{1000e^{\sqrt{x}-rx}}_{\text{positive}}\underbrace{\left(\frac{1}{2\sqrt{x}} - r\right)}_{\text{negative}}$$

$\frac{dy}{dx}$ is the product of a positive factor and a negative factor and is thus negative. Since $\frac{dy}{dx}$ goes from positive to negative, it follows from the first derivative test that

$$y = 1000e^{\sqrt{x}-rx}$$

has a maximum value at $x = 1/4r^2$. Therefore the higher the interest rate is, the shorter the optimal waiting time will be.

Thus, for example, if interest is accumulated at the rate of 10% compounded continuously, the optimal time to sell the timber is

$$x = \frac{1}{4(0.10)(0.10)} = 25 \text{ years}$$

after its acquisition. Its value after 25 years is

$$y = 1000e^{\sqrt{25}-0.12(25)} = e^{2.5} = 1000(12.1825) = \$12{,}182.50.$$

We should also note that while our formulation of the sale-time optimization problem takes interest considerations into account, it does not take storage or maintenance costs into account.

EXERCISES

1. A wine merchant has purchased a quantity of wine which is to be aged and resold at a later time. The function

$$S = 500e^{\frac{1}{2}\sqrt{x}}$$

expresses S, the dollar selling price of the wine, as a function of x, the number of years after its acquisition. Interest accumulates at the rate of 10% compounded continuously. Determine the optimal storage time of the wine.

2. An investor has purchased timber-bearing land which is to be resold at a later time. The function

$$S = 2000e^{\sqrt{x}-0.1x}$$

expresses the selling price of the land as a function of the number of years after its acquisition. For interest accumulating at the rate of 10% compounded continuously, determine the optimal time to sell the timber.

CHAPTER 8

Curve Sketching

8.1. TOOLS FOR CURVE SKETCHING

The **graph of *a* function *f***, we recall, is the graph of the equation $y = f(x)$, where x takes on only those values in its domain of definition. What may be termed crude graph sketching is based on the technique of letting the independent variable x take on a number of values, calculating the corresponding values of $y = f(x)$, plotting the points $(x, f(x))$, and joining these points with a smooth curve. There are a number of function characteristics that often play an important role in accurate graph sketching. Some of these characteristics, such as continuity, limit behavior as $x \to \infty$ or $x \to -\infty$, extreme values, and concavity behavior, have been considered in other connections. In this section we shall explore the use of these and other properties for graph sketching.

Continuity and Differentiability

In joining plotted points with a smooth curve, one must be careful of jumps, breaks, and gaps in the graph; thus the question of continuity arises. If $y = f(x)$ is continuous at a value, then its graph has no gaps, jumps, or breaks at the value; if $y = f(x)$ is not continuous at a value, then proceed with great caution near the value, since an irregularity is indicated. Let us also recall from Section 5.2 that, if the derivative of $y = f(x)$ exists at a value, then the function is continuous at the value. In fact, wherever the derivative of the function exists, its graph is not only in one piece, but is smooth as well. On the other hand, let us also keep in mind that a function that does not have a derivative at a value may still be continuous at the value. To settle the continuity question at values where the derivative does not exist, one must use the definition of continuity (see Section 3.3).

To illustrate, let us consider, $f(x) = 1/x$. This function is defined for all nonzero values of x. It is not defined at zero and is thus not continuous at zero. This signals that caution should be exercised near zero. The derivative exists for all nonzero values of x, $f'(x) = -1/x^2$, and thus the graph of $f(x) = 1/x$ is in one piece and is smooth for all nonzero values of x. Plotting points and connecting those to the right of the y-axis with a smooth curve and those to the left of the y-axis with a smooth curve yields the graph shown in Figure 8.1.

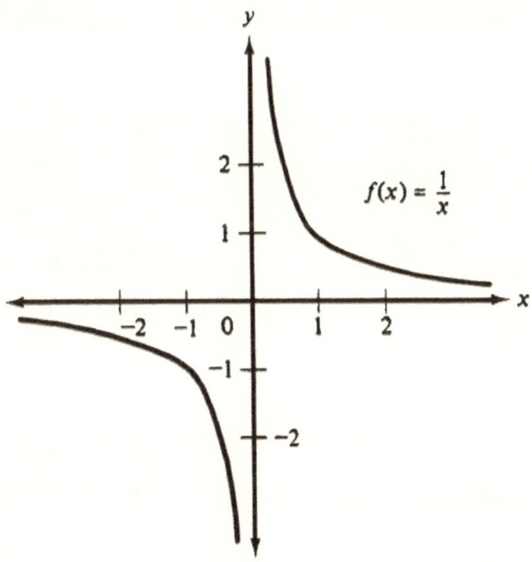

Figure 8.1

As another example, consider the function

$$f(x) = \frac{1}{3}x^3 - \frac{1}{2}x^2 - 6x + 1$$

which is defined for all real numbers. Since $f'(x)$ is also defined for all real numbers, the graph of $f(x)$ has no gaps, jumps, or breaks and is smooth.

Increasing and Decreasing Behavior

Figures 8.2 and 8.3 illustrate functions that are increasing over an interval, and Figures 8.4 and 8.5 illustrate functions that are decreasing over an interval. Formally stated, **$y = f(x)$ is increasing in an interval** if, for any two values a and b in the interval such that $a < b$, $f(a) < f(b)$. If $a < b$ and $f(a) > f(b)$, **$y = f(x)$ is said to be**

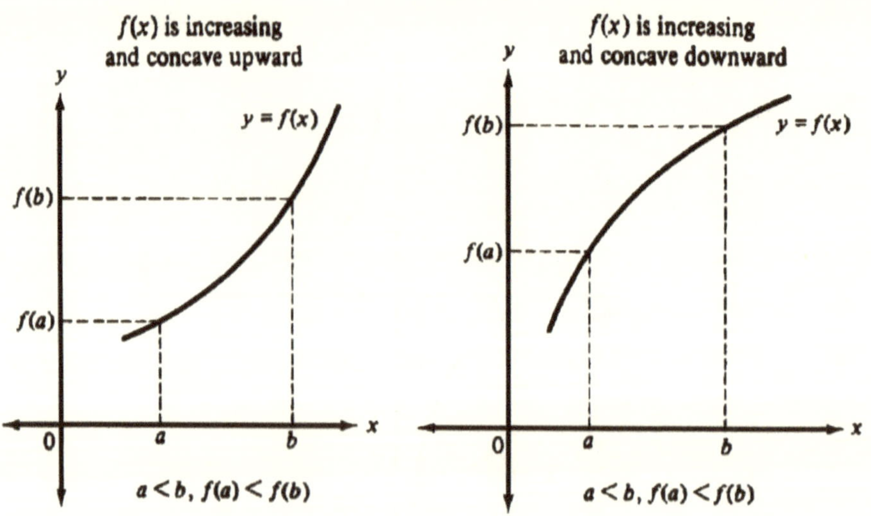

Figure 8.2

Figure 8.3

decreasing in the interval. In addition to intervals, these definitions hold for half-lines and the entire x-axis.

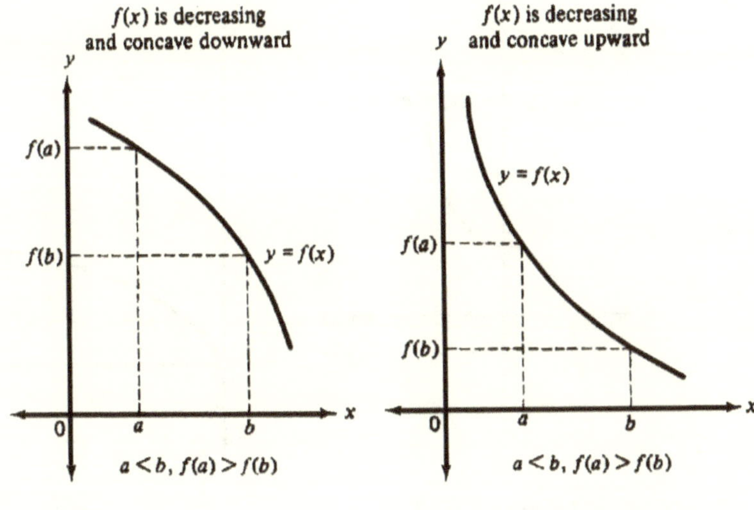

Figure 8.4 *Figure 8.5*

First Derivative Test

The increasing or decreasing behavior of $y = f(x)$ is shown by the sign of its first derivative.

> If $f'(x) > 0$ in an interval, then $y = f(x)$ is **increasing in the interval**. If $f'(x) < 0$ in an interval, then $y = f(x)$ is **decreasing in the interval**. Such a result holds for $y = f(x)$ defined on a half-line or the entire x-axis.

To illustrate, consider the functions $f(x) = x^3 + 1$, $f(x) = \sqrt{x}$, and $f(x) = 1/x^2$ shown in Figures 8.6 through 8.8, respectively. $f(x) = x^3 + 1$ is increasing over the entire x-axis and $f'(x) = 3x^2$, which is positive for all x. $f(x) = \sqrt{x}$ is increasing for $x \geq 0$ and $f'(x) = \frac{1}{2}\sqrt{x}$, which is positive for $x > 0$. $f(x) = 1/x^2$ is increasing for $x < 0$ and decreasing for $x > 0$; $f'(x) = -2/x^3$ is positive for $x < 0$ and negative for $x > 0$.

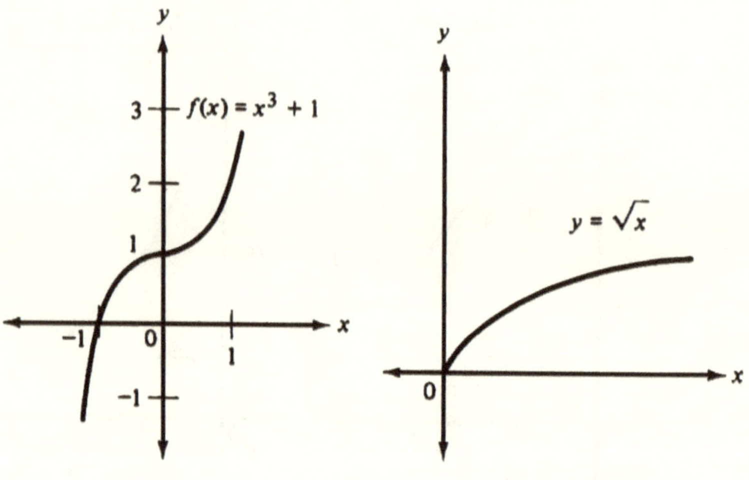

Figure 8.6

Figure 8.7

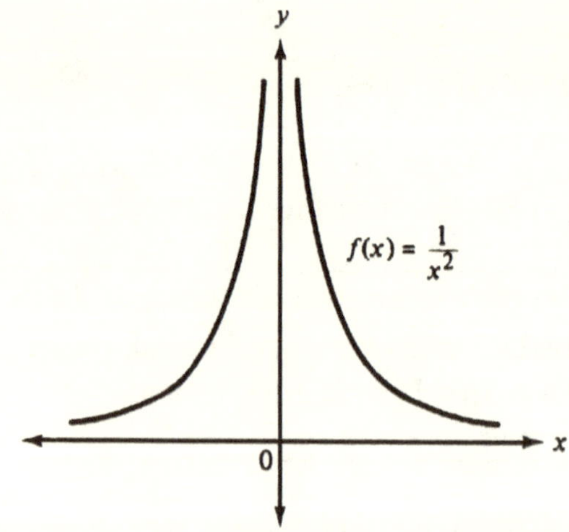

Figure 8.8

To further illustrate, let us return to the function

$$f(x) = \frac{1}{3}x^3 - \frac{1}{2}x^2 - 6x + 1$$

The derivative $f'(x)$ is

$$f'(x) = x^2 - x - 6$$
$$= (x+2)(x-3)$$

To determine where $f'(x)$ is positive and where it is negative, it is useful to first find where it is zero. Setting $f'(x)$ equal to zero yields

$$(x+2)(x-3) = 0$$
$$x = -2, \quad x = 3$$

Consider three cases: $x < -2$, $-2 < x < 3$, and $x > 3$.

For $x < -2$, $f'(x) = \underbrace{(x-3)}_{\text{neg.}}\underbrace{(x+2)}_{\text{neg.}}$ is positive.

For $-2 < x < 3$, $f'(x) = \underbrace{(x-3)}_{\text{neg,}}\underbrace{(x+2)}_{\text{pos.}}$ is negative.

For $x > 3$, $f'(x) = \underbrace{(x-3)}_{\text{pos.}}\underbrace{(x+2)}_{\text{pos.}}$ is positive.

Thus $f(x)$ is increasing for $x < -2$, decreasing for x between -2 and $3\,(-2 < x < 3)$, and increasing for $x > 3$.

Extreme Values

Extreme values, considered in Section 7.1, play an important role in graph sketching. Returning to

$$f(x) = \frac{1}{3}x^3 - \frac{1}{2}x^2 - 6x + 1$$

we see from our discussion of its increasing-decreasing behavior that $f'(-2) = 0$ and $f'(x)$ goes from positive to negative. Thus

$f(-2) = 8.3$ is a local maximum value. Since $f'(3) = 0$ and $f'(x)$ goes from negative to positive, $f(3) = -12.5$ is a local minimum value.

Concavity Behavior

The graph of the increasing functionshown in Figure 8.2 is concave upward, whereas the graph of the increasing function shown in Figure 8.3 concave downward; the graph of the decreasing function shown in Figure 8.4 is concave downward, whereas the graph of the decreasing function shown in Figure 8.5 is concave upward (a formal definition of concavity is given on p. 225). Thus it is not sufficient to know where a function is increasing and decreasing, since this by itself does not determine the concavity behavior of the graph. Let us also recall that the concavity behavior of the graph of $y = f(x)$ is determined by the sign of the second derivative $f''(x)$ in the following way.

Second Derivative Test for Concavity

If $f''(x) > 0$ for all x in an interval, then the graph of $y = f(x)$ is **concave upward in the interval.** If $f''(x) < 0$ for all x in an interval, then the graph of $y = f(x)$ is **concave downward in the interval**. This test also holds for half-lines and the entire x-axis.

If a point $P(c, f(c))$ is such that $f''(c) = 0$ and $f''(x)$ goes from positive to negative, or negative to positive, as we go from x less than c to x greater than c, then P, called an **inflection point of the graph**, separates concave upward and concave downward portions of the graph.

To illustrate, consider the graph of $f(x) = x^3 + 1$ (shown in Figure 8.6), which is concave downward for $x < 0$ and concave upward for $x > 0$.

$$f'(x) = 3x^2, \quad f''(x) = 6x$$

To determine inflection points, we begin by setting $f''(x)$ equal to zero and solving for x. In this case, we obtain

$$6x = 0, \qquad x = 0$$

Thus $f''(0) = 0$. For $x < 0$, $f''(x) = 6x$ is negative, and for $x > 0$, $f''(x) = 6x$ is positive. Thus, $(0, f(0)) = (0, 1)$ is an inflection point.

The graph of $f(x) = 1/x^2$ (shown in Figure 8.8) is concave upward for $x < 0$ concave upward for $x > 0$. The first and second derivatives are

$$f'(x) = -\frac{2}{x^3}, \quad f''(x) = \frac{6}{x^4}$$

$f''(x)$ is positive for $x < 0$ and positive for $x > 0$. There are no inflection points since $f''(x) \neq 0$.

The function

$$f(x) = \frac{1}{3}x^3 - \frac{1}{2}x^2 - 6x + 1$$

has first and second derivatives

$$f'(x) = x^2 - x - 6, \quad f''(x) = 2x - 1$$

Setting $f''(x)$ equal to zero and solving for x yields

$$2x - 1 = 0, \quad x = \frac{1}{2}$$

Since $f''(\frac{1}{2}) = 0$ and $f''(x) = 2x - 1$ is negative for $x < \frac{1}{2}$ and positive for $x > \frac{1}{2}$, the point $(\frac{1}{2}, f(\frac{1}{2})) = (\frac{1}{2}, -2.1)$ is an inflection point, and the graph of $f(x)$ is concave downward for $x < \frac{1}{2}$ and concave upward for $x > \frac{1}{2}$.

Limit Behavior as $x \to \infty$ or $x \to -\infty$

It is sometimes useful to examine the limit behavior of $y = f(x)$ as $x \to \infty$ or $x \to -\infty$. For

$$f(x) = \frac{1}{3}x^3 - \frac{1}{2}x^2 - 6x + 1$$

the term of highest power, $\frac{1}{3}x^3$, dominates the behavior of $f(x)$ as $x \to \infty$ and $x \to -\infty$. Thus we have

$$\lim_{x \to \infty} f(x) = \infty \qquad \lim_{x \to -\infty} f(x) = -\infty$$

EXERCISES

Determine where the following functions are increasing and decreasing, the points of inflection, and the concavity.

1. $f(x) = x^2 + 1$
2. $f(x) = \sqrt{x}$
3. $f(x) = -2x^2 + 4$
4. $f(x) = \frac{1}{x}$
5. $f(x) = -\frac{1}{x^2}$
6. $f(x) = 2x^3 + 3x^2$
7. $f(x) = \frac{1}{3}x^3 - \frac{1}{2}x^2 - 2x + 1$
8. $f(x) = 3x^4 - 4x^3$
9. $f(x) = \frac{1}{4}x + 5 + \frac{1}{x}$
10. $f(x) = e^{-x^2}$
11. $f(x) = xe^x$
12. $f(x) = x + 10 + \frac{100}{x}$

8.2. GRAPHS OF FUNCTIONS

In summary, then, knowledge of the following may be of considerable value in sketching the graph of a function:

1. Continuity and differentiability.
2. Extreme values.
3. Regions in which the function is increasing and decreasing.
4. Inflection points and concavity.
5. Limit behavior as $x \to \infty$ and $x \to -\infty$.

The determination of these characteristics is not intended as a replacement for, but rather as a supplement to, the practice of assigning selected values to the independent variable x, calculating the corresponding values of $f(x)$, and plotting the points $(x, f(x))$. Knowledge of these properties reduces the number of points that must be plotted and gives us a more accurate picture of what's happening in between the points plotted.

EXAMPLE 1

Sketch the graph of $f(x) = \frac{1}{3}x^3 - \frac{1}{2}x^2 - 6x + 1$.

In our discussion of graph sketching, the following data were accumulated about this function.

1. Continuity and differentiability. The derivative of $f(x)$ exists for all values of x. Thus the graph of $f(x)$ has no gaps, jumps, or breaks, and furthermore is smooth.
2. Extreme values. $(-2, 8.3)$ is a maximum point and $(3, -12.5)$ is a minimum point.
3. Increasing-decreasing behavior. $f(x)$ is increasing for $x < -2$, decreasing in the interval $-2 < x < 3$, and increasing for $x > 3$.

4. Points of inflection and concavity. $(\frac{1}{2}, -2.1)$ is an inflection point, and the graph of $f(x)$ is concave downward for $x < \frac{1}{2}$ and concave upward for $x > \frac{1}{2}$.
5. Limit behavior. $\text{limit}_{x \to \infty} f(x) = \infty$, $\text{limit}_{x \to -\infty} f(x) = -\infty$.

In Table 8.1, values of $f(x)$ are given for selected values of x. We thus obtain the graph shown in Figure 8.9.

Table 8.1

x	-4	-3	-1	0	1	2	4	5	6
$f(x)$	-4.3	5.5	6.2	1	-5.2	-10.3	-9.7	0.2	19

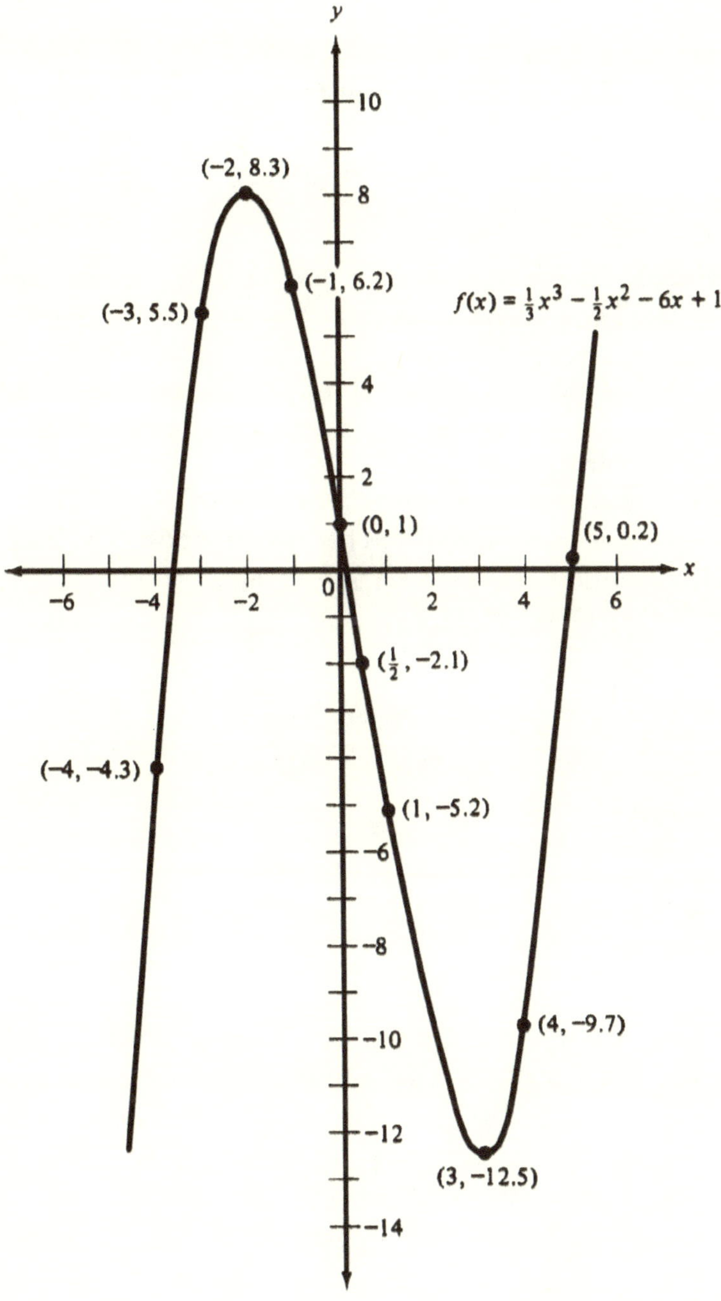
y
x
10
8
4
2
−6
−8
−10
−12
−14
−6
−4
−2
0
2
4
6
(−2, 8.3)
(−1, 6.2)
(−3, 5.5)
$f(x) = \frac{1}{3}x^3 - \frac{1}{2}x^2 - 6x + 1$
(0, 1)
(5, 0.2)
($\frac{1}{2}$, −2.1)
(−4, −4.3)
(1, −5.2)
(4, −9.7)
(3, −12.5)

Figure 8.9

EXAMPLE 2

Sketch the graph of the area function

$$A(x) = 102 + 3x + \frac{192}{x}$$

where $x > 0$, which arises from the type-plate-dimensions problem discussed in Section 7.1.

1. Continuity and differentiability. Since $A'(x)$ exists for all $x > 0$, the graph of $A(x)$ has no gaps, jumps, or breaks and is smooth.
2. Extreme values. There are neither end-point critical values nor values at which $A'(x)$ does not exist. To obtain type 1 critical values, we must determine $A'(x)$, set it equal to zero, and solve for x.

$$A'(x) = 3 - \frac{192}{x^2}$$

where $x > 0$.

$$3 - \frac{192}{x^2} = 0, \quad 3x^2 = 192, \quad x^2 = 64, \quad x = 8$$

Thus $x = 8$ is a critical value. To determine if $A(8)$ is an extreme value, we use the second-derivative test.

$$A''(x) = \frac{384}{x^3}, \quad A''(8) = \frac{384}{8^3} > 0$$

Since $A'(8)$ is positive, $A(8) = 150$ is a minimum value.

3. Increasing-decreasing behavior. To determine where $A(x)$ is increasing and decreasing, we must determine where $A'(x)$ is positive and negative.

$$A'(x) = 3 - \frac{192}{x^2} = \frac{3x^2}{x^2} - \frac{192}{x^2}$$

$$= \frac{3x^2 - 192}{x^2} = \frac{3(x-8)(x+8)}{x^2}$$

For $0 < x < 8$, $A'(x) = \dfrac{3\overbrace{(x-8)}^{\text{neg.}}\overbrace{(x+8)}^{\text{pos.}}}{\underbrace{x^2}_{\text{pos.}}}$ is negative.

For $x > 8$, $A'(x) = \dfrac{\overbrace{3(x-8)}^{\text{pos.}}\overbrace{(x+8)}^{\text{neg.}}}{\underbrace{x^2}_{\text{pos.}}}$ is positive.

Thus $A(x)$ is decreasing for $0 < x < 8$ and increasing for $x > 8$.

4. Points of inflection and concavity. The first step in finding points of inflection is to find where the second derivative is zero.

$$A''(x) = \frac{384}{x^3}$$

is positive for $x > 0$. Thus there are no inflection points, and the graph of $A(x)$ is concave upward for $x > 0$.

5. Limit behavior. As $x \to \infty$, $A(x) = 102 + 3x + 192/x \to \infty$. Since $A(x)$ has a term with x in the denominator, the behavior of $A(x)$ as x approaches 0 may be of interest. Let us observe that as x approaches 0 through positive values of x $(x = \frac{1}{10}, \frac{1}{100}$, and so on), $A(x) \to \infty$.

In summary, then, the graph of $A(x)$ has no gaps, jumps, or breaks and is smooth. $A(8) = 150$ is a minimum value; $A(x)$ is decreasing for x between 0 and 8 and increasing for $x > 8$. There are no inflection points, and the graph of $A(x)$ is concave upward for $x > 0$. As $x \to \infty$, $A(x) \to \infty$, and as $x \to 0$ through positive values, $A(x) \to \infty$.

Table 8.2 gives values of $A(x)$ for selected values of x. We thus obtain the graph shown in Figure 8.10.

Table 8.2

x	$\frac{1}{10}$	$\frac{1}{2}$	1	2	4	12	16	24	32
$A(x)$	2022	488	297	204	162	154	162	182	204

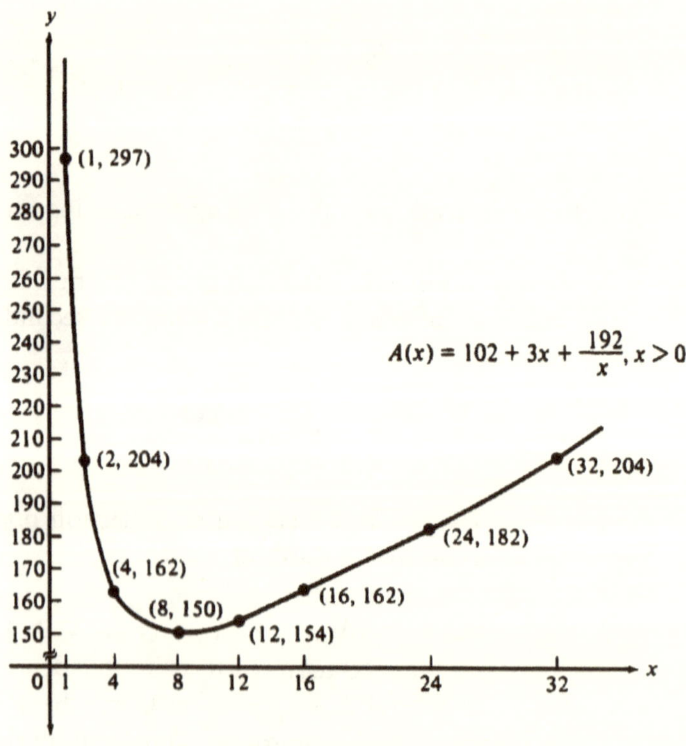

Figure 8.10

EXAMPLE 3

Sketch the graph of $f(x) = \frac{1}{\sqrt{2\pi}} e^{-x^2/2}$.

1. Continuity and differentiability. Since $f'(x)$ exists for all x, the graph of $f(x)$ has no gaps, jumps, or breaks and is smooth.
2. Extreme values. There are neither end-point critical values nor values at which $f'(x)$ does not exist. To obtain type 1 critical values, we must determine $f'(x)$, set it equal to zero, and solve for x. To determine $f'(x)$, we use the chain rule.

$$\begin{aligned} f'(x) &= \frac{1}{\sqrt{2\pi}} D_x(e^{-x^2/2}) \\ &= \frac{1}{\sqrt{2\pi}} \left[e^{-x^2/2} D_x \left(\frac{-x^2}{2} \right) \right] \\ &= \frac{1}{\sqrt{2\pi}} [e^{-x^2/2}(-x)] \\ &= \frac{-x}{\sqrt{2\pi}} e^{-x^2/2} \end{aligned}$$

Setting $f'(x)$ equal to zero and solving for x yields

$$= \frac{-x}{\sqrt{2\pi}} e^{-x^2/2} = 0$$

$$-x = 0, \quad x = 0$$

Thus 0 is a critical value.

For $x < 0$, $f'(x) = \frac{-x}{\underbrace{\sqrt{2\pi}}_{\text{pos.}}} \underbrace{e^{-x^2/2}}_{\text{pos.}}$ is positive.

For $x > 0$, $f'(x) = \underbrace{\frac{-x}{\sqrt{2\pi}}}_{\text{neg.}} \underbrace{e^{-x^2/2}}_{\text{pos.}}$ is negative.

Since $f'(x)$ goes from positive to negative, it follows from the first-derivative test that $f(0) = 1/\sqrt{2\pi}$ is a maximum value.

3. Increasing-decreasing behavior. Since $f'(x)$ is positive for $x < 0$ and negative for $x > 0$, $f(x)$ is increasing for $x < 0$ and decreasing for $x > 0$.
4. Points of inflection and concavity. To determine points of inflection, we must first determine $f''(x)$. By the product theorem and chain rule, we obtain

$$\begin{aligned} f''(x) &= \frac{-x}{\sqrt{2\pi}} D_x(e^{-x^2/2}) + e^{-x^2/2} D_x\left(\frac{-x}{\sqrt{2\pi}}\right) \\ &= \frac{-x}{\sqrt{2\pi}}\left[e^{-x^2/2} D_x\left(\frac{-x^2}{2}\right)\right] + e^{-x^2/2}\left(\frac{-x}{\sqrt{2\pi}}\right) \\ &= \frac{-x}{\sqrt{2\pi}} e^{-x^2/2}(-x) - \frac{1}{\sqrt{2\pi}} e^{-x^2/2} \\ &= \frac{x^2}{\sqrt{2\pi}} e^{-x^2/2} - \frac{1}{\sqrt{2\pi}} e^{-x^2/2} \end{aligned}$$

Factoring out $\frac{1}{\sqrt{2\pi}} e^{-x^2/2}$ yields

$$\begin{aligned} f''(x) &= \frac{1}{\sqrt{2\pi}} e^{-x^2/2}(x^2 - 1) \\ &= \frac{1}{\sqrt{2\pi}} e^{-x^2/2}(x-1)(x+1) \end{aligned}$$

Setting $f''(x)$ equal to zero and solving for x yields

$$x=-1, \quad x=1$$

For $x<-1$, $f''(x)=\underbrace{\frac{1}{\sqrt{2\pi}}}_{\text{pos.}}e^{-x^2/2}\underbrace{(x-1)}_{\text{neg.}}\underbrace{(x+1)}_{\text{neg.}}$ is positive.

For $-1<x<1$, $f''(x)=\underbrace{\frac{1}{\sqrt{2\pi}}}_{\text{pos.}}e^{-x^2/2}\underbrace{(x-1)}_{\text{neg.}}\underbrace{(x+1)}_{\text{pos.}}$ is negative.

For $x>1$, $f''(x)=\underbrace{\frac{1}{\sqrt{2\pi}}}_{\text{pos.}}e^{-x^2/2}\underbrace{(x-1)}_{\text{pos.}}\underbrace{(x+1)}_{\text{pos.}}$ is positive.

Thus $(-1, f(-1))=(-1,\ 1/\sqrt{2\pi e})$ and $(1, f(1))=(1,\ 1/\sqrt{2\pi e})$ are inflection points, and the graph of $f(x)$ is concave upward for $x<-1$, concave downward for x between -1 and 1, and concave upward for $x>1$.

5. Limit behavior. As $x\to\infty$,

$$e^{-x^2/2}=\frac{1}{e^{x^2/2}}\to 0$$

so that

$$f(x)=\frac{1}{\sqrt{2\pi}}e^{-x^2/2}\to 0$$

Also, as $x\to-\infty$, $f(x)\to 0$.

6. Other points of interest. The graph of $f(x)$ is symmetric with respect to the y-axis; that is, the behavior of the graph for positive x is duplicated for corresponding negative x. Since $e^{-x^2/2}$ is positive for all x, the graph of

$f(x)$ is entirely above the x-axis. The graph of $f(x)$ is shown in Figure 8.11. This curve is of particular importance in probability and statistics and is called the standard normal curve.

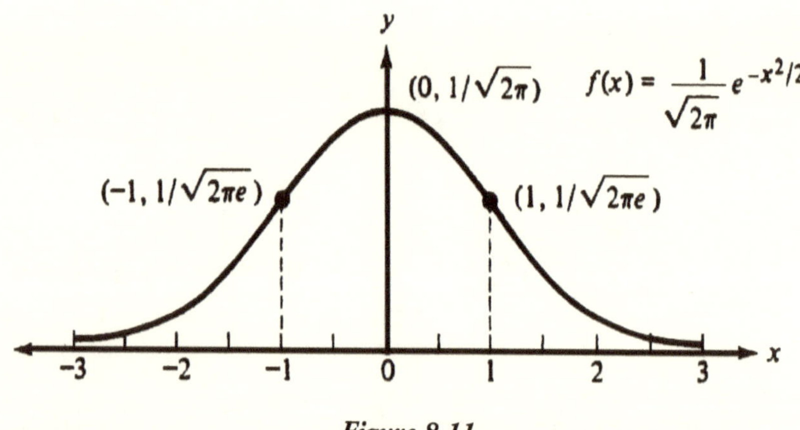

Figure 8.11

EXERCISES

Sketch the graphs of the following functions.

1. $f(x)=3x^2+6x+4$
2. $f(x)=-x^2-4x+2$
3. $f(x)=2x^3+3x^2$
4. $f(x)=2x^3-3x^2-12x+4$
5. $f(x)=\frac{1}{4}x+5+\frac{1}{x}$
6. $f(x)=3x^4-4x^3$
7. $f(x)=xe^x$
8. $f(x)=x^3-3x$
9. $f(x)=x+10+\frac{100}{x}$
10. $f(x)=\frac{1}{3}x^3-\frac{1}{2}x^2-2x+1$
11. $f(x)=x+\frac{1}{x}$
12. $f(x)=e^x-e^{-x}$
13. $f(x)=x^2e^{-x}$
14. $f(x)=xe^{-x^2}$

SELF-TESTS FOR CHAPTERS 1-8

Allow 100 or so minutes for each self-test. Go over each one before proceeding to the next.

Self-Test 1

In questions (1) through (5) find, if it exists, the limit required. If the limit does not exist or is infinity or minus infinity, so state. Explain the basis for your conclusion in each case.

1. $\lim\limits_{x\to 2} f(x)$, where $f(x)=\begin{cases} 4x+2, & x<2 \\ 12, & x=2. \\ 5x-2, & x>2 \end{cases}$
2. $\lim\limits_{x\to 2} \dfrac{5}{x-2}$
3. $\lim\limits_{t\to\infty} \left(1+\dfrac{7}{t}\right)^t$
4. $\lim\limits_{x\to 3} \dfrac{4}{(x-3)^2}$
5. $\lim\limits_{x\to 3} \dfrac{2x-6}{x-3}$
6. Is the function $f(x)$ defined in question (1) continuous at 2? Explain the basis for your conclusion in terms of the definition of continuity.
7. State the definition of the derivative of $y = f(x)$.
8. By using the definition of derivative find, if it exists, $f'(3)$ for $f(x)=2x^2+3x-1$. If $f'(3)$ does not exist, so state and explain.
9. Is the function $f(x)$ defined in question (1) differentiable at 2? Explain.

In questions (10) through (13) find the derivatives of the functions stated.

10. $y = 2x^4 e^x$
11. $y = \ln(4x^3 + 3x)$
12. $y = 5e^{-x^2+3x}$
13. $y = \sqrt{5x^2 + 2x + 4}$
14. Find the tangent line to the graph of $f(x) = 2x^2 + 4$ at the point where $x = 3$.
15. For $f(x) = \ln x$ find $f''(2)$.
16. Find the critical values of $f(x) = x^3 + x^2 - x + 5$, where $0 \leq x \leq 2$.
17. A producer has revenue and cost functions

$$R(x) = 1236x - 2x^2, \quad c(x) = x^2 + 6x + 5000,$$

where output is in tons per week. Find the optimal output level.
18. With respect to the conditions described in (17), what would the optimal output level be if a lump-sum tax of $7000 were imposed?
19. With respect to the conditions described in (17), what would the optimal output level be if an output tax of $480 per ton were imposed?
20. With respect to the conditions described in (17), suppose that an output tax of t dollars per ton were imposed. State the function that describes tax revenue in terms of t.
21. With respect to (20), find t for which the tax revenue function is maximized.

Self-Test 2

1. For $f(x) = 2x + \frac{1}{x}$ find $f''(3)$.
2. For $f(x) = \ln(3x^2 + 4x)$ find $f'(x)$.
3. For $f(x) = 3x + x^2 \cdot \ln x$ find $f'(x)$.
4. Find the inverse of $y = 2x^3$.

5. A copper producer has revenue and cost functions

$$R(x) = 8056x - 3x^2, \quad c(x) = \frac{1}{2}x^2 + 6x + 3000$$

Where output is in tons per week. Find the optimal output level.

6. With respect to the conditions described in (5), what would the optimal output level be if a 30% of profit tax were imposed?
7. With respect to the conditions described in (5), what would the optimal output level be if an output tax of $350 per ton were imposed?
8. With respect to the conditions described in (5), suppose that an output tax of t dollars per ton were imposed. State the function that describes tax revenue in terms of t.
9. With respect to (8), find t for which the tax revenue function is maximized.
10. Find the slope of the tangent line to the graph of $y = 3/x$ at $P(6, \frac{1}{2})$.
11. List the critical values of $f(x) = 3x^2 - 2x + 10$, defined for $-1 \leq x \leq 2$.
12. State the definition of derivative for $y = f(x)$ at 3; at 0.
13. By using the definition of derivative find, if it exists, $f'(0)$ for

$$f(x) = \begin{cases} x, & \text{for } x \geq 0 \\ -x, & \text{for } x < 0 \end{cases}.$$

If $f'(x)$ does not exist, so state and explain.
14. Is $f(x)$ defined in (13) continuous at 0? Explain.
15. Find, if it exists, $f''(-1)$ for $f(x) = \ln x$. If $f''(-1)$ does not exist, so state and explain.
16. Find the derivative of the inverse of $y = 3x^2 + 1$.

17. Find $f'(x)$ for $f(x)=\dfrac{4x^3-2x}{3x^2+x}$
18. Find $f'(x)$ for $f(x)=e^{5x+2}$
19. Find $f'(x)$ for $f(x)=\sqrt{2x^3+2x-1}$.
20. Define *e*.
21. Sketch the graph of $\overline{c}(x)=x+2+\dfrac{100}{x}$, $x > 0$, the average cost function of a chocolate producer.

Self-Test 3

In questions (1) through (7) find, if it exists, the limit required. If the limit does not exist or is infinity or $-\infty$, so state. Explain the basis for your conclusion in each case.

1. $\operatorname*{limit}_{x\to 3} f(x)$, where $f(x)=\begin{cases}4x+1, & x<3\\ 10, & x=3.\\ 5x-2, & x>3\end{cases}$

2. $\operatorname*{limit}_{x\to 2} f(x)$, where $f(x)$ is defined in (1).

3. $\operatorname*{limit}_{x\to\infty}\dfrac{3x^3+4x+1}{4x^3-x^2-3}$

4. $\operatorname*{limit}_{x\to 3}\dfrac{6x-18}{x-3}$

5. $\operatorname*{limit}_{x\to 2}\dfrac{6}{(x-2)^2}$

6. $\operatorname*{limit}_{x\to 1}\dfrac{4}{(x-1)^3}$

7. $\operatorname*{limit}_{x\to\infty}\left(6-\dfrac{1}{x}+\dfrac{5}{x^2}\right)$

8. Is $f(x)$ defined in (1) continuous at 3? Explain in terms of the definition of continuity.
9. Is $f(x)$ defined in (1) differentiable at 3? Explain.
10. Sketch the graph of $f(x)=\begin{cases} 3x-1, & x\geq 2 \\ -2x+1, & x<2 \end{cases}$
11. What is the domain of definition of $f(x)=\dfrac{x-1}{3x+4}$? Explain.
12. Find the effective rate corresponding to a rate of 6% per annum compounded continuously.
13. What does the result obtained in answer to (12) mean?
14. Define instantaneous velocity and marginal cost and explain their points of similarity and difference.
15. The income tax paid in Vorex City is \$0.00 if taxable income is less than or equal to \$10,000, and \$200 plus 5% of the excess over \$10,000 if taxable income is over \$10,000. Express tax to be paid as a function of taxable income.
16. A pipe line to transport oil is to connect cites A and B (shown in Figure 1) which are located on opposite sides of a river 5 miles wide. Site A is located 20 miles downstream from point C which is directly across the river from B.

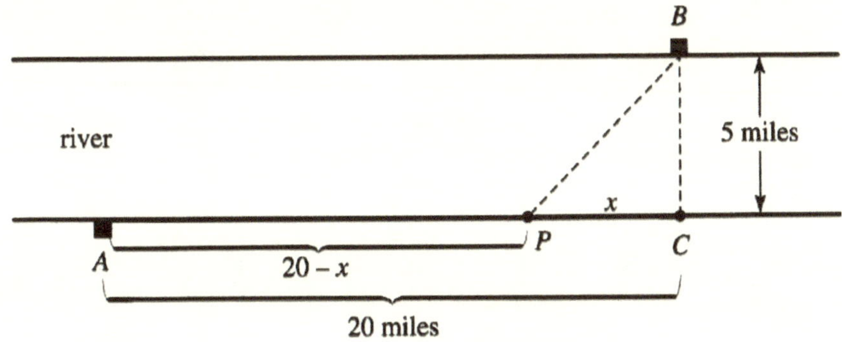

Figure 1

Constructing the pipe line along the line from A to C on the bank of the river would cost \$4000 per mile, while laying pipe across the river costs \$5000 per mile. What is the most economical route that can be used in constructing the pipe line? That is, at what point P on the bank should the pipe be laid across the river to B after having been laid along the river bank from A to P if the total cost is to be minimized? What is the minimum cost?

17. Would it be correct to say that if a function $y = f(x)$ is not differentiable at a value c, then it cannot be continuous at c?

18. Find $f'(8)$, where $f(x) = \sqrt[3]{x}$.

19. Find the maximum and minimum values of $f(x) = 4x^5 - 5x^4 + 2$.

20. Is the second derivative test applicable to the type 1 critical values that arise for $f(x)$ in (19)? Explain.

CHAPTER 9

Integral Calculus: Indefinite Integrals

9.1. INDEFINITE INTEGRALS

Thus far our work has taken us down the road concerned with finding the derivative of a given function $f(x)$. Also of great importance is the road that takes us in the opposite direction, so to speak. That is, given a function $f(x)$, is $f(x)$ the derivative of some other function $F(x)$? Find, if it exists, a function $F(x)$ whose derivative $F'(x)$ is $f(x)$. To illustrate, consider

$$f(x) = 2x$$

There are many functions whose derivative is $f(x) = 2x$; for example,

$$F_1(x) = x^2, \quad F_2(x) = x^2 - 3, \quad F_3(x) = x^2 + \frac{1}{2}$$

More generally,

$$\begin{aligned} F(x) &= F_1(x) = C \\ &= x^2 + C \end{aligned}$$

where C is any constant, has the property that $F'(x) = 2x$. Each such function whose derivative is $f(x) = 2x$ is called an indefinite integral, or antiderivative, of $f(x) = 2x$.

More generally, if $f(x)$ is a function, then any function $F(x)$ with the property that $F'(x) = f(x)$ is called an **indefinite integral**, or **antiderivative**, of $f(x)$. As we have observed in connection with $f(x) = 2x$, since the derivative of a constant is zero, if $F(x)$ is one indefinite integral of $f(x)$, then the function obtained by adding a constant to $F(x)$ is another indefinite integral of $f(x)$. Moreover, all indefinite integrals of $f(x)$ can be obtained by adding constants to a given indefinite integral. That is, if $F(x)$ and $G(x)$ are indefinite integrals, then they differ by a constant; there is a constant C such that $F(x) = G(x) + C$. The notation

$$\int f(x)dx$$

is used to denote an indefinite integral of $f(x)$. The symbol $\int$ is called an **integral sign**, and the role of dx is to identify x as the independent variable. If $F(x)$ is an indefinite integral of $f(x)$, then we write

$$\int f(x)dx = F(x) + C$$

to indicate that every indefinite integral can be obtained by adding a constant to $F(x)$. Thus the indefinite integrals of $f(x) = 2x$ are expressed by writing

$$\int 2x\, dx = x^2 + C$$

The process of finding the indefinite integrals of $f(x)$ is called **integrating** $f(x)$.

By definition, then, the operations of differentiation and integration are inverses of each other in the sense that

$$D_x \int f(x)dx = f(x). \quad \int D_x f(x)dx = f(x) + C$$

Some Basic Integration Formulas

From differential calculus let us recall the following results:

$$D_x x = 1, \quad D_x(2x) = 2$$

and more generally,

$$D_x(kx) = k$$

where k is any constant. Stated in integration terms, we have the following:

$$\int 1\,dx = x + C \quad \text{or} \quad \int dx = x + C$$

$$\int 2\,dx = 2x + C$$

and, more generally,

$$\int k\,dx = kx + C$$

where k is any constant.

To differentiate such functions as $f(x) = x^2$, $f(x) = x^3$, and, more generally, $f(x) = x^n$, we take the exponent as a multiplier and drop the exponent by 1, thus yielding the well-known $f'(x) = 2x$, $f'(x) = 3x^2$, and $f'(x) = nx^{n-1}$. To integrate a function of this sort, it is thus clear that we must at the very least raise the exponent by 1. Consider, for example,

$$f(x) = x^3$$

If we raise the exponent by 1 to obtain $F_1(x) = x^4$, we have $F'(x) = 4x^3$, so that an unwanted 4 enters the scene. To prevent this unwanted 4 from entering the scene, we must divide x^4 by 4, thus obtaining

$$F(x) = \frac{1}{4}x^4$$

Now we have

$$F'(x) = \frac{1}{4}(4x^3) = x^3$$

More generally, then, to integrate $f(x) = x^n$, raise the exponent by 1, yielding $n + 1$, and divide by the new exponent. That is,

$$\int x^n \, dx = \frac{1}{n+1} x^{n+1} + C$$

where n is any real number except for -1. When $n = -1$, we have

$$\int x^{-1} \, dx = \int \frac{dx}{x} = \ln x + C$$

since the derivative of $\ln x$ is $1/x$, where $x > 0$.

As further illustrations, consider the following:

$$\int \frac{dx}{x^3} = \int x^{-3} dx = \frac{1}{-3+1} x^{-3+1} + C = -\frac{1}{2} x^{-2} + C$$

$$\int x^4 dx = \frac{1}{4+1} x^{4+1} + C = \frac{1}{5} x^5 + C$$

$$\int \sqrt{x} \, dx = \int x^{1/2} dx = \frac{1}{\frac{1}{2}+1} x^{1/2+1} + C = \frac{2}{3} x^{3/2} + C$$

$$\int x^{1/4}dx = \frac{1}{\frac{1}{4}+1}x^{1/4+1} = \frac{4}{5}x^{5/4} + C$$

From differential calculus, let us recall that the derivative of the sum, or difference, of two functions is the sum, or difference, of the derivatives of the component parts, and the derivative of a constant times a function is the constant times the derivative of the function. The integration counterparts of these theorems are the following:

$$\int [f(x) + g(x)]\, dx = \int f(x)\, dx + \int g(x)\, dx$$

$$\int kf(x)\, d(x) = k \int f(x)\, dx$$

To integrate a sum, or difference, of functions, integrate the components and add, or subtract. This last result tells us that constants can be brought across the integral sign and back. Thus, for example,

$$\int (x^3 + x^2 - 1)\, dx = \int x^3 dx + \int x^2 dx - \int 1\, dx = \frac{1}{4}x^4 + \frac{1}{3}x^3 - x + C$$

$$\int \left(3 - \frac{1}{x}\right) dx = \int 3dx - \int \frac{dx}{x} = 3x - \ln x + C$$

$$\int 3x^6\, dx = 3 \int x^6\, dx = 3\left(\frac{1}{7}x^7\right) + C = \frac{3}{7}x^7 + C$$

$$\int \left(4x^2 - \frac{2}{x}\right) dx = 4 \int x^2\, dx - 2 \int \frac{dx}{x} = \frac{4}{3}x^3 - 2\ln x + C$$

In connection with exponential functions, we have,

$$\int e^x\, dx = e^x + C$$

EXERCISES

Find the following integrals and check your results by differentiation.

1. $\int x^5\,dx$
2. $\int x^8\,dx$
3. $\int x^{12}\,dx$
4. $\int \frac{dx}{x^2}$
5. $\int \frac{dx}{x^4}$
6. $\int \frac{dx}{x^8}$
7. $\int x^{3/2}\,dx$
8. $\int x^{1/3}\,dx$
9. $\int \frac{dx}{x^{1/4}}$
10. $\int \frac{2}{x^{1/2}}\,dx$
11. $\int 5x^{1/5}\,dx$
12. $\int (4x^6+3x)\,dx$
13. $\int (5x^2-3x+1)\,dx$
14. $\int \left(2x^{1/3}-x+\frac{4}{x}\right)dx$
15. $\int (3x^4-3\sqrt{x}-1)\,dx$
16. $\int (3e^x-\sqrt{x})\,dx$
17. $\int \left(\frac{3}{x^2}-4x^2\right)dx$
18. $\int \left(4\sqrt{x}-\frac{3}{x}\right)dx$
19. $\int (5x^7-3x^3+4)\,dx$
20. $\int \left(\frac{5}{x^{1/3}}-3\right)dx$
21. $\int (6x^2-3e^x)\,dx$
22. $\int \left(\frac{1+x}{x^2}\right)dx$
23. $\int \left(\frac{x^2+1}{x}\right)dx$
24. $\int (4x^3-x^{-1/2})\,dx$

9.2. THE CHAIN RULE IN THE WORLD OF INTEGRALS

To obtain a basic integration tool, which has the chain rule for derivatives as its basis, suppose that $f(x)$ is a function, and consider the function

$$y=\frac{1}{n+1}[f(x)]^{n+1}+C$$

where $n \neq -1$ and C are constants. To differentiate this function, we use the chain rule (see Section 6.3, p. 173) to obtain

$$D_x\left(\frac{1}{n+1}[f(x)]^{n+1}+C\right)=\frac{1}{n+1}(n+1)[f(x)]^{(n+1)-1}f'(x)+0$$
$$=[f(x)]^n f'(x)$$

By definition of indefinite integral, this means

$$\int [f(x)]^n f'(x)dx=\frac{1}{n+1}[f(x)]^{n+1}+C$$

To successfully apply this theorem, one must have the differentiation theorems under good control so that the structure under the integral sign can easily be recognized. To illustrate our integral theorem, consider the following examples.

EXAMPLE 1

Find $\int 6x(3x^2+4)^5\,dx$.

If we take $f(x) = 3x^2 + 4$, then $f'(x)=6x$. Thus the needed structure is present ($n = 5$ here), and we have

$$\int 6x(\underbrace{3x^2+4})^5\,dx=\frac{1}{6}(3x^2+4)^6+C$$

$$\downarrow \qquad \downarrow$$

$$f'(x) \quad f(x)$$

EXAMPLE 2

Find $\int 3x^2\sqrt{x^3+2}\,dx$.

If we take $f(x) = x^3 + 2$, then $f'(x) = 3x^2$. Here too the needed structure is present $(n = \frac{1}{2})$, and we have

$$\int \underbrace{3x^2}_{f'(x)} \underbrace{(x^3+2)}_{f(x)}{}^{1/2}\, dx = \frac{1}{\frac{1}{2}+1}(x^3+2)^{1/2+1} + C$$

$$= \frac{2}{3}(x^3+2)^{3/2} + C$$

EXAMPLE 3

Find $\int x^2(4x^3+2)^6\, dx$.

If we take $f(x) = 4x^3 + 2$, then $f'(x) = 12x^2$. Under the integral sign we have something close to $f'(x)$, that is, x^2, but we need $12x^2$. Fortunately, what we need differs from what we have by a constant, 12, and, as we have seen, constants can be moved back and forth across the integral sign. Thus to get the structure we need we simply insert the needed 12 under the integral sign and at the same time insert its reciprocal, $\frac{1}{12}$, so as to not change the value of the integral. In effect, we are multiplying x^2 under the integral sign by $12(\frac{1}{12})$, which is 1. This yields

$$\int (\frac{1}{12}) 12x^2(4x^3+2)^6\, dx$$

We don't want the $\frac{1}{12}$ value under the integral sign, since it disturbs the structure we need. Thus we move it in front of the integral to obtain

$$\frac{1}{12}\int 12x^2(4x^3+2)^6\,dx$$

Now with $f(x) = 4x^3 + 2$ and $f'(x) = 12x^2$, we have the structure we need under the integral sign. With $n = 6$, we obtain

$$\frac{1}{12}\int(\underbrace{12x^2}_{f'(x)})(\underbrace{4x^3+2}_{f(x)})^6\,dx = \frac{1}{12}\left[\frac{1}{7}(4x^3+2)^7\right]+C$$

$$= \frac{1}{84}(4x^3+2)^7+C$$

The problem of course is to see what choice of $f(x)$ will do the job, and this is where a good foundation in differentiation is helpful.

> ***Caution***. In some student circles it has become popular to insert variables and their reciprocals under the integral sign and move one or the other across the integral sign. In a word, **disastrous**! Keep in mind that only constants can be handled in this way.

Another way of looking at the theorem

$$\int[f(x)]^n f'(x)dx = \frac{1}{n+1}[f(x)]^{n+1}+C$$

is from the point of view of finding a helping function $u = f(x)$ such that the function being integrated takes the form

$$\int u^n D_x u\,dx, \quad \text{where } u = f(x)$$

Then

$$\int u^n D_x u\,dx = \frac{1}{n+1}u^{n+1}+C$$

where $u = f(x)$. We also know that, in terns of u itself (without reference to functions of x),

$$\int u^n du = \frac{1}{n+1} u^{n+1} + C$$

Thus the problem of integrating a function of x with the structure

$$\int u^n D_x u \, dx$$

where $u = f(x)$, can be reduced to integrating the function of u

$$\int u^n du$$

provided that after this integration has been done, yielding

$$\frac{1}{n+1} u^{n+1} + C$$

u is replaced by the initially introduced helping function $u = f(x)$. To illustrate, let us consider from this point of view the problem taken up in Example 1.

EXAMPLE 4

Find $\int 6x(3x^2 + 4)^5 \, dx$.

We take as our helping function $u = 3x^2 + 4$. Then $D_x u = 6x$. Thus we have

$$\int \underbrace{(6x)}_{D_x u} \underbrace{(3x^2 + 4)}_{u}{}^5 \, dx = \int u^5 du = \frac{1}{6} u^6 + C$$

Replacing u by what it is in terms of x, $3x^2 + 4$, yields

$$\int (6x)(3x^2+4)^5\,dx = \frac{1}{6}(3x^2+4)^6 + C$$

Thus we obtain the same result by using the same theorem, but looked at from a different point of view. This point of view emphasizes the idea of reducing an integration problem in terms of x to a simpler integration problem in terms of u by introducing a suitable helping function, integrating the function of u, and then replacing u by the helping function introduced to obtain the originally sought after integral. The virtue of this point of view is its more general applicability. The following examples illustrate.

EXAMPLE 5

Find $\int e^{2x+1}\,dx$

We take $u = 2x + 1$ as our helping function. Then $D_x u = 2$. We don't have 2 under the integral sign, so we introduce 2 and its reciprocal $\frac{1}{2}$, yielding

$$\frac{1}{2}\int \underset{\substack{\downarrow \\ D_x u}}{2}\, e^{\overset{2x+1}{\underset{u}{\downarrow}}}\,dx$$

This structure allows us to transfer the problem to the world of u as follows:

$$\frac{1}{2}\int \underset{\substack{\downarrow \\ D_x u}}{2}\, e^{\overset{2x+1}{\underset{u}{\downarrow}}}\,dx = \frac{1}{2}\int e^u\,du = \frac{1}{2}e^u + C$$

Replacing u by $2x + 1$ yields

$$\int e^{2x+1}dx = \frac{1}{2}e^{2x+1} + C$$

EXAMPLE 6

Find $\int \frac{2}{4x+3}dx$.

We take $u = 4x + 3$ as our helping function. Then $D_x u = 4$. To obtain the needed 4 under the integral sign, we introduce 2 and its reciprocal $\frac{1}{2}$ into the scene, thus yielding

$$\frac{1}{2}\int \frac{\overset{\overset{D_x u}{\downarrow}}{4}}{\underbrace{4x+3}_{u}}dx$$

This structure allows us to transfer the problem to the world of u as follows:

$$\frac{1}{2}\int \frac{4}{4x+3}dx = \frac{1}{2}\int \frac{du}{u} = \frac{1}{2}\ln u + C$$

Replacing u by $4x + 3$ yields

$$\int \frac{2}{4x+3}dx = \frac{1}{2}\ln(4x+3) + C$$

EXERCISES

Find the following integrals.

1. $\int 8x(4x^2+4)^4\,dx$
2. $\int 10x(5x^2-1)^5\,dx$
3. $\int 3(3x+8)^8\,dx$
4. $\int 15x^2(5x^3-2)^2\,dx$
5. $\int x(3x^2-1)^9\,dx$
6. $\int x^3(x^4-1)^5\,dx$
7. $\int 4x\sqrt{x^2+4}\,dx$
8. $\int 3x^2\sqrt{2x^3-7}\,dx$
9. $\int e^{4x-1}\,dx$
10. $\int \frac{dx}{8x-1}$
11. $\int xe^{x^2}\,dx$
12. $\int \frac{8x}{(4x^2+1)^2}\,dx$
13. $\int \frac{4x^2}{(4x^3+1)^4}\,dx$
14. $\int \frac{2x+3}{\sqrt{x^2+3x}}\,dx$
15. $\int \frac{5x^4}{x^5+1\,dx}$
16. $\int 3e^{2x}\,dx$
17. $\int \frac{f'(t)}{f(t)}\,dt$
18. $\int \frac{3x^2+4}{x^3+4x+1}\,dx$
19. $\int e^{-3x+1}\,dx$
20. $\int \frac{2x}{\sqrt{5x^2+1}}\,dx$

9.3. INTEGRATION BY PARTS

From the product theorem for derivatives, we have:

$$D_x f(x)g(x) = f(x)g'(x) + g(x)f'(x)$$

Integrating both sides yields

$$f(x)g(x) = \int f(x)g'(x)dx + \int g(x)f'(x)dx$$

which we rewrite as

$$\boxed{\int f(x)g'(x)dx = f(x)g(x) - \int g(x)f'(x)dx \qquad (9.1)}$$

This result, called the integration-by-parts formula, allows us to reduce one integration problem to another based on the idea of choosing $f(x)$ and $g'(x)$ so that $g(x)$ is easily obtained by integration, and $\int g(x)f'(x)\,dx$ is easier to evaluate than the given integral $\int f(x)g'(x)\,dx$.

EXAMPLE 1

Find $\int xe^x\,dx$

Choose $f(x)$ and $g'(x)$ as follows:

$$f(x) = x \qquad \text{and } g'(x) = e^x$$

Then $f'(x)$ and $g(x)$ are

$$f'(x) = 1 \quad \text{and } g(x) = e^x$$

From the integration-by-parts formula we obtain the following results. (Note that we wait until we have the final expression for the integral before adding the constant C.)

$$\int xe^x\,dx = xe^x - \int e^x\,dx = xe^x - e^x + C$$

The function notation employed in the preceding analysis makes clear the logic of the integration-by-parts theorem, but the application of the result is often made simpler by casting it in the following form.

Let

$$u = f(x), \qquad v = g(x)$$

and

$$du = f'(x)dx, \qquad dv = g'(x)dx.$$

Then (9.1) can be written in the form:

$$\int u\,dv = uv - \int v\,du \qquad (9.2)$$

To illustrate (9.2) we return to the problem of determining $\int xe^x\,dx$.

Take $u = x$, $dv = e^x dx$. Then:

$$du = 1dx, \quad v = \int e^x\,dx = e^x$$

From (9.2) we have:

$$\int xe^x\,dx = xe^x - \int e^x\,dx = xe^x - e^x + C$$

The key to success in employing integration-by-parts is in making a "good" choice of u and dv so that the problem of integrating dv is easier than that of integrating the function we started with.

EXAMPLE 2

Find $\int \ln x\,dx$

Take $u = \ln x$, $dv = 1dx$. Then:

$$du = \frac{1}{x}dx, \quad v = \int 1dx = x$$

From (9.2) we have:

$$\int \ln x\,dx = x\ln x - \int x\frac{1}{x}dx$$

$$= x\ln x - x + C$$

EXAMPLE 3

Find $\int (2200-100t)e^{-0.1t}\,dt$

Take $u=2200-100t,\quad dv=e^{-0.1t}$. Then:

$$du=-100dt,\ v=\int e^{-0.1t}\,dt=-10e^{-0.1t}$$

We have:

$$\begin{aligned}\int (2200-100t)e^{-0.1t}\,dt &= -10(2200-100t)e^{-0.1t} \\ &\quad -\int 1000e^{-0.1t}\,dt \\ &= -22,000e^{-0.1t}+1000te^{-0.1t} \\ &\quad +\frac{1000}{0.1}e^{-0.1t}+C \\ &= -12,000e^{-0.1t}+1000te^{-0.1t}+C\end{aligned}$$

EXERCISES

Determine the following integrals.

1. $\int xe^{-x}\,dx$
2. $\int xe^{3x}\,dx$
3. $\int (1800-50t)e^{-0.1t}\,dt$
4. $\int (1000-20t)e^{-0.05t}\,dt$
5. $\int xe^{-2x}\,dx$
6. $\int (100+3t)e^{-0.1(5-t)}\,dt$
7. $\int (500-10t)e^{-0.08t}\,dt$
8. $\int \frac{x^3}{\sqrt{1+x^2}}\,dx$
9. $\int x\ln x\,dx$
10. $\int x^2e^{-x}\,dx$

CHAPTER 10

Introduction to Mathematical Modeling

In this chapter we take up the nature of mathematical modeling in preparation for a discussion of mathematical models based on derivatives of functions to be determined—called differential equation models—which are considered in the next chapter.

10.1. MATH MODELS FOR A VACATION TRIP

Ann's Model

Members of the Adams family were recently engaged in planning their trips from home in Brooklyn, New York to the vacation town of Kennebunkport, Maine. Ann planned to make the trip in late July with a major stop at Putnam, Connecticut for the annual picnic held there. Ann's problem was to set up a mathematical representation or model of the situation which would enable her to predict the total time required for the journey.

The setting of any such problem presents numerous features and characteristics, many of which are irrelevant or unessential to the focus of the problem. In developing her model Ann had to sort this out and decide on which features were negligible. This required discretion and judgment, the most controversial aspect of the math model development process; one person's essential might be another's irrelevancy.

Ann examined a map and laid out a route. She made assumptions about the traffic flow to be expected along various points, speeds that would be possible, and the number of rest stops to be made and their duration. These considerations led her to a model consisting of two line segments joining points representing Brooklyn, Putnam, and Kennebunkport, the sum of whose lengths is 350 miles, and the problem of determining how long it would take an object moving at an average speed of 50 miles an hour to travel this distance. She envisioned an approximate 2 hour stay at the picnic in Putnam before continuing to Kennebunkport.

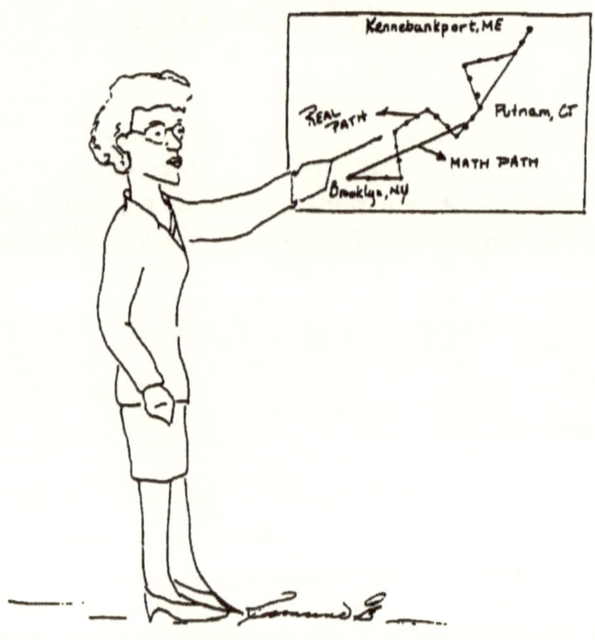

Ann's mathematical model is an idealized, abstract rendering of the real situation involving a trip from Brooklyn to Kennebunkport with a stop at Putnam. It is intended to capture the main features involved in taking such a trip and reflects these features as she sees them and the assumptions that she was led to make. It is possible that someone else planning such a trip would see things in another light and make very different assumptions.

By employing the mathematical operation division we obtain the valid conclusion that an object moving along the idealized path of Ann's model at an average speed of 50 miles per hour would take 350/50 = 7 hours to make the journey. If we add 2 hours for the picnic, we have a total of 9 hours for the travel time. This conclusion is a valid one with respect to Ann's model, valid in the sense that it is an inescapable consequence of the assumptions she was led to make. If we accept Ann's model as a starting point, then we must accept this conclusion as following from it in a deductive logical sense; it is inherent in the model. Mathematical methods, whether simple or the ultimate in technical sophistication, give valid conclusions from the mathematical models we set up. Valid conclusions obtained from a mathematical model are called theorems and the assumptions underlying the model are called postulates or axioms.

The interpretation of Ann's theorem is that if the trip were made under conditions realistically described by her postulates, it would take around 9 hours to complete the journey. What is the acid test of the accuracy of Ann's theorem—accuracy in the sense of how closely it describes the journey in the real world? Take the trip and see how long it takes. This is just what Ann did. It took her 9 hours and 10 minutes, which is close to the projected 9 hours for the journey. This established that her theorem is realistic in this case and, by reflection, was evidence in support of the realism of her model for the trip.

Andy's Model

Ann's son Andrius, called Andy by his friends, was planning his own trip to Kennebunkport to be taken in August. Andy did not intend to stop at Putnam and laid out a different model. Andy's assumptions about traffic flow, speeds that would be possible, and the duration of the one rest stop that he envisioned led him to a model consisting of a line 330 miles long joining points representing Brooklyn and Kennebunkport and the problem of determining how long it would take an object moving at an average speed of 55 miles per hour to cover this distance.

By again employing the mathematical operation division we obtain from Andy's model the valid conclusion that it would take 330/55 = 6 hours. This conclusion is valid with respect to Andy's model in the sense that it follows as an inescapable consequence of his model; it is inherent in Andy's model.

The interpretation of Andy's theorem is that if the trip were made under conditions realistically described by the postulates of his model, it would take around six hours to complete the journey. Andy took the trip in the middle of August and it took him 5 hours and 55 minutes, which is close to the projected time stated by his theorem. This established that his theorem is realistic in this case and, by reflection, was evidence in support of the realism of his model for the trip.

Rasa's Trip: Who's Right?

Andy's sister Rasa was planning to take a brief vacation trip to Kennebunkport during Labor Day weekend. Andy's model looked reasonable to her and she decided to follow the route it prescribed, expecting the journey to take around 6 hours.

Rasa took the trip on Labor Day weekend, but it took her 7 hours and 20 minutes. This actual trip time differs considerably from the projected 6 hour trip time of Andy's theorem so that something clearly went wrong; but what? "Your theorem stinks," Rasa shouted at her brother in a somewhat agitated manner; "it's wrong; it's not valid," she continued.

Rasa's experience proved Andy's theorem wrong in terms of reality, but not validity. The distinction is a fundamental one. In the confrontation between what actually happened and what the theorem says should happen, what actually happened—reality—wins. Andy's theorem is a false statement as a description of the travel time to Kennebunkport on a Labor Day weekend, but it remains a theorem. It is an inescapable consequence of Andy's model—more specifically, his assumptions or postulates—and this is what makes a theorem a theorem. Mathematical methods, in this case the division of 330 by 55 yielding 6, did what it is capable of doing; it yielded a valid conclusion with respect to Andy's model. That valid conclusion may or may not be realistic. Validity is not the same as reality, but they are linked in this way: If the postulates of a model are realistic, so will be the valid conclusions obtained from it. If a valid conclusion is found to be false, this sends us a signal that some of the postulates of the model are unrealistic and require modification.

Andy's model, Rasa's experience showed, is not realistic for travel to Kennebunkport on a Labor Day weekend. In reexamining Andy's postulates we find that they do not realistically take into account unusually heavy traffic delays around the tollgates of the Whitestone Bridge characteristic of holiday weekends. Further examination of Rasa's actual trip shows that this is precisely where she had the difficulty.

We should keep in mind too that Andy did not design his model with a holiday weekend trip in mind. Major lesson: Look before you leap! That is, look at a mathematical model's assumptions before you use it.

Ann's and Andy's experiences showed that their theorems are realistic for their trips. By reflection, this is evidence in support of the realism of their models for the envisioned trips. This evidence does not, however, establish that their models are completely in accord with reality, as Rasa distressingly discovered when she employed her brother's model.

A mathematical model gives us a picture of a situation or phenomenon, but we cannot view the model as complete or the only possible picture. Other models are possible as well.

EXERCISES

1. Rasa's cousin Asta was planning to take a vacation trip from home in Woodhaven, Queens, to Kennebunkport, Maine, at the end of July. Based on data she collected, Audre formulated the following postulates:

Asta's Postulates

P1. A line segment of length 300 miles joining points representing Woodhaven and Kennebunkport is to be taken as an idealized representation for the actual path of the trip.

P2. And object moving at an average speed of 60 miles per hour from the point representing Woodhaven to the point representing Kennebunkport along the line segment is to be taken as an idealized representation of the car trip itself.

P. Factors not considered in the formulation of P1 and P2 have a negligible impact on the vacation trip.

(a) What theorem can be deduced from Asta's postulates concerning the time it would take to make the trip?

(b) Audre took the trip at the end of July as planned, but the journey took her six hours. Does the discrepancy between the projected trip time obtained from her postulates and the time it took her to make the trip establish that her conclusion is not valid? Explain.

(c) If your answer to (b) in no, what does this discrepancy establish? Explian.

10.2. THE MATHEMATICAL MODEL BUILDING PROCESS

The development of a mathematical model for a real world problem or phenomenon basically consists of the following steps.

1. Specify the real world problem or phenomenon to be investigated.

Let us reconsider, for example, Andy's desire to travel from home in Brooklyn, New York, to the vacation town of Kennebunkport, Maine. His problem is to determine how long the trip would take.

2. Collect relevant data about the problem or phenomenon being studied and formulate an idealized representation for it, called a **mathematical model** (or more properly, the **hypothesis of a mathematical model**), from which valid conclusions, called **predictions**, can be obtained by mathematical methods.

The setting of a problem or phenomenon contains numerous features and characteristics, many of which are irrelevant or unessential to the focus of the study. In developing a model assumptions must be made as to which features of the setting are essential to the study and which are negligible. This calls for insight and discretion and is the most controversial aspect of the model development process; one person's essential might be another's irrelevancy.

Moreover, a delicate balance between realism and mathematical manageability must be struck. If we adopt a play-it-safe approach which favors incorporating into the model as many features as possible in the belief that we would thereby be less likely to leave out something important, then not only do we run the risk of clogging up the model with gibberish, but we might make it mathematically intractable; that is, we might not be able to apply existing mathematical methods to obtain useful theorems from the hypothesis of the model. Let us note that a **mathematical model** consists of its assumptions (or postulates) together with its theorems, although we will sometimes use the term model synonymously with its assumptions. If we dismiss too many features of the problem or phenomenon, or the wrong ones, as unessential to our study, then our model diverges too greatly from the phenomenon it is to represent and thus becomes unsuitable in this regard.

Concerning Andy's Brooklyn to Kennebunkport trip, for example, he began the data collection phase by examining a map and laying out a route. Estimates were obtained on the nature of the traffic flow along various parts of the route and the speeds that are possible. He collected data about the distances between various points on the route and introduced assumptions about the number of rest stops to be made and their duration. All such considerations led him to a model consisting of a line 330 miles long joining points representing Brooklyn and Kennebunkport and the problem of determining how long it takes an object going an average speed of 55 miles an hour to travel this distance. This mathematical model is an abstraction of the real problem of determining how long a trip from Brooklyn to Kennebunkport would take. How suitable this model is depends on how realistic are the assumptions on which the model is based.

> 3. Apply mathematical methods to obtain valid conclusions (theorems) with respect to the hypothesis of the model.

For Andy's model for the Brooklyn to Kennebunkport trip the mathematical operation division yields the valid conclusion that a point going an average speed of 55 miles an hour along a straight line path 330 miles long will travel this distance in 330/55 = 6 hours.

> 4. Interpret the theorems obtained from the hypothesis of the model in terms of the real world phenomenon the model represents.

For Andy's Brooklyn to Kennebunkport trip we would expect the journey to take somewhere in the neighborhood of 6 hours, provided that the conditions under which our actual trip takes place are realistically represented in our model.

> 5. Test the accuracy of the model; that is, compare the interpreted conclusions of the model with the results obtained from reality within some desired degree of accuracy.

If the interpreted conclusions can be shown to closely approximate reality within some degree of accuracy, that is, to be true within some desired degree of accuracy (by means of observation which may be coupled with experimentation), then this increases our confidence in the mathematical model as a realistic description of the phenomenon under study.

If in making the trip from Brooklyn to Kennebunkport Andy found that the time taken is close to 6 hours, then this would serve as an acid test which confirms the model as a realistic representation of his undertaking.

However, a model developed for a phenomenon may not be the only representation for the phenomenon. Other models are possible as well; we cannot entertain uniqueness in model representation. Andy, for example, may well have put together another model for his envisioned trip based on another route and assumptions.

Suppose alternative models are available for a phenomenon; how do we choose between them? The decisive verdict is rendered by reality. The crucial question is, are the model's theorems in accord with reality? When some of the model's theorems, suitably interpreted, are found to be in disagreement with the findings of experimentation and observation, this tells us that some part of the model's hypothesis is incomplete or unrealistic and that the model must be refined or abandoned as a description of the phenomenon in question. A mathematical model lives as a description of a phenomenon as long as it avoids head on collision with some contrary fact of observation or experiment.

A theorem found to be false (or true for that matter) does not lose its status as a theorem. It remains a valid consequence of the hypothesis of the model, correct in the sense of validity, incorrect in the sense of reality. It is unfortunate that many writers on science and mathematics use the terms valid and true synonymously. (Sometimes this is done out of sloppiness, sometimes out of ignorance.)

Suppose, as Rasa found, much to her displeasure, that the Brooklyn to Kennebunkport trip, made on Labor Day weekend, took 7 hours and 20 minutes instead of the 6 hours stated in the model's theorem. The theorem remains a theorem of Andy's model, but it now acquires the status of false theorem for travel on a holiday weekend. His model is unsuitable for holiday weekend travel and its hypothesis would have to be modified for such a purpose.

The following characteristics are also important for mathematical models. A model should bring in only observable entities connected with the phenomenon it is to describe. Although a model has its origins in data and observations, it must go beyond them in leading to significant new predictions and results; the theorems of the model should provide us with significant new insights into the phenomenon in question.

6. Refine the model as evidence obtained from experimentation and observation make necessary.

The model Rasa employed for her Labor Day weekend trip from Brooklyn to Kennebunkport proved unsatisfactory because its predicted travel time of 6 hours differed considerably from the actual 7 hours and 20 minutes needed for the journey. In reexamining the hypothesis of Andy's model she found that it did not take into account unusually heavy traffic delays around the tollgates of the Whitestone bridge characteristic of holiday weekends. This would lead her to suitably refine the hypothesis of Andy's model to make it applicable to holiday weekend trips.

Figure 10.1 summarizes the basic steps in the development of a mathematical model in diagramatic form.

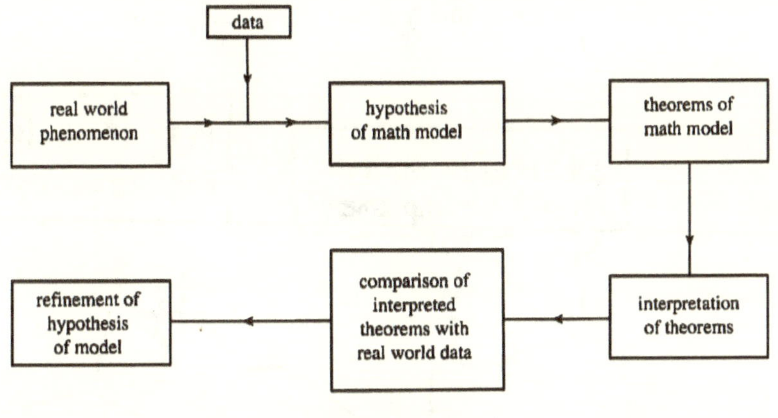

Figure10.1

EXERCISES

1. The Gaja Company makes shoes. The Company plans to introduce two new styles, designated by A-18 and A-21, for the fall season and management wants to know how to set its monthly production schedule so as to maximize profit. Two

consulting firms were hired to study this problem and to make recommendations. Each consulting firm formulated a mathematical model, designated by M1 and M2 for convenience. A conclusion obtained from M1 states that to maximize profit 50,000 A-18 pairs and 35,000 A-21 pairs should be produced monthly with an anticipated monthly profit of $230,000. A conclusion obtained from M2 states that to maximize profit 40,000 A-18 and 50,000 A-21 pairs should be produced monthly with an anticipated profit of $200,000.

(a) If mathematics is the precise subject it is reputed to be, should there not be one solution to the problem of developing a production schedule to maximize profit instead of two? Explain.
(b) Since two solutions emerge, does it follow that not both are valid? Explain.
(c) Would you implement model M1 because it projects a larger monthly profit than M2? Explain.

(d) It came to pass that M1 was implemented. Subsequently, the Gaja Company sustained a consistent monthly loss on the A-18 and A-21 styles and discontinued them. How can one explain these developments?

2. While on your way to your Aunt Alice's birthday celebration, her hundredth, it has been suggested, the world seemed to be in conspiracy against you. It was snowing, traffic was bumper to bumper, and then someone in an oldsmobile sideswiped your new buick and took off. You finally arrive at your aunt's celebration, but, needless to say, not in very good humor. Before you know it you find yourself in a social group being "educated" by Uncle George. George thinks he knows everything about everything and on this occasion his subject is mathematics, specifically geometry. "Since their truth was established by the precise mathematical reasoning for which the ancient Greek mathematicians are famous, the truth of the theorems of Euclidean geometry is beyond question," bellows George in his most authoritative sounding tone. Usually your attitude toward George is one of toleration, but this time you're already for bear. What reply would you give to his "profound observation"?

Because of precise mathematical reasoning the truth of Euclidean geometry is beyond question!
Sounds good after a drink!
Sounds better after two drinks!
Sounds like nonsense before a drink!

CHAPTER 11

Differential Equation Models

11.1. DIFFERENTIAL EQUATIONS

A **differential equation** is an equation that involves derivatives of an unknown function. For example,

$$\frac{dy}{dx} = 2x + 1$$

is a differential equation in which y is an unknown function of x. Given such an equation, the problem that arises is to find a function $y = f(x)$ that satisfies the equation. Such a function is called a solution of the equation. Solving the equation

$$\frac{dy}{dx} = 2x + 1$$

involves simple integration. We obtain

$$y = \int (2x+1)dx = x^2 + x + C$$

Since C can be any given value, there are many functions that satisfy our differential equation. A **particular solution** is obtained when a specific value is given to, or determined for, the constant C. Sometimes a differential equation arises and we also know the value of the unknown function for a particular value of the independent variable. Such a condition permits determination of the constant of integration C. For example, if we were given the differential equation

$$\frac{dy}{dx} = 2x + 1$$

and we were also told that $y = 4$ for $x = 1$, then by integrating we would obtain

$$y = x^2 + x + C$$

as before. Substituting 4 for y and 1 for x in this function yields

$$4 = (1)^2 + 1 + C$$
$$4 = 2 + C$$
$$2 = C$$

Thus the particular solution

$$y = x^2 + x + 2$$

satisfies the given differential equation with the condition, called a **boundary condition**, that $y = 4$ for $x = 1$.

EXAMPLE 1

Find the curves whose tangent line at any point (x,y) has slope $2x$. Determine the curve which contains the point $(1, -1)$.

Suppose $y = f(x)$ has as its graph one of the curves we seek; then $\frac{dy}{dx} = 2x$. By integrating we obtain the following result.

$$y = \int 2x\,dx = x^2 + C$$

$y = x^2 + C$ describes a family of parabolas. A specific member of this family is determined by giving C a specific value. For $C = 0$ we obtain

$$y = x^2.$$

For $C = 1$ we obtain

$$y = x^2 + 1.$$

For $C = -2$ we obtain

$$y = x^2 - 2.$$

All of the members of this family of parabolas have slope $2x$ at any point (x,y) (see Figure 11.1).

To determine the member of this family of parabolas which contains the point $(1,-1)$, we must determine C so that $y = x^2 + C$ is satisfied by $(1,-1)$. To determine C, substitute 1 for x and -1 for y in $y = x^2 + C$ and solve for C.

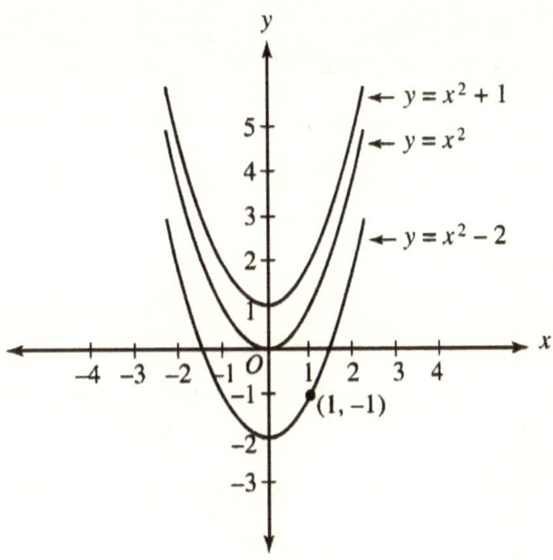

Figure 11.1

$$-1 = (1)^2 + C$$
$$C = -2$$

Thus $y = x^2 - 2$ has slope $2x$ at (x,y) and contains the point $(1,-1)$.

EXAMPLE 2

The marginal cost function of a sugar refinery is $c'(x) = \frac{2}{10}x + 2$, where x is output in tons per week. If the sugar refinery incurs a fixed cost of $200 per week, determine the total cost function $c(x)$.

To determine $c(x)$ we integrate $c'(x) = \frac{2}{10}x + 2$.

$$c(x) = \int (\frac{2}{10}x + 2)dx = \frac{1}{10}x^2 + 2x + C$$

Since the constant factor in the total function represents fixed costs, we have $C = 200$. Thus the total cost function is

$$c(x) = \frac{1}{10}x^2 + 2x + 200.$$

EXERCISES

In Exercises 1 through 5, find the solution of the given differential equation that satisfies the prescribed boundary condition.

1. $f'(x) = 5x + 4, \quad f(x) = \frac{3}{2}$ 2. $f'(x) = 4x^2 + 2x - 1, \quad f(2) = \frac{2}{3}$

3. $\frac{dy}{dx} = -4x + 2 + \frac{1}{x}, \quad y = 5 \textit{ for } x = 1$

4. $f''(x) = \frac{1}{x^2} + 1, \quad f'(1) = 3, f(3) = 0$

5. $f'(x) = -3f(x), \quad f(0) = 10$

6. The tangent line to the graph of $y = f(x)$ has slope $3x^2 - 1$ at the point (x, y), and the curve passes through the point (1, 5). Find $y = f(x)$.

7. The marginal-cost function of a tea producer is $f(x) = \frac{1}{10}x + 5$, where x is output in tons per week. The tea producer incurs a fixed cost of $1000 per week. Find the total cost function.

8. Find the equation of the curve whose tangent line at any point (x, y) has slope $x/(x^2 + 1)$, where the curve passes through the point (0, 5).

9. The instantaneous velocity at time t of a particle moving in a straight line path is $v(t) = 4t + 5$. If the particle moves 33 feet in 3 seconds, find the time-distance function that describes the motion of the particle.

10. The marginal revenue function of a firm is $f(x) = 500 - 4x$ for an output of x tons per day. Determine the total revenue function.

11.2. EXPONENTIAL GROWTH AND DECAY MODELS

In many situations of interest a mathematical model is formulated based on the assumption that the rate of change $\frac{dy}{dt}$ of the amount y of a quantity with respect to time t is proportional to the amount of y that is present at time t. In differential-equation terms, this says that

$$\frac{dy}{dt} = ky$$

or, equivalently,

$$f'(t) = kf(t)$$

where k is the constant of proportionality, and $y = f(t)$ is the function to be determined. We shall solve this equation and then discuss some applications. Dividing both sides of $f'(t) = kf(t)$ by $f(t)$ yields

$$\frac{f'(t)}{f(t)} = k$$

Thus

$$\int \frac{f'(t)}{f(t)} dt = \int k\, dt \qquad (11.1)$$
$$\ln f(t) = kt + C$$

Let N_0 denote the amount of y that is present when $t = 0$ (the beginning of the process). Substituting N_0 for $f(t)$ and 0 for t in (11.1) yields

$$\ln N_0 = C$$

Replacing C by $\ln N_0$ in (11.1) gives us

$$\ln f(t) = kt + \ln N_0$$

Subtracting $\ln N_0$ from both sides yields

$$\ln f(t) - \ln N_0 = kt$$

Since the logarithm of a quotient is the difference of the logarithms of the components, we can write the above as follows:

$$\ln \frac{f(t)}{N_0} = kt$$

To say that $\log_a M = w$ means, by definition, that $a^w = M$. Using this here with $M = f(t)/N_0$, $w = kt$, and $a = e$ gives us

$$e^{kt}\frac{f(t)}{N_0}$$

or

$$f(t) = N_0 e^{kt} \qquad (11.2)$$

where k is the constant of proportionality (determined from data intrinsic to the nature of the process) and $N_0 = f(0)$, the value of the function at the beginning of the process ($t = 0$).

Two interesting applications of this model type are to radiocarbon dating and population growth.

11.3. RADIOCARBON DATING

Earth is continually bombarded by sub-atomic particles, termed cosmic radiation, which are emitted by the Sun. These highly energetic particles react with atoms in the atmosphere to produce neutrons, which then collide with nitrogen atoms in the atmosphere to produce a radioactive form of carbon called carbon-14 or radiocarbon. Carbon-14 decays spontaneously, giving off an electron, and changing to nitrogen. This decay process is such that it takes about 5,730 years for half the amount of carbon-14 in a substance to disintegrate. This value is called the half-lie of carbon-14. Carbon-14 behaves chemically in the same way as ordinary carbon-12.

Living plants and animals absorb carbon-14 along with carbon-12. When they die this absorption ceases, but the radioactive disintegration of carbon-14 continues, slowly at a fixed known rate. The simple idea that emerges, at least in theory, is based on measuring the proportion of carbon-14 left in a sample whose age is to be determined. Since the initial proportion of carbon-14 in the substance is known, to a good approximation, when it was living, we should be able to calculate how long the radioactive decay process had been going on, which gives us the age of the material.

Shortly after World War II Willard F. Libby proposed a way of employing the radioactive decay of carbon-14 to estimate the age of substances containing carbon-14, particularly organic remains. This includes charcoal, wood, cloth, limestone, bones, hair and soil. Libby's carbon-14 dating technique, for which he was awarded the 1960 Nobel Prize in chemistry, is based on the following **assumptions**:

1. Carbon-14 is produced in the atmosphere at an approximately constant rate.
2. Except for recent geologic time (the last half-century), there is a constant concentration of carbon-14 in all living things, which is about 1 carbon-14 atom to 1 trillion carbon-12 atoms.
3. The rate at which carbon-14 decreases is proportional to the amount present.

These **assumptions** lead to an exponential decay model

$$\frac{dy}{dt} = ky,$$

where $y = f(t)$ expresses the weight (in suitable units) of carbon-14 in a given portion of matter and t is time measured in years.

If we let $N_0 = f(0)$ denote the amount of carbon-14 present at time $t = 0$ (the beginning of the process), then from (11.2) we obtain the function

$$y = N_0 e^{kt}. \tag{11.3}$$

If the value of k is known, then $y = N_0 e^{kt}$ e enables us to compute the value of y for each given t, or the value of t for each $y > 0$. The value of k can be obtained by using the fact that the half-life of carbon-14 is 5730 years; that is, the length of time it takes half of the carbon-14 nuclei in a sample to decay is 5730 years. In other words, when $t = 5730$, $y = N_0/2$. Substitution into (11.3) gives us:

$$\frac{N_0}{2} = N_0 e^{5730k}$$

$$\frac{1}{2} = e^{5730k}$$

In logarithm form this yields:

$$5730k = \ln (1/2) \simeq -0.69315$$

$$k \simeq -0.0001209$$

Replacing k by -0.0001209 in (11.3) yields

$$y = N_0 e^{-0.0001209t} \tag{11.4}$$

as the function describing the amount of carbon-14 in a substance in terms of time. The graph of (11.4) is shown in Figure 11.2.

This is what the model says, but what does reality say? Libby put his method to the test in 1955 when he obtained radiocarbon dates for a number of samples, mostly from Egypt, whose ages had been determined by other means. The radiocarbon dates were close to the established dates, and a powerful new dating tool was made available to archeologists. This tool made possible decisive advances. Colin Renfrew notes that one of the greatest contributions of the first radiocarbon revolution was in making possible the study of world prehistory. Developments throughout the world may now be studied on a comparative basis with a sound framework of dates. [1]

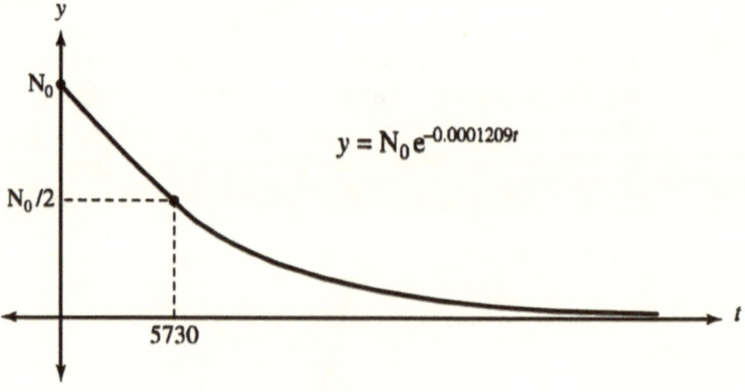

Figure 11.2

EXAMPLE 1

The skin, bone, and clothing of an adult mummy found in a cave near Lake Winnemucca, Nevada, was found to contain 74 percent of the original carbon-14. What is the age of the mummy?

Substituting $0.74\ N_0$ for y in (11.4) yields:

$$0.74N_0 = N_0e^{-0.0001209t}$$
$$0.74 = e^{-0.0001209t}$$

By definition of natural logarithm, this means:

$$\ln 0.74 = -0.0001209t$$

$$t = \frac{\ln 0.74}{-0.0001209}$$

$$\simeq -\frac{0.3010}{-0.0001209}$$

$$= 2490$$

Thus the mummy is approximately 2500 years old.

The basic radiocarbon dating model has undergone a number of refinements. Libby's second assumption, for one, is open to question. The amount of carbon-14 in the atmosphere has not been constant with time. It has varied by as much as ±5% because of changes in solar activity and Earth's magnetic field. In recent years contamination from the burning of fossil fuels and testing of nuclear weapons has resulted in significant changes in the amount of carbon-14 in the atmosphere. Studies of the bristlecone pine, a tree that grows in the White Mountains of California and lives for up to 5000 years, has allowed scientists to develop calibration curves for carbon-14 dates to correct for changes in the level of carbon-14 over time. Accurate tree ring records of age are available for a period as far back as 9000 years and scientists have sought other indicators of age against which carbon-14 dates can be compared. One such indicator is a uranium-thorium dating technique based on the decay of uranium-234 to thorium-230. Recently conducted studies at the Lamont-Doherty Geological Laboratory of Columbia University indicate that age estimates using carbon dating and uranium-thorium dating were in basic agreement for the period from 9000 years ago to the present. For earlier times the carbon-14 dates were substantially younger than those obtained by uranium-thorium analysis. The largest deviation, 3500 years, was obtained for samples that were approximately 20,000 years old. These results make clear some of the limitations of carbon dating and may lead to revisions in the age estimates that have been assigned by carbon dating to a number of "older" samples. [2]

A Tale of Fine Art Forgery

On May 29, 1945 the Dutch painter H.A. Van Meegeren was arrested on a charge of collaboration with the Nazis in having sold the priceless Vermeer painting "Woman Taken in Adultery" to Herman Goering. Van Meegeren claimed that he was not guilty of aiding the enemy in acquiring priceless Dutch art because he himself had painted "Woman Taken in Adultery," the famous "Disciples at Emmaus," as well as other paintings attributed to Vermeer and the less famous de Hooghs. The art world was understandably skeptical, and to prove his point Van Meegeren began, while in prison, to forge the Vermeer painting "Jesus Amongst the Doctors". When the work was nearly completed the charge of collaboration was changed to the less serious charge of forgery. Van Meegeren then refused to finish and age the painting in the

hope of thwarting the investigation. An international panel of experts that was appointed to investigate the matter concluded that the alleged Vermeers were forgeries. Van Meegeren was vindicated and was sentenced to a year in prison for forgery. While in prison he died of a heart attack on December 30, 1947. In spite of the evidence gathered by the panel of experts, many people refused to believe that "Disciples at Emmaus" was a Vermeer forgery. It did not seem possible that such a masterpiece could have been forged by a painter considered third rate. More conclusive proof was demanded, and in 1967 scientists at Carnegie Mellon University took up the problem.

Their analysis was based on the radioactive decay of lead-210 and radium-226, small amounts of which are found in the widely used pigment lead oxide. Lead-210 has a comparatively short half-life of 22 years and if the paint used in painting is very old in comparison to this short half-life, the amount of radioactivity from the lead-210 in the paint will be approximately equal to the amount of radioactivity due to the radium; if the painting is recent, the amount of radioactivity from the lead-210 will be much greater than that generated by the radium. The Carnegie Mellon group made this precise by employing a mathematical model similar to, but more complex than, the one employed for carbon-14 decay. They showed that the amount of radioactivity from lead-210 is much greater than that from radium-226.

In the end van Meegeren was in a sense vindicated; he was not a Nazi collaborator and he did paint a great work that had been attributed to Vermeer. [3]

Reality Strikes Back

Before too deep a feeling of self-satisfaction sets in, we should look more closely at the Libby exponential decay model because it seems to be in contradiction to the basic nature of the radioactive decay process. The weight y of the carbon-14 sample suddenly decreases when a particle is radiated and otherwise remains constant. This feature leads to a graph of $y = N(t)$ of the sort shown in Figure 11.3 which

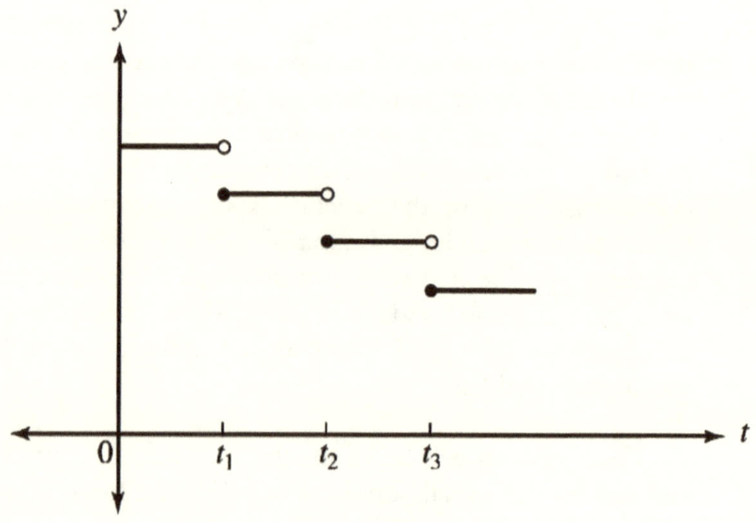

Figure 11.3

is that of a step function discontinuous at those instances t_1, t_2, . . . at which a particle is radiated and otherwise constant.

In a sense this fact proves the first description unrealistic, but it would be appropriate to view the new model as a suitable replacement for the exponential decay model when we are concerned with far greater detail than the exponential decay model could provide us with. The function (11.4), suitably refined, is accurate for macroscopic predictions, that is, for predictions involving time intervals which go back as far as 10,000 years; its accuracy falls off for time periods exceeding 10,000 years and gives poor results in connection with microscopic predictions involving a "small number" of atoms of carbon-14.

REFERENCES

1. C. Renfrew, *Before Civilization: The Radiocarbon Revolution and Prehistoric Europe* (New York: Knopf, 1973).
2. M. Browne, "Errors are Feared in Carbon Dating," *The New York Times*, May 31, 1990.
3. P. Coremans, *Van Meegeren's Faked Vermeers and DeHooghs* (Amsterdam: Meulenhoff, 1949).

EXERCISES

1. Artifacts obtained in an unearthed settlement are found to contain 60 percent of the original amount of carbon-14. Find the date of the artifacts. Note: $\ln 0.6 \simeq -0.5108$.
2. Radium is an unstable element which undergoes radioactive decay. Let $y = N(t)$ denote the number of grams of radium in a given portion of matter in terms of time t, in years. Set up a differential equation model for $y = N(t)$ based on the assumption that the rate at which radium decreases is proportional to the amount present. Taking 1,600 years as the half-life of radium, determine $y = N(t)$.
3. Bones found in a tomb are found to contain 80% of the original amount of carbon-14. Based on carbon-14 dating, approximately how old are the bones? Note: $\ln 0.8 \simeq -0.22314$.

11.4. POPULATION GROWTH

Let $N = N(t)$ denote a function which describes the size of a population in terms of time. The problem we set ourselves here is to determine the nature of $N = N(t)$ under various assumed conditions.

If we assume that there are no constraints on food and space, then it seems reasonable to assume that at any time the rate of increase of the population is proportional to the size of the population at that time. This assumption is expressed in mathematical model form by the differential equation

$$\frac{dN}{dt} = bN$$

where b is a positive constant.

To solve this differential equation we follow in the footsteps used to solve its brother which arose in our analysis of radioactive decay. This yields

$$N = N_0 e^{b(t-t_0)},$$

where $N_0 = N(t_0)$ is the size of the population at some time t_0. This function is called the *Malthusian growth function* after Thomas Malthus (1766-1834), who put forth an exponential growth function for human population growth in his *An Essay on the Principle of Population as it Affects the Future Improvement of Society* (1798).

A Malthusian model, with suitable value of b, has given surprisingly accurate predictions for the Earth's population in recent times (see Exercise 3). For suitable values of b it has also given accurate estimates for the population growth of some mammalian species under certain conditions and time periods (see Exercise 4). It is not realistic when applied to the Earth's population in the distant future (see Exercise 3*c*) or to bacterial growth under laboratory conditions over extended periods of time.

Consider the single bacterium in its culture flask. If we enrich the culture medium with glucose and other elements required for bacterial growth, the population grows exponentially at first, levels off, briefly remains stable, and then starts to decline rapidly and soon dies out. For the bacteria glucose meant food and energy. As the population expanded the glucose was used up and as supplies began to dwindle so did the basis for growth. When any essential resource becomes in short supply, it becomes a limiting factor for population growth. Even if nutrients continued to be supplied, the population would decline and die out unless we removed the increasing amount of metabolic wastes produced by the expanding population which would poison the environment.

The Verhulst-Pearl-Reed Model

A limited growth model which takes into account environmental resistance to further growth was first introduced by the Dutch mathematical biologist P.F. Verhulst (1804-1849) in 1837 and independently by the Americans R. Pearl and L.J. Reed in 1920. Let K denote the maximal population that can exist in the given microcosm with the given level of food, resources, etc. Following Gause in his classic work on the biomathematics of populations and competition between them, *The Struggle for Existence* [1; pp. 34-35], we have:

The difference between the maximally possible and the already accumulated population $(K-N)$, taken in a relative form, i.e., divided by the maximal population, $(K-N)/K$, shows the relative number of 'still vacant places' for definite species in a given microcosm at a definite moment of time. According to the number of the still vacant places only a definite part of the potential rate of increase can be realized. At the beginning of the population growth when the relative number of unoccupied places is considerable the potential increase is realized to a great extent, but when the already accumulated population approaches the maximally possible or saturating one, only an insignificant part of the biotic potential will be realized (Fig. 11.4). Multiplying

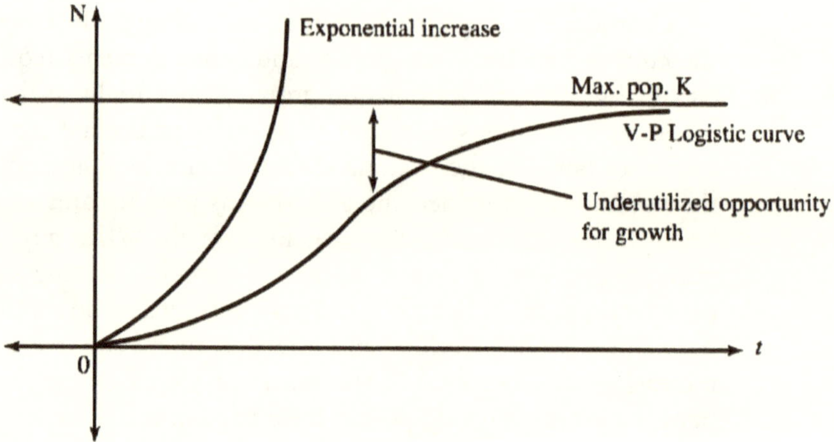

Figure 11.4

the biotic potential of the population bN by the relative number of still vacant places or its "degree of realization" $(K-N)/K$, we shall have the increase of population per infinitesimal unit of time:

$$\begin{bmatrix}\text{Rate of}\\ \text{growth}\\ \text{or increase}\\ \text{per unit}\\ \text{time}\end{bmatrix} = \begin{bmatrix}\text{Potential}\\ \text{increase of}\\ \text{population}\\ \text{per unit}\\ \text{of time}\end{bmatrix} \text{x} \begin{bmatrix}\text{Degree of realization}\\ \text{of the potential}\\ \text{increase. Depends}\\ \text{on the number of}\\ \text{still vacant places}\end{bmatrix}$$

In mathematical terms we have the **Verhulst-Pearl-Reed differential equation model** stated by

$$\frac{dN}{dt} = bN\frac{(K-N)}{K},$$

whose solution gives us the Verhulst-Pearl-Reed logistic curves:

$$N(t) = \frac{K}{1+de^{-bt}}$$

K is the maximum population level or saturation level, b is a proportionality constant—"the potential number of offspring which the organism can produce per unit time," as Gause puts it—and d is the integration constant whose value is determined by a specified initial condition.

Gause [1, pp. 36-39] placed five *Paramecium caudatum* in a small test tube containing 0.5 cm^3 of a nutritive medium and observed and counted them for six days. Gause found the saturating population to be $K = 375$. In fitting a logistic curve to his observations he obtained

$$N(t) = \frac{375}{1 + e^{5.69\ -2.309t}}$$

or equivalently,

$$N(t) = \frac{375}{1 + 74e^{-2.309t}}$$

The "biotic potential" of one *Paramecium* b was found to be 2.309, which "means that per unit of time (one day) under our conditions of cultivation every *Paramecium* can potentially give 2.309 new *Paramecia* [1, p. 37]. Gause's logistic curve was in excellent agreement with his empirical observations.

In their study of population growth in the United States Pearl and Reed [2] obtained a logistic growth function based on census figures for 1790, 1850 and 1910 which can be stated by:

$$N(t) = \frac{197{,}273{,}522}{1 + e^{-0.03134(t-1914.32)}}$$

Table 11.1 compares the predictions obtained from the Pearl-Reed logistic growth function with census figures. There is excellent agreement between observed and predicted values through 1950, but the discrepancy widens significantly after that (see Exercise 5).

Table 11.1

Year	Observed Population	Predicted Population	Percent Error
1790	3,929,000	3,929,000	0.0
1800	5,308,000	5,336,000	0.5
1810	7,240,000	7,227,000	–0.2
1820	9,638,000	9,756,000	1.2
1830	12,866,000	13,108,000	1.9
1840	17,069,000	17,505,000	2.6
1850	23,192,000	23,191,000	–0.0
1860	31,443,000	30,410,000	–3.3
1870	38,558,000	39,370,000	2.1
1880	50,156,000	50,175,000	0.0
1890	62,948,000	62,767,000	–0.3
1900	75,995,000	76,867,000	1.1
1910	91,972,000	91,970,000	–0.0
1920	105,711,000	107,393,000	1.6
1930	122,755,000	122,396,000	–0.3
1940	131,699,000	136,317,000	3.5
1950	150,697,000	148,677,000	–1.3
1960	179,323,000	159,230,000	–11.2
1970	203,212,000	167,943,000	–17.4
1980	226,505,000	174,941,000	–22.8

REFERENCES

1. G.F. Gause. *The Struggle for Existence*, The Williams and Wilkens Co., 1934. Reprinted by Dover, 1971.
2. R. Pearl and L.J. Reed. "On the Rate of Growth of the Population of the United States since 1790 and its Mathematical Representation," *Proceedings of the National Academy of Sciences*, 1920.

EXERCISES

1. The growth function of a population is an integer valued function, but the Malthusian growth model is defined by a

differential equation. Doesn't the former condition make a Malthusian model impossible? Explain.

2. Establish that if $dN/dt = bN$, where $N_0 = N(t_0)$, then $N = N_0 e^{b(t-t_0)}$.
3. Let $N = N(t)$ denote the human population of Earth at time t (in years). Assume that Earth's population has been increasing at the rate of 2% per year and that the rate of change of this population is proportional to the number of people on Earth at any time. It has been estimated that Earth's human population in 1961 was 3,060 million.

   (a) Express the human population as a function of time.
   (b) Earth's human population for 1950 was estimated as 2,510 million. Determine the predicted population value obtained from the growth function obtained in answer to part (a) and compare this with the observed value. How close is the argument? What is the percent error?
   (c) Use the growth function obtained in part (a) to estimate the human population 2510. Does the value seem reasonable? Explain.

4. *Microtus Arvallis Pall* is a small rodent which reproduces very quickly. Take the unit of time to be a month, assume the population is increasing at a rate of 40% per month, and that there are two rodents initially at time t = 0.

   (a) Express the rodent population as a function of time.
   (b) Use the growth function to estimate the size of the rodent population at the end of 2, 6 and 10 months.
   (c) The number of rodents observed at the end of 2, 6 and 10 months were 5, 20 and 109, respectively. How do the predicted values compare with the observed values.

5. Concerning the Pearl-Reed logistic function for the population of the United States, what factors might explain the significant discrepancy between observed and predicted values past 1950?

11.5. MONEY GROWTH

Let $y = a(t)$ denote the amount of money available at time t and consider the condition that the rate of growth of y with respect to t is proportional to the amount of money available at time t (in years), where the proportionality constant is the annual interest rate r.

In differential equation terms this point of view is expressed by

$$\frac{dy}{dt} = ry$$

which yields the asset-growth function

$$a(t) = Ae^{rt}.$$

A is the principal on deposit when the growth process begins, which is time $t = 0$. $a(t)$ expresses the principal on hand t years after A dollars has been deposited and the growth process initiated.

It is interesting to note that it is the same asset-growth function obtained as a consequence of continuous compounding of interest developed in Section 4.2 (p. 124).

CHAPTER 12

Integral Calculus: Definite Integrals

12.1. AN AREA DEFINITION PROBLEM

In this section we take up a problem in the world of geometry that leads to the concept of definite integral, the central idea of integral calculus. Since sums of quantities often make an appearance in this connection and it is useful to have a notation for sums, we begin our discussion with a few comments on summation notation.

The Greek letter Σ (sigma) is used in mathematics to denote sums. For example, consider the sum

$$1^2 + 2^2 + 3^2 + 4^2 + 5^2$$

First, we look for a pattern in the way in which the terms in the sum change. In this situation the exponent is the same for all terms, but the base increases by 1 as we go from term to term. We next introduce a variable, k, say, to denote the quantity that is changing from term to term. Letting k denote the base value, we see that a typical term in the sum can be denoted by k^2. In terms of Σ notation, the sum is indicated by

$$\sum_{k=1}^{k=5} k^2$$

which is understood to mean that k is given integer values between 1 and 5 inclusive, thus yielding 1^2, 2^2, 3^2, 4^2, and 5^2, and that these terms are added.

which is understood to mean that k is given integer values between 1 and 5 inclusive, thus yielding 1^2, 2^2, 3^2, 4^2, and 5^2, and that these terms are added.

The sum

$$f(m_1)x_1 + f(m_2)x_2 + f(m_3)x_3 + f(m_4)x_4$$

is expressed by

$$\sum_{k=1}^{k=4} f(m_k)x_k$$

and the notation

$$\sum_{k=1}^{k=3} f(m_k)(x_k - x_{k-1})$$

expresses the sum

$$f(m_1)(x_1 - x_0) + f(m_2)(x_2 - x_1) + f(m_3)(x_3 - x_2)$$

An Area Problem

In geometry, the area $A(R)$ of a rectangle R with height h and base with length b is defined as the product $h \cdot b$. The area of R serves as a numerical measure of the amount of space occupied by R. Two basic properties of area are the following:

1. If r is a rectangle that is contained in R, then $A(r) \leq A(R)$.
2. If R is partitioned into two subrectangles R_1 and R_2 (see Figure 12.1), then $A(R) = A(R_1) + A(R_2)$.

In basic geometry, the concept of area is extended from rectangles to polygons in such a way that these fundamental properties are preserved. Polygons are bounded by line segments, but how is an area concept to be defined for a region not all of whose sides are line segments in such a way that the fundamental properties of area as defined in basic geometry are maintained? This is the question we now take up and, to be specific, let us consider the region S bounded below by the x-axis, on the sides by the lines $x = 3$ and $x = 12$, and above by the graph of the function $f(x) = x^2$ (see Figure 12.2). The problem is to define the area $A(S)$ of S in a way that is consistent with our experience with area from basic geometry.

Let m and M denote values in the base interval [3, 12] at which $f(x) = x^2$ takes on its absolute minimum and absolute maximum values, respectively. From the nature of the region S, it is clear that $m = 3$, $M = 12$, and that the minimum and maximum values are 9 and 144, respectively. Let r denote the rectangle with base [3, 12] and height $f(m) = f(3) = 9$. Since r is contained in region S (see Figure 12.3), we shall refer to r as an **inner rectangle**. Let R denote the rectangle with base [3, 12] and height $f(M) = f(12) = 144$. Since R contains region S (see Figure 12.4), we shall refer to R as an **outer rectangle**. If we are to be consistent with the concept of area developed in basic geometry for rectangles and polygons, then, however $A(S)$ is defined, $A(S)$ should neither be less than $A(r)$ nor greater than $A(R)$. $A(r) = 9(9) = 81$ and $A(R) = 144(9) = 1296$, we have

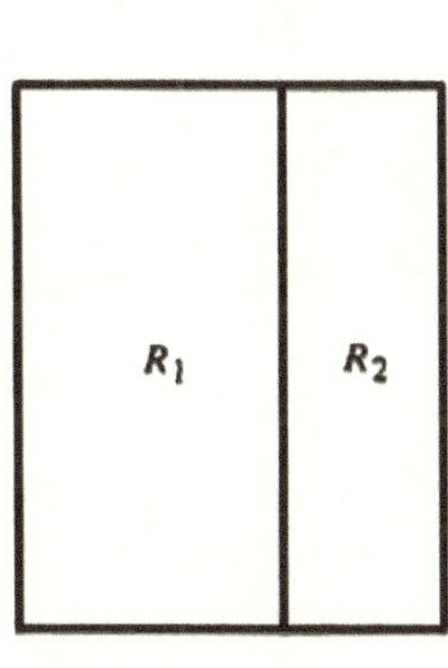

Figure 12.1

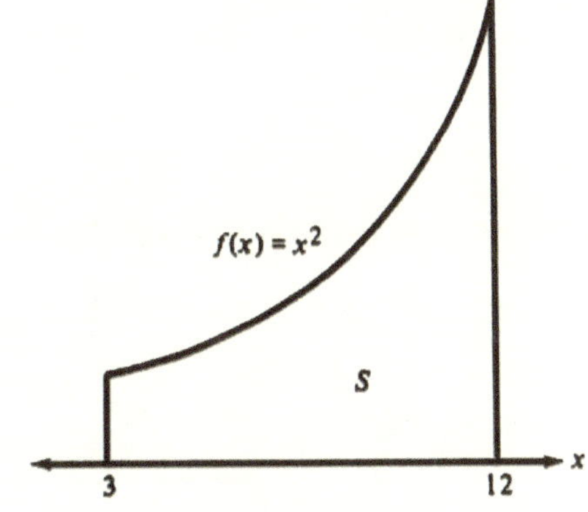

Figure 12.2

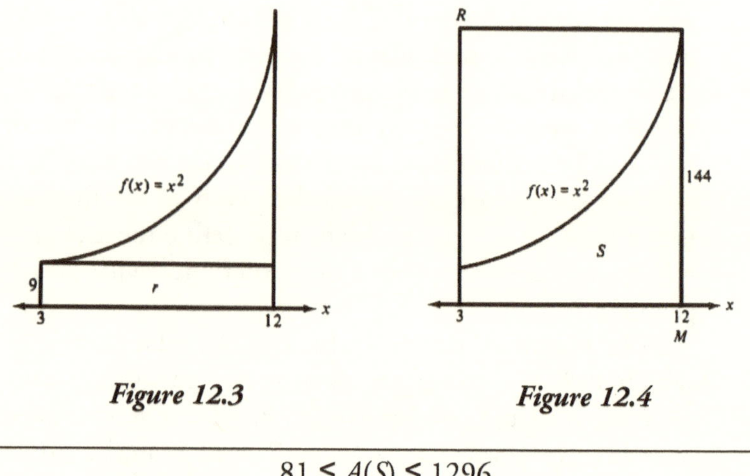

Figure 12.3 *Figure 12.4*

$$81 \leq A(S) \leq 1296$$

There's a lot of room between 81 and 1296, and to narrow the gap we partition S into two subregions, S_1 and S_2, and use the same approach on each one. To partition S into two subregions, we choose a value, 8, say, in the interval [3, 12]. By choosing 8, we partition the interval [3, 12] into two subintervals, [3, 8] and [8, 12] (see Figure 12.5). Let m_1 and M_1 denote values in [3, 8] at which $f(x) = x^2$ takes on its absolute minimum and absolute maximum values, respectively ($m_1 = 3$, $M_1 = 8$); let m_2 and

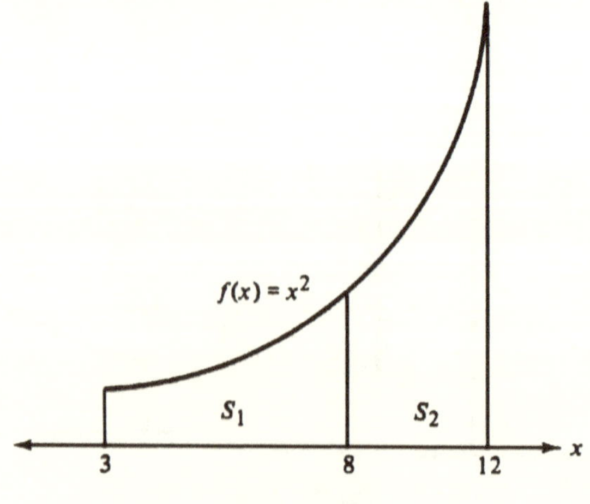

Figure 12.5

M_2 denote values in [8, 12] at which $f(x) = x^2$ takes on its absolute minimum and absolute maximum values, respectively ($m_2 = 8$, $M_2 = 12$). The minimum values themselves are $f(m_1) = 9$ and $f(m_2) = 64$, and lead to inner rectangles r_1 and r_2, which are contained in S_1 and S_2, respectively (see Figure 12.6). The maximum values are $f(M_1) = 64$ and $f(M_2) = 144$, and lead to outer rectangles R_1 and R_2, which contain S_1 and S_2, respectively (see Figure 12.7). Since the areas of these subregions, $A(S_1)$ and $A(S_2)$, should neither be less than the areas of their respective inner rectangles, $A(r_1)$ and $A(r_2)$, nor greater than the areas of their respective outer rectangles, $A(R_1)$ and $A(R_2)$, we emerge with the conditions

$$A(r_1) \leq A(S_1) \leq A(R_1)$$

$$A(r_1) \leq A(S_2) \leq A(R_2)$$

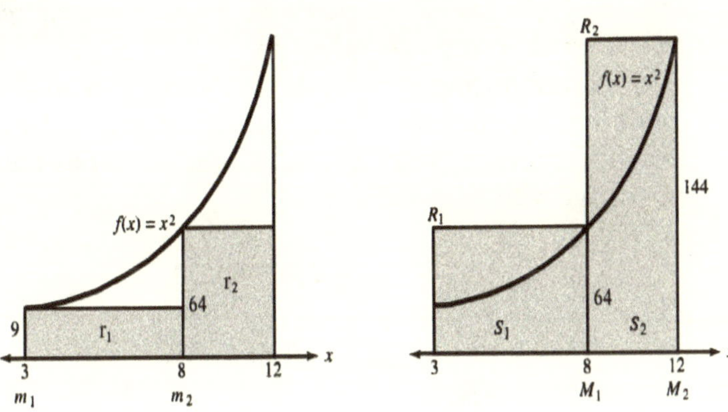

Figure 12.6 *Figure 12.7*

For the areas of these inner and outer rectangles, we have

$$f(3)(8-3) \leq A(S_1) \leq f(8)(8-3)$$
$$f(8)(12-8) \leq A(S_2) \leq f(12)(12-8)$$

which yields

$$45 \leq A(S_1) \leq 320$$

$$256 \leq A(S_2) \leq 576$$

By adding, we obtain

$$301 \leq A(S_1) + A(S_2) \leq 896$$

Since S_1 and S_2 were obtained by partitioning S into two nonoverlapping pieces, it is reasonable to impose on $A(S)$ the condition $A(S) = A(S_1) + A(S_2)$. Thus we have

$$\boxed{301 \leq A(S) \leq 896}$$

This raises our lower bound for $A(S)$ from 81 to 301, and lowers our upper bound from 1296 to 896.

The gap is still wide, but we have developed a method for narrowing the gap. If we partition S into three subregions S_1, S_2, and S_3 by choosing two values in [3, 12], 5 and 8, for example, then by introducing inner and outer rectangles and imposing the condition that the area of each subregion neither be less than the area of its inner rectangle nor exceed the area of its outer rectangle, we emerge with the conditions

$$18 \leq A(S_1) \leq 50$$

$$75 \leq A(S_2) \leq 192$$

$$256 \leq A(S_3) \leq 576$$

Adding and imposing the condition $A(S) = A(S_1) + A(S_2) + A(S_3)$ yields

$$\boxed{349 \leq A(S) \leq 818}$$

thereby raising the lower bound on $A(S)$ from 301 to 349, and lowering the upper bound from 896 to 818.

The general idea illustrated by these specific partitions of S into subregions is that lower and upper bounds for $A(S)$ can be obtained by partitioning S into subregions, introducing inner and outer rectangles for each subregion as indicated, taking the sum of the areas of the inner rectangles to obtain a lower bound for $A(S)$, and taking the sum of the areas of the outer rectangles to obtain an upper bound for $A(S)$. To express this idea in a precise way, we must introduce appropriate notation. More often than not, it is the notation and not the basic idea that disturbs some people. Keep in mind the simplicity of the idea and the fact that the notation is the servant of the idea, not vice versa.

To partition S into a number of subregions, n, denoted by $S_1, S_2, \ldots, S_n$, we introduce a sequence of partition points denoted by $3 = x_0, x_1, x_2, \ldots, x_{n-1}, x_n = 12$ (see Figure 12.8). Subregion S_1 has base $[x_0, x_1]$, S_2 has base $[x_1, x_2]$; more generally, S_i has base $[x_{i-1}, x_i]$, and finally S_n base $[x_{n-1}, x_n]$. The inner rectangles $r_1, r_2, \ldots, r_n$ which are contained in $S_1, S_2, \ldots, S_n$, respectively, are shown in Figure 12.9, and the outer rectangles $R_1, R_2, \ldots, R_n$, which contain $S_1, S_2, \ldots, S_n$, respectively, are shown in Figure 12.10. The requirement that the area of each subregion neither be less than

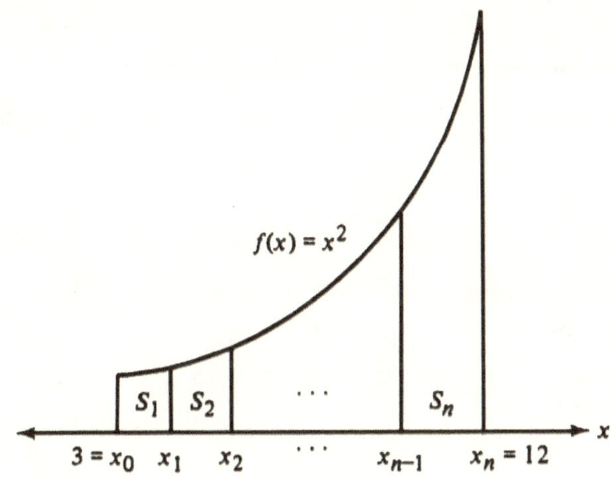

Figure 12.8

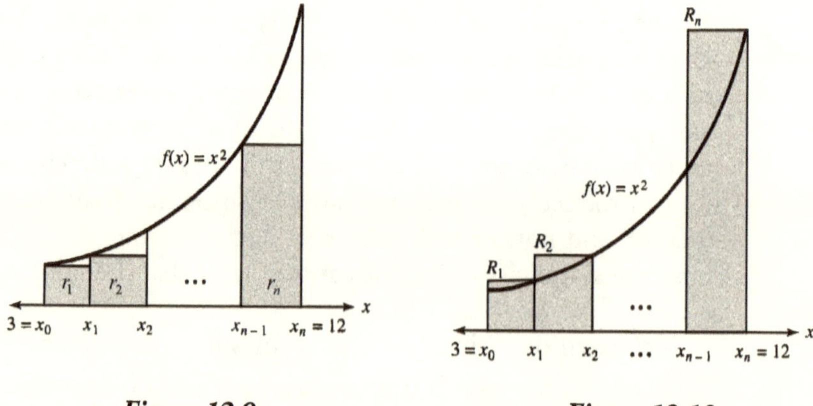

Figure 12.9 *Figure 12.10*

the area of its inner rectangle nor exceed the area of its outer rectangle is expressed by the conditions

$$A(r_1) \le A(S_1) \le A(R_1)$$
$$A(r_2) \le A(S_2) \le A(R_2)$$
$$\vdots$$
$$A(r_n) \le A(S_n) \le A(R_n)$$

Adding and requiring that $A(S)$ be equal to the sum of the areas of the components $S_1, S_2, \ldots, S_n$, leads us to the condition

$$\underbrace{A(r_1)+\ldots+A(r_n)}_{\text{sum of areas of inner rectangles}} \le A(S) \le \underbrace{A(R_1)+\ldots+A(R_n)}_{\text{sum of areas of outer rectangles}}$$

The area of each of these rectangles is equal to the product of its dimensions. To express this precisely, let $m_1, m_2, \ldots, m_n$ denote values in the respective subintervals $[x_0, x_1], [x_1, x_2], \ldots, [x_{n-1}, x_n]$ at which $f(x) = x^2$ takes on its absolute minimum value in these respective subintervals. (In this case, $m_1 = x_0$, $m_1 = x_1, \ldots m_n = x_{n-1}$.) Then the height of r_1 is $f(m_1)$, and the length of its base is $(x_1 - x_0)$, so that

$$A(r_1) = f(m_1)(x_1 - x_0)$$

More generally,

$$A(r_2) = f(m_2)(x_2 - x_1), \ldots, A(r_n) = f(m_n)(x_n - x_{n-1})$$

so that the sum of the areas of the inner rectangles

$$A(r_1) + A(r_2) + \ldots + A(r_n)$$
$$= f(m_1)(x_1 - x_0) + f(m_2)(x_2 - x_1) + \ldots + f(m_n)(x_n - x_{n-1})$$

which is expressed in summation notation by

$$\sum_{k=1}^{k=n} f(m_k)(x_k - x_{k-1})$$

Let M_1, M_2, . . . , M_n denote values in the respective subintervals $[x_0, x_1]$, $[x_1, x_2]$, . . . , $[x_{n-1}, x_n]$ at which $f(x) = x^2$ takes on its absolute maximum value in these respective subintervals. (In this case $M_1 = x_1$, $M_2 = x2$, . . . , $M_n = x_n$.) Then the height of R_1 is $f(M_1)$, and the length of its base is $x_1 - x_2$, so that

$$A(R_1) = f(M_1)(x_1 - x_0)$$

More generally,

$$A(R_2) = f(M_2)(x_2 - x_1), \ldots, f(M_n)(x_n - x_{n-1})$$

so that the sum of the areas of the outer rectangles

$$A(R_1) + A(R_2) + \ldots + A(R_n)$$
$$= f(M_1)(x_1 - x_0) + f(M_2)(x_2 - x_1) + \ldots + f(M_n)(x_n - x_{n-1})$$

which is expressed in summation notation by

$$\sum_{k=1}^{k=n} f(M_k)(x_k - x_{k-1})$$

In summary, then, the condition that $A(S)$ neither be less than the sum of the areas of the inner rectangles nor exceed the sum of the areas of the outer rectangles is expressed in algebraic terms as follows:

$$\underbrace{\sum_{k=1}^{k=n} f(m_k)(x_k - x_{k-1})}_{\text{sum of areas of inner rectangles}} \leq A(S) \leq \underbrace{\sum_{k=1}^{k=n} f(M_k)(x_k - x_{k-1})}_{\text{sum of areas of outer rectangles}} \qquad (12.1)$$

Such lower and upper bounds for $A(S)$ are obtained for every sequence of partition values $3 = x_0, x_1, \ldots, x_n = 12$.

Closing the Gap

By partitioning S into more and smaller subregions, we narrow the gap between the lower and upper sums. The problem is to close it entirely. One way of approaching this problem is to examine the behavior of the lower and upper sums

$$\sum_{k=1}^{k=n} f(m_k)(x_k - x_{k-1}) \text{ and } \sum_{k=1}^{k=n} f(M_k)(x_k - x_{k-1})$$

as we choose sequences of partition values $3 = x_0, x_1, \ldots, x_n = 12$ with more and more values $(n \to \infty)$ in such a way that the distance between consecutive values $(x_1 - x_0,\ x_2 - x_1$, and, more generally, $x_k - x_{k-1}$ approaches zero $(x_k - x_{k-1}) \to 0$ for all k).

The situation we face is analogous to preparing to run a videotape which features two friends—call them Jack and Jill—walking towards what seems to be a common point of interest. Each frame in the videotape shows Jack and Jill, and when we run the tape we see them in motion. Each partition of [3, 12] may be thought of as a frame in a mathematical videotape which shows a lower sum and upper sum. As we run the mathematical tape, which is accomplished by taking partitions with more and more points in such a way that the difference between consecutive values goes to zero, do the mathematical friends—the lower and upper sums—approach the same numerical value? If they do approach a common value, call it I (see Figure 12.11), then it would make sense, I think we would agree, to define $A(S)$—the area of S—to be I.

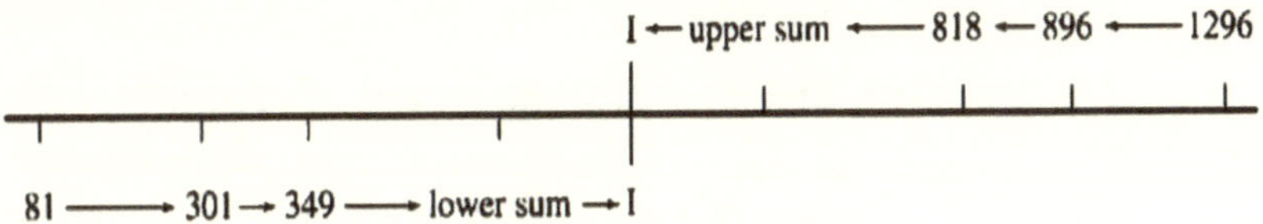

Figure 12.11

Finding the Value of I

While this approach to closing the gap between the lower and upper sums is natural, it raises some important and difficult questions. Under what conditions do the lower and upper sums approach the same number I, and how can the value of I be determined? In answer to the first question, it suffices to note that the continuity of $f(x) = x^2$ over the interval [3, 12] is sufficient to guarantee that $A(S)$ is well defined, that is, that the lower and upper sums converge to the same number I.

To determine I, we refer to the regions V and W shown in Figure 12.12, and

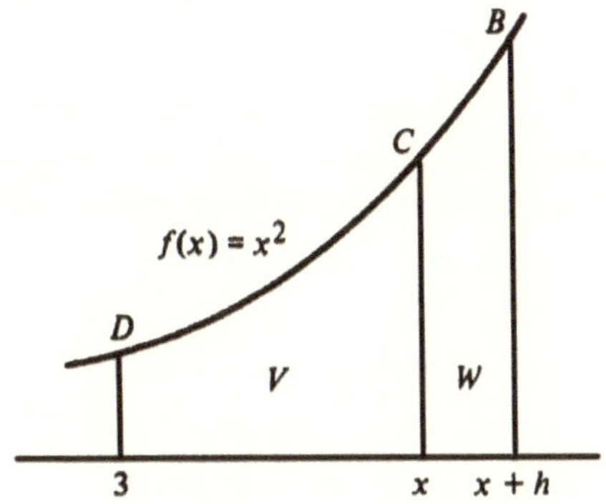

Figure 12.12

define an area function $H(x)$ as follows:

$$H(x) = \text{area of region } V \text{ (with vertices } 3xCD)$$
$$H(3) = 0$$

Our first task is to show that the derivative of $H(x)$ is $f(x)$.

$H(x + h)$ is the area of V plus the area of W. Thus.

$$H(x + h) - H(x) = \text{area of } W$$

With respect to region W, we introduce inner and outer rectangles r and R (see Figure 12.13) with areas $f(m)h$ and $f(M)h$, respectively, where m and M are values in $[x, x + h]$

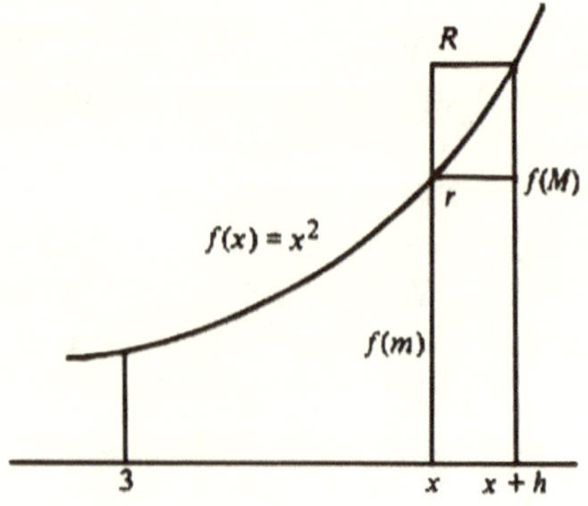

Figure 12.13

at which $f(x) = x^2$ takes on its absolute minimum and maximum values, respectively ($m = x$ and $M = x + h$). Since the area of W, which is $H(x + h) - H(x)$, is between the areas of the inner and outer rectangles of W, we have

$$f(m)h \le H(x+h) - H(x) \le f(M)h$$

Since $m = x$ and $M = x + h$, we obtain

$$f(x)h \le H(x+h) - H(x) \le f(x+h)h$$

Dividing by h, which is positive, yields

$$f(x) \leq \frac{H(x+h)-H(x)}{h} \leq f(x+h)$$

Let h approach 0; $f(x)$ remains constant, and $f(x + h)$ approaches $f(x)$. Thus

$$\frac{H(x+h)-H(x)}{h}$$

is trapped between $f(x)$ and a quantity that approaches $f(x)$. Therefore,

$$\operatorname*{limit}_{h \to 0} \frac{H(x+h)-H(x)}{h} = f(x)$$

In other words, the derivative of $H(x)$ is $f(x)$,

$$H'(x) = f(x)$$

which means that $H(x)$ is an indefinite integral of $f(x)$.

Thus if $F(x)$ is any other integral of $f(x) = x^2$, then $F(x)$ differs from $H(x)$ by some constant C.

$$H(x) = F(x) + C$$

Our problem is to determine the area of the region bounded on the sides by $x = 3$ and $x = 12$, which, in terms of the area function $H(x)$, is expressed by $H(12)$. We have

$$H(12) = F(12) + C$$
$$0 = H(3) = F(3) + C$$

Subtracting yields

$$H(12) = F(12) - F(3)$$

This tells us that to obtain $H(12)$ all we have to do is find an indefinite integral of $f(x)=x^2$ [take $F(x)=\frac{1}{3}x^3$], evaluate $F(x)$ at 12 and at 3, and subtract:

$$F(12)=\frac{1}{3}(12)^3=576, \quad F(3)=\frac{1}{3}(3)^3=9$$
$$H(12)=576-9=567$$

The area of the region S bounded above by the graph of $f(x) = x^2$, on the sides by the lines $x = 3$ and $x = 12$, and below by the x-axis (the value I approached by the lower and upper sums) is 567.

Extension of Area to a More General Region

More generally, let S denote a region bounded below by the x-axis, on the sides by the lines $x = a$ and $x = b$, and above by the graph of a function $y = f(x)$, which is continuous for all x satisfying $a \le x \le b$ (see Figure 12.14). By choosing a sequence of values $a = x_0, x_1, x_2, \ldots, x_{n-1}, x_n = b$, we partition S into subregions $S_1, S_2, \ldots, S_n$ (see Figure 12.15). For each subregion, we introduce inner and outer rectangles (see Figures 12.16 and 12.17).

The requirement that the area of each subregion neither be less than the area of its inner rectangle nor exceed the area of its outer rectangle is expressed by the conditions

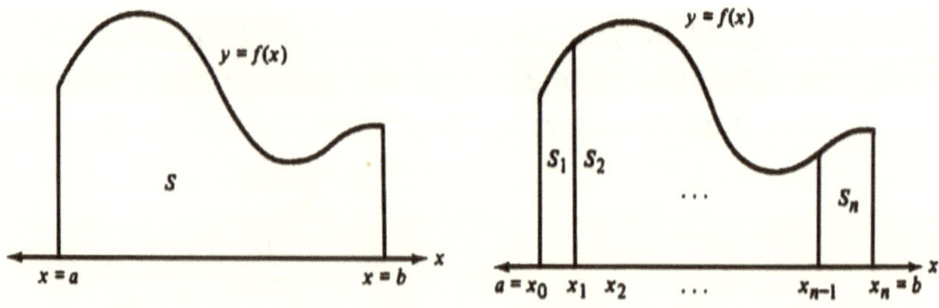

Figure 12.14 *Figure 12.15*

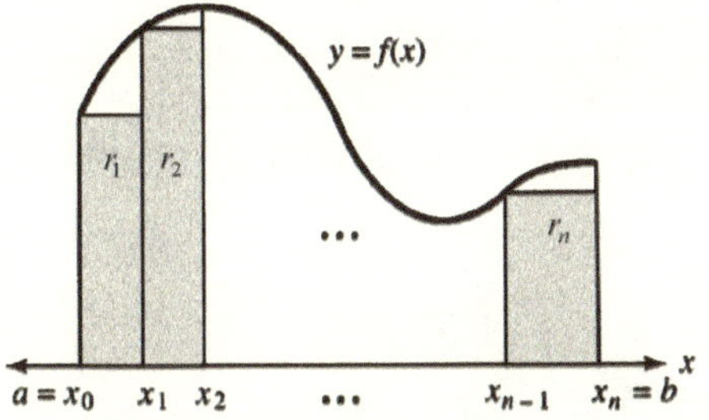

Figure 12.16

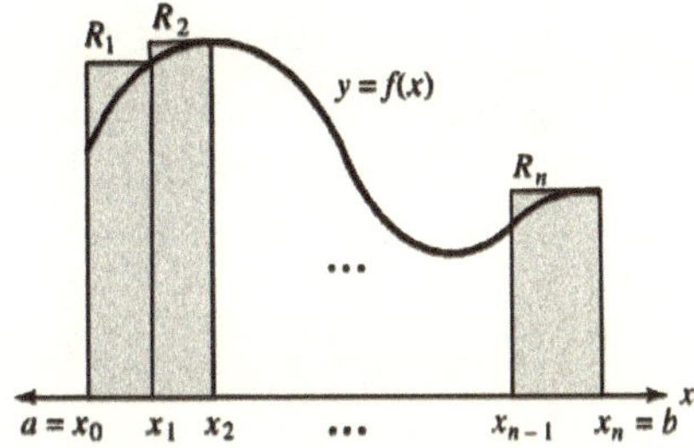

Figure 12.17

$$A(r_1) \le A(S_1) \le A(R_1)$$
$$A(r_2) \le A(S_2) \le A(R_2)$$
$$\vdots \qquad \vdots \qquad \vdots$$
$$A(r_n) \le A(S_n) \le A(R_n)$$

Adding and requiring that $A(S)$ be equal to the sum of the areas of the component subregions S_1, S_2, . . . , S_n, leads us to the condition

$$\underbrace{A(r_1)+\ldots+A(r_n)}_{\text{sum of areas of inner rectangles}} \le A(S) \le \underbrace{A(R_1)+\ldots+A(R_n)}_{\text{sum of areas of outer rectangles}}$$

Since the area of each inner and outer rectangle is the product of its height and the length of its base, we have

$$\begin{array}{ll} A(r_1)=f(m_1)(x_1-x_0) & A(R_1)=f(M_1)(x_1-x_0) \\ A(r_2)=f(m_2)(x_2-x_1) & A(R_2)=f(M_2)(x_2-x_1) \\ \vdots \qquad \vdots & \vdots \qquad \vdots \\ A(r_n)=f(m_n)(x_n-x_{n-1}) & A(R_n)=f(M_n)(x_n-x_{n-1}) \end{array}$$

where m_1, m_2, . . . , m_n are values in the subintervals $[x_0, x_0]$, $[x_1, x_2]$, . . . , $[x_{n-1} - x_n]$ at which $y = f(x)$ takes on its absolute minimum value in these respective subintervals, and M_1, M_2, . . . ,

M_n are values in these subintervals at which $y = f(x)$ takes on its absolute maximum value in these respective subintervals.

Thus, as bounds for $A(S)$ we obtain the now well recognized lower and upper sums:

$$f(m_1)(x_1 - x_0) + ... + f(m_n)(x_n - x_{n-1}) \le A(S)$$
$$\le f(M_1)(x_1 - x_0) + ... + f(M_n)(x_n - x_{n-1})$$

In summation notation, we have

$$\underbrace{\sum_{k=1}^{k=n} f(m_k)(x_k - x_{k-1})}_{\text{sum of areas of inner rectangles}} \le A(S) \le \underbrace{\sum_{k=1}^{k=n} f(M_k)(x_k - x_{k-1})}_{\text{sum of areas of outer rectangles}}$$

If, as we choose sequences of partition values $a = x_0, x_1, \ldots, x_n = b$ with more and more values in such a way that the distance between consecutive values approaches zero ($n \to \infty$ and $x_k - x_{k-1} \to 0$ for all k), the corresponding lower sums approach a single number I,

$$\sum_{k=1}^{k=n} f(m_k)(x_k - x_{k-1}) \to I$$

and the corresponding upper sums also approach I,

$$\sum_{k=1}^{k=n} f(M_k)(x_k - x_{k-1}) \to I$$

then we **define $A(S)$ to be I.**

To **evaluate** $A(S)$, find an indefinite integral $F(x)$ of $f(x)$ and determine $F(a)$ and $F(b)$. Then

$$A(S) = F(b) - F(a)$$

EXAMPLE 1

Find the area of the region S bounded by the x-axis, the lines $x = -1$ and $x = 1$, and the graph of $f(x) = x^3 + 1$ (see Figure 12.18).

$F(x) = \frac{1}{4}x^4 + x$ is an indefinite integral of $f(x) = x^3 + 1$. Thus

$$\begin{aligned} A(S) &= F(1) - F(-1) \\ &= \frac{5}{4} - \left(-\frac{3}{4}\right) \\ &= 2 \end{aligned}$$

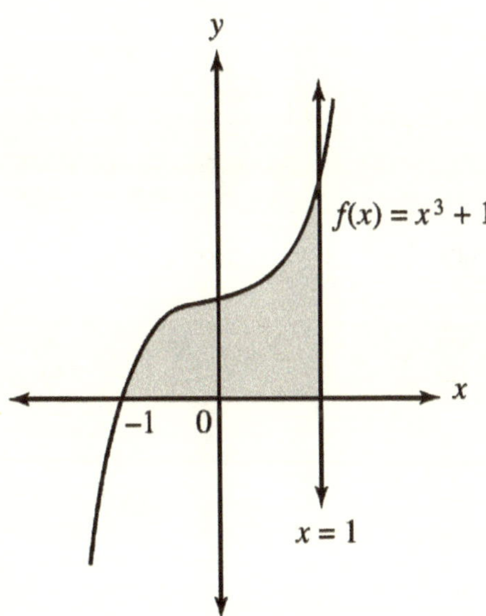

Figure 12.18

EXERCISES

1. Express the following sums in summation notation.

   (a) $3^1 + 3^2 + 3^3 + 3^4 + 3^5$

   (b) $x_1 2^2 + x_2 2^3 + x_3 2^4 + x_4 2^5$

(c) $(x_1)^2 y_0 + (x_2)^2 y_1 + (x_3)^2 y_2 + (x_4)^2 y_3$

(d) $2(x_1 + x_0) + 3(x_2 + x_1) + 4(x_3 + x_2) + 5(x_4 + x_3)$

2. Write the following expressions without summation notation.

(a) $\sum_{k=0}^{k=4} (x+1)^3$ (b) $\sum_{k=1}^{k=4} x_k (x_{k+1} - 3)$ (c) $\sum_{k=1}^{k=4} x_k (x_{k-1})^2$

(d) $\sum_{k=1}^{k=5} 3^k k^2$

3. Given that

$$x_1 = 1, \ x_2 = -2, \ x_3 = -4, \ x_4 = 2$$
$$y_1 = 3, \ y_2 = -1, \ y_3 = -3, \ y_4 = 5$$

evaluate the following expressions:

(a) $\sum_{k=1}^{k=4} 2x_k$ (b) $\sum_{k=1}^{k=4} (x_k + y_k)$ (c) $\sum_{k=1}^{k=4} (x_k)^2 y_k$

(d) $\sum_{k=1}^{k=4} (3x_k - y_k)^2$

4. Consider the region *S* bounded below by the *x*-axis, on the left and right by the lines $x = 1$ and $x = 10$, and above by the graph of $f(x) = 3x^2 - 1$.

   (a) Find lower and upper bounds for *A*(*S*) on the basis of the sequence of values 1, 5, 10 that partitions *S* into two subregions.
   (b) Find lower and upper bounds for *A*(*S*) on the basis of the sequence of values 1, 5, 8, 10 that partitions *S* into three subregions.

(c) Find lower and upper bounds for $A(S)$ on the basis of the sequence of values 1, 3, 5, 8, 10 that partitions S into four subregions.

(d) Find $A(S)$.

In Exercises 5 through 10, find the area of the region S described by the following bounds. Sketch the graph of S.

5. x-axis, $x = 1$, $x = 3$, $f(x) = x^2 + 4$

6. x-axis, $x = -1$, $x = 2$, $f(x) = 2x^2 + 3$

7. x-axis, $x = -2$, $x = 3$, $f(x) = -x^2 + 9$

8. x-axis, $x = -1$, $x = 2$, $f(x) = 2x^3 + 4$

9. x-axis, $x = -2$, $x = 4$, $f(x) = 3x^2 + 2x + 1$

10. x-axis, $x = 1$, $x = 9$, $f(x) = \sqrt{x} + 1$

12.2. THE DEFINITE INTEGRAL AND ITS DETERMINATION

To define the concept of definite integral we follow in the footsteps of the development of the concept of area.

Let $y = f(x)$ denote a function that is defined and continuous at each value of an interval $[a, b]$. Consider a sequence of values $a = x_0, x_1, x_2, \ldots, x_{n-1}, x_n = b$ that partitions the interval $[a, b]$ into subintervals $[x_0, x_1], [x_1, x_2], \ldots, [x_{n-1} - x_n]$. From each such sequence, we obtain a lower sum

$$\sum_{k=1}^{k=n} f(m_k)(x_k - x_{k-1}) = f(m_1)(x_1 - x_0) + \ldots + f(m_n)(x_n - x_{n-1})$$

and an upper sum

$$\sum_{k=1}^{k=n} f(M_k)(x_k - x_{k-1}) = f(M_1)(x_1 - x_0) + \ldots + f(M_n)(x_n - x_{n-1})$$

where $m_1, m_2, \ldots, m_n$ are values in $[x_0, x_1], [x_1, x_2], \ldots, [x_{n-1} - x_n]$ at which $y = f(x)$ takes on its absolute minimum value in these respective subintervals, and $M_1, M_2, \ldots, M_n$ are values in $[x_0, x_1]$, $[x_1, x_2], \ldots, [x_{n-1} - x_n]$ at which $y = f(x)$ takes on its absolute maximum value in these respective subintervals.

If as we choose sequences $a = x_0, x_1, \ldots, x_{n-1}, x_n = b$ with more and more values $(n \to \infty)$ in such a way that the distance between consecutive values approaches zero $(x_k - x_{k-1} \to 0)$, the corresponding lower sums approach a single number I,

$$\sum_{k=1}^{k=n} f(m_k)(x_k - x_{k-1}) \to I$$

and the corresponding upper sums approach I,

$$\sum_{k=1}^{k=n} f(M_k)(x_k - x_{k-1}) \to I$$

then I is defined as the **definite integral of $y = f(x)$ with respect to interval $[a, b]$**, and is denoted by

$$\int_a^b f(x)dx$$

The definition of definite integral is abstract in the sense that in its formulation no interpretation is given to the underlying function $y = f(x)$, interval $[a, b]$, and the lower and upper sums

$$\sum_{k=1}^{k=n} f(m_k)(x_k - x_{k-1}) \text{ and } \sum_{k=1}^{k=n} f(M_k)(x_k - x_{k-1})$$

The benefit of this abstract formulation is in its clarity and flexibility, clarity in the sense that all unessentials have been stripped away, allowing us to see more clearly the structural essentials, and flexibility in the sense that whatever we learn about definite integrals can be applied to any concept with the definite-integral structure, no matter what world of application it arises from.

The area concept developed in the preceding section is the definite integral of a function plus the interpretation given the components due to the nature of the setting from which the area concept emerged. The function $y = f(x)$ is viewed as defining the upper boundary of a region S of interest, the interval $[a, b]$ is the lower boundary of the region, and the lower and upper sums express the sum of the areas of the inner and outer rectangles, respectively.

Alternative Formulation of the Definite Integral

The definition of definite integral can be stated in a slightly different way that is worthy of note, and we now turn our attention to this equivalent formulation. As we have seen, the introduction of a sequence of values $a = x_0, x_1, \ldots, x_{n-1}, x_n = b$ partitions the interval $[a, b]$ into subintervals $[x_0, x_1]$, $[x_1, x_2], \ldots,$ $[x_{n-1}, x_n]$.

An alternative to forming lower and upper sums is to arbitrarily choose values $c_1, c_2, \ldots, c_n$ in $[x_0, x_1]$, $[x_1, x_2], \ldots,$ $[x_{n-1}, x_n]$, respectively, form the more general sum

$$\sum_{k=1}^{k=n} f(c_k)(x_k - x_{k-1}) = f(c_1)(x_1 - x_0) + \ldots + f(c_n)(x_n - x_{n-1})$$

and examine its behavior as we choose sequences $a = x_0, x_1, \ldots,$ $x_{n-1}, x_n = b$ with more and more values $(n \to \infty)$ in such a way that the distance between consecutive values approaches zero $(x_k - x_{k-1} \to 0)$. If, under these conditions, our general sum approaches one value I.

$$\sum_{k=1}^{k=n} f(c_k)(x_k - x_{k-1}) \to I$$

then I is the **definite integral of $y = f(x)$ with respect to interval $[a, b]$**.

To establish that this formulation of the definite integral in terms of general sums is equivalent to the formulation in terms of lower and upper sums, we must show that, if the lower and upper sums approach I, then the general sum approaches I, and also that, if the general sum approaches I, then the lower and upper sums approach I. Although it may sound complicated, it is actually very simple to do.

First, suppose that the lower and upper sums approach I; that is,

$$\sum_{k=1}^{k=n} f(m_k)(x_k - x_{k-1}) \to I$$

and

$$\sum_{k=1}^{k=n} f(M_k)(x_k - x_{k-1}) \to I$$

Our original sum

$$\sum_{k=1}^{k=n} f(c_k)(x_k - x_{k-1})$$

is between the lower and upper sums, since the lower sum, being based on minimum values of the function, $f(m_1), \ldots, f(m_n)$, is the smallest sum of this kind, and the upper sum, being based on maximum values of the function, $f(M_1), f(M_2), \ldots, f(M_n)$, is the largest sum of this kind. Since the lower and upper sums are approaching I, the general sum, which is trapped between them, must also approach I (see Figure 12.19). This is what you might call a mathematics squeeze play.

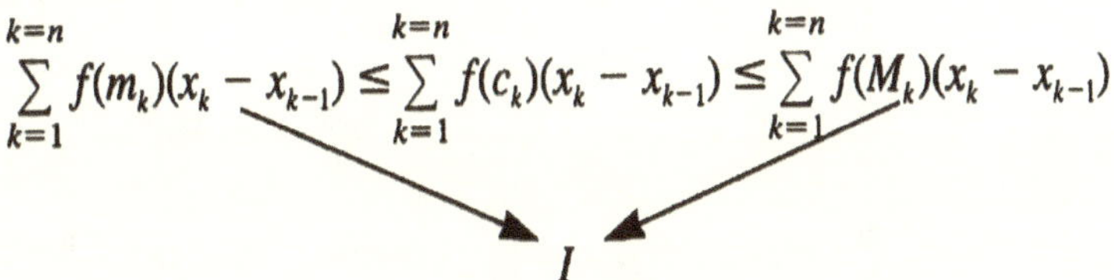

Figure 12.19

Now suppose that the general sum approaches I,

$$\sum_{k=1}^{k=n} f(c_k)(x_k - x_{k-1}) \to I$$

where $c_1, c_2, \ldots, c_n$ are *any* values in $[x_0, x_1]$, $[x_1, x_2]$, . . . , $[x_{n-1}, x_n]$, respectively. If we take $c_1 = m_1$, $c_2 = m_2, \ldots, c_n = m_n$, then we obtain

$$\sum_{k=1}^{k=n} f(m_k)(x_k - x_{k-1}) \to I$$

That is, the lower sums approach I. If we take $c_1 = M_1$, $c_2 = M_2, \ldots, c_n = M_n$, then we obtain

$$\sum_{k=1}^{k=n} f(M_k)(x_k - x_{k-1}) \to I$$

That is, the upper sums approach I.

The definition of $\int_a^b f(x)dx$ presupposes that $a < b$. If $a > b$, then $\int_a^b f(x)dx$ is defined as $-\int_a^b f(x)dx$. Thus, for example, $\int_2^1 f(x)dx = -\int_1^2 f(x)dx$. If $a = b$, then $\int_a^b f(x)dx = \int_a^a f(x)dx$ is defined to be zero.

Properties of the Definite Integral

The following are three basic properties of definite integrals:

1. $\int_a^b k \cdot f(x)dx = k\int_a^b f(x)dx$, where k is any constant.
2. $\int_a^b [f(x)+g(x)]dx = \int_a^b f(x)dx + \int_a^b g(x)dx$. The integral of a sum is the sum of the integrals of the component parts, provided that the integrals of the component parts exist. The result holds for a sum of any number of functions.
3. $\int_a^b f(x)dx + \int_b^c f(x)dx = \int_a^c f(x)dx$, where a, b, and c are any values.

If b is between a and c, then geometrically speaking this theorem says that the area under the region defined by $y = f(x)$ from a to c is the sum of the areas under $y = f(x)$ from a to b and b to c (see Figure 12.20).

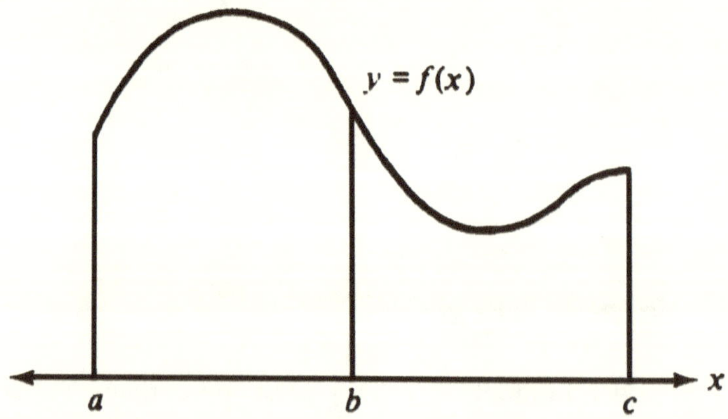

Figure 12.20

Insofar as the existence of the definite integral is concerned, we have the following basic result:

If $y = f(x)$ is continuous at each value of the interval $[a, b]$, then the definite integral $\int_a^b f(x)dx$ exists; that is, the lower and upper sums approach the same value, which is denoted by $\int_a^b f(x)dx$.

Because of the close connection between definite integrals and areas, it is not unexpected that the procedure we developed for determining area values applies to definite integrals in general.

Fundamental Theorem of Integral Calculus

If $y = f(x)$ is **continuous for** $\boldsymbol{a \le x \le b}$, and $F(x)$ is any indefinite integral of $f(x)\,[F'(x) = f(x)$ for $a \le x \le b]$, then

$$\int_a^b f(x)dx = F(b) - F(a)$$

That is, to obtain the value of $\int_a^b f(x)dx$, determine an indefinite integral, $F(x)$, of $f(x)$, evaluate $F(x)$ at a and b, and subtract $F(a)$ from $F(b)$. The difference $F(b) - F(a)$ is often denoted by $F(x)]_a^b$

Very Important Point: Watch the Hypothesis

If $y = f(x)$ is **not** continuous for $a \le x \le b$, it may turn out that $\int_a^b f(x)dx$ does not exist, in which case $\int_a^b f(x)dx = F(b) - F(a)$ would be nonsense.

EXAMPLE 1

Determine $\int_{-1}^3 (3x^2 + 2x - 5)dx$.

Our first task is to obtain an indefinite integral of $f(x) = 3x^2 + 2x - 5$. We have

$$F(x) = \int (3x^2 + 2x - 5)dx = x^3 + x^2 - 5x$$

$$F(3) = (3)^3 - 5(3) = 21$$
$$F(-1) = (-1)^3 + (-1)^2 - 5(-1) = 5$$

Thus

$$\int_{-1}^{3} (3x^2 + 2x - 5)dx = F(3) - F(-1)$$
$$= 21 - 5$$
$$= 16$$

EXAMPLE 2

Determine $\int_1^2 x^2(1+x^3)^2\,dx$.

Our first problem is to obtain an indefinite integral

$$F(x) = \int x^2(1+x^3)^2\,dx.$$

If we introduce the helping function $u = 1 + x^3$, then $D_x u = 3x^2$. Under the integral sign we have x^2. To obtain $3x^2$, we introduce 3 under the integral sign and compensate with $\frac{1}{3}$ outside the integral sign. Thus we obtain

$$\int x^2(1+x^3)^2\,dx = \frac{1}{3}\int \underbrace{(3x^2)}_{D_x u}\underbrace{(1+x^3)^2}_{u}\,dx$$
$$= \frac{1}{3}\int u^2 du$$
$$= \frac{1}{9}u^3$$

Replacing u by $1 + x^3$ yields

$$\int x^2(1+x^3)^2\,dx = \frac{1}{9}(1+x^3)^3$$

Thus

$$\int_1^2 x^2(1+x^3)^2 dx = (\frac{1}{9}(1+x^3)^3\Big]_1^2$$
$$= \frac{729}{9} - \frac{8}{9} = \frac{721}{9}$$

EXERCISES

Determine the following definite integrals.

1. $\int_2^5 (x^2+1)dx$
2. $\int_1^4 (5x+3)dx$
3. $\int_{-1}^3 5x^2 dx$
4. $\int_{-2}^2 (4x^3+2)dx$
5. $\int_1^5 (x^2+2x-1)dx$
6. $\int_{-1}^4 (2x^3-x+2)dx$
7. $\int_1^3 \frac{dx}{x}$
8. $\int_1^3 (4x^3-3x-1)dx$
9. $\int_1^2 4x(3+2x^2)^2 dx$
10. $\int_0^{12} \sqrt{4+x}dx$
11. $\int_0^4 8x(5-4x^2)^3 dx$
12. $\int_0^2 3x^2\sqrt{1+x^3}dx$
13. $\int_0^4 10e^{0.5x}dx$
14. $\int_0^{10} 5e^{0.2x}dx$
15. $\int_0^3 \frac{x}{\sqrt{x^2+16}}dx$
16. $\int_1^3 x^2\sqrt{x^3+3}dx$
17. $\int_1^2 \frac{2x}{x^2}dx$
18. $\int_{-1}^4 \sqrt{5+x}dx$
19. $\int_1^3 \frac{dx}{2x-1}$
20. $\int_0^5 10e^{0.1x}dx$
21. $\int_0^3 300e^{0.06x}dx$
22. $\int_0^5 600e^{0.09x}dx$
23. $\int_0^5 2000e^{0.08(5-t)}dt$
24. $\int_0^2 1000e^{0.06(2-t)}dt$

12.3. IMPROPER INTEGRALS

From our discussion of integration we have that

$$\int_1^t \frac{dx}{x^2} = \int_1^t x^{-2}dx = -1x^{-1}\Big]_1^t = -\frac{1}{x}\Big]_1^t = -\frac{1}{t}+1.$$

Let us further observe that

$$\underset{t\to\infty}{\text{limit}} \int_1^t \frac{dx}{x^2} = \underset{t\to\infty}{\text{limit}}(-\frac{1}{t}+1) = 1.$$

This sequence of operations defines the **improper integral** denoted by

$$\int_1^\infty \frac{dx}{x^2}$$

which is said to have value 1.

Geometrically speaking, the region S bounded above by the graph of $f(x) = 1/x^2$, below by the x-axis, and on the left by the line $x = 1$ is said to have area 1 (see Figure 12.21. S is an unbounded region since it is not bounded on the right, and the concept of improper integral that we have introduced provides us with a means for extending, in a natural way, the definition of area to regions like S.

Let us also note that the "proper" integral

$$\int_1^t \frac{dx}{x^2} = -\frac{1}{t}+1$$

defines the area of the region S_t bounded above by the graph of $f(x) = 1/x^2$, below by the x-axis, on the left by the line $x = 1$, and on the right by the line $x = t$ (see Figure 12.22). Thus, geometrically speaking, the area of S, defined by the improper integral $\int_1^\infty \frac{dx}{x^2}$, is the limit as $t \to \infty$ of the area of S_t defined by the "proper" integral $\int_1^t \frac{dx}{x^2}$.

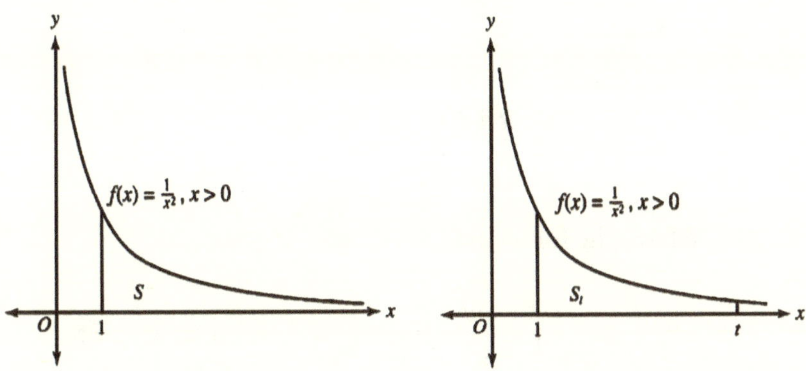

Figure 12.21 ***Figure 12.22***

More generally, the ***improper integrals***

$$\int_{-\infty}^{\infty} f(x)dx, \quad \int_{-\infty}^{a} f(x)dx, \quad \text{and} \quad \int_{-\infty}^{\infty} f(x)dx$$

are defined as follows.

1. $\int_{a}^{\infty} f(x)dx = \underset{t\to\infty}{\text{limit}} \int_{a}^{t} f(x)dx$, if this limit exists.
2. $\int_{-\infty}^{a} f(x)dx = \underset{t\to-\infty}{\text{limit}} \int_{t}^{a} f(x)dx$, if this limit exists.
3. $\int_{-\infty}^{\infty} f(x)dx = \int_{-\infty}^{0} f(x)dx + \int_{0}^{\infty} f(x)dx$, if
   $\int_{-\infty}^{0} f(x)dx = \underset{t\to-\infty}{\text{limit}} \int_{t}^{0} f(x)dx$ and
   $\int_{0}^{\infty} f(x)dx = \underset{t\to\infty}{\text{limit}} \int_{0}^{t} f(x)dx$ provided that both of these limits exist.

In definition 1 it is assumed that $y = f(x)$ is continuous for $x \geq a$; in definition 2 it is assumed that $y = f(x)$ is continues for $x \leq a$; and in definition 3 it is assumed that $y = f(x)$ is continuous for all real numbers.

EXAMPLE 1

Find, if it exists, the value of $\int_{-\infty}^{-1} \frac{dx}{x^2}$.

Since

$$\int_t^{-1}\frac{dx}{x^2}=-1x^{-1}\Big]_t^{-1}=-\frac{1}{x}\Big]_t^{-1}=1+\frac{1}{t}$$

we have the following.

$$\int_{-\infty}^{-1}\frac{dx}{x^2}=\underset{t\to-\infty}{\text{limit}}\int_t^{-1}\frac{dx}{x^2}=\underset{t\to-\infty}{\text{limit}}(1+\frac{1}{t})=1$$

EXAMPLE 2

Find, if it exists, the value of $\int_1^{\infty}\frac{dx}{x}$.

Since

$$\int_1^{t}\frac{dx}{x}=\ln t-\ln 1=\ln t-0=\ln t$$

we have the following.

$$\int_1^{\infty}\frac{dx}{x}=\underset{t\to\infty}{\text{limit}}\int_t^{t}\frac{dx}{x}=\underset{t\to\infty}{\text{limit}}\ln t=\infty$$

Thus the improper integral $\int_1^{\infty}\frac{dx}{x}$ does not exist.

EXERCISES

Find the value of each of the following improper integrals, if it exists.

1. $\int_1^{\infty}\frac{dx}{x^3}$

2. $\int_{-\infty}^{-1}\frac{dx}{x^4}$

3. $\int_4^{\infty} \frac{dx}{\sqrt{x}}$

4. $\int_0^{\infty} e^{-x} dx$

5. $\int_0^{\infty} xe^{-x^2} dx$

6 $\int_{-\infty}^{3} \frac{dx}{(4-x)^2}$

7. $\int_0^{\infty} \frac{x}{1+x^2} dx$

8. $\int_{\infty}^{0} e^x dx$

12.4. ANOTHER LOOK AT THE DEFINITE INTEGRAL

The standard formulation of the definite integral concept requires us to examine the behavior of lower, middle, and upper sums— $\sum_{k=1}^{k=n} f(m_k)(x_k - x_{k-1})$, $\sum_{k=1}^{k=n} f(c_k)(x_k - x_{k-1})$, and $\sum_{k=1}^{k=n} f(M_k)(x_k - x_{k-1})$, respectively— as we take partitions with more and more points $(n \to \infty)$ in such a way that the difference between consecutive partition values goes to zero $(x_k - x_{n-1} \to 0)$. There is another approach to this based on the concepts of least upper bound and greatest lower bound of a set of numbers which we consider here.

Upper and Lower Bounds

Let B denote a set of numbers. A number k is said to be an **upper bound** of set B if every number in B is less than or equal to k.

The **least upper bound** of B (LUB of B) is a number I such that

1. I is an upper bound of B;
2. no number less than I is an upper bound of B, that is, I is the smallest of the upper bounds of B.

If $B = \{\frac{1}{2}, 3, \sqrt{10}, 15, 27\}$, then 27 is an upper bound of B and any number greater than 27 is an upper bound of B. 27 = LUB of B. If B is the set of numbers between 0 and 2, then 2 is an upper bound of B, as is any number greater than 2. LUB of B = 2.

A number c is said to be a **lower bound** of set B if every number in B is greater than or equal to c.

The **greatest lower bound** of B (GLB of B) is a number J such that

1. J is a lower bound of set B;
2. no number greater than J is a lower bound of B, that is, J is the largest of the lower bounds of B.

If $B = \{\frac{1}{2}, 3, 10, 15, 27\}$, then $\frac{1}{2}$ is a lower bound of B, as is any number less than $\frac{1}{2}$. GLB of $B = \frac{1}{2}$. If B is the set of numbers between 0 and 2, then 0 is a lower bound of B, as is any number less than 0. GLB of B = 0.

EXERCISES

1. In a discussion of the set of numbers between 2 and 5 it was said that the least upper bound of this set is 4.99 and the greatest lower bound is 1.99. Do you agree? Explain.
2. Determine the least upper bound of the set of numbers between 1 and 9. What is the basis for your conclusion?
3. Determine the greatest lower bound of the set of numbers between 1 and 9. What is the basis for your conclusion?
4. Determine the least upper bound and greatest lower bound of the set $\left\{1, \frac{1}{2}, \frac{1}{3}, \frac{1}{4}, \ldots, \frac{1}{n}, \ldots\right\}$. What is the basis for your conclusions?

Return to the Problem of Closing the Gap

In considering the problem of defining the area of the region S bounded below by the x-axis, on the sides by the lines $x = 3$ and $x = 12$ and above by the graph of $f(x) = x^2$ (Sec. 12.1, pp. 336-342) the crucial point in the analysis comes when we arrive at condition (12.1) for $A(S)$, the area of S which is to be defined (p. 336).

$$\underbrace{\sum_{k=1}^{k=n} f(m_k)(x_k - x_{k-1})}_{\text{sum of areas of inner rectangles}} \leq A(S) \leq \underbrace{\sum_{k=1}^{k=n} f(M_k)(x_k - x_{k-1})}_{\text{sum of areas of outer rectangles}} \qquad (12.1)$$

Geometrically speaking, the left and right sides of (12.1) are the sums of the areas of the inner and outer rectangles shown in Figures 12.23 and 12.24, respectively.

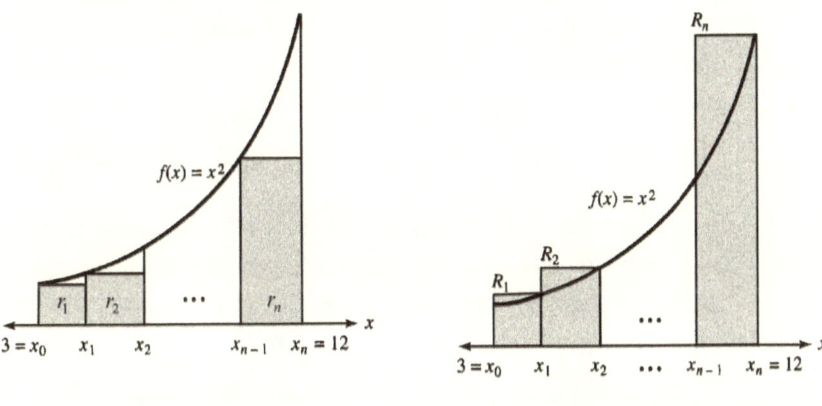

Figure 12.23 ***Figure 12.24***

As we have seen, the lower sums increase and the upper sums decrease as we partition region S into more and smaller subregions. Another approach to obtaining a common meeting point for these sums is based on considering the least upper bound and greatest lower bound of these sums.

Let

$$I = \text{LUB of} \left\{ \sum_{k=11}^{n} f(m_k)(x_k - x_{k-1}) \right\}, \quad J = \text{GLB of} \left\{ \sum_{k=1}^{n} f(M_k)(x_k - x_{k-1}) \right\}$$

The lower sums move up towards I and the upper sums move down towards J (see Figure 12.25). If ***I* and *J* are the same value,** then the gap between the lower and upper sums will, in a sense, have been closed and it makes sense to define $A(S)$ to be their common value.

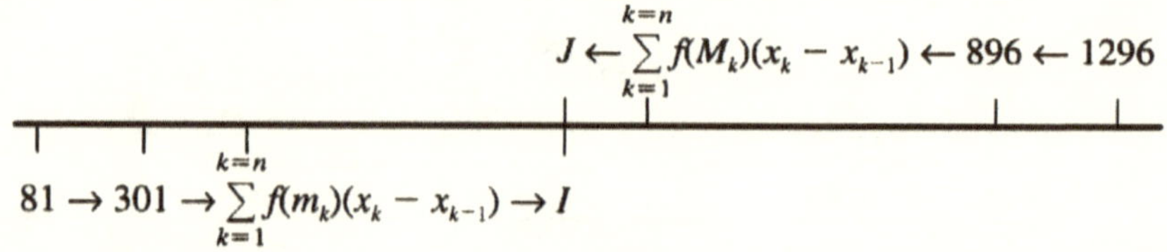

Figure 12.25

Doing this yields results consistent with the limit of a sums approach so that we may conclude that the value 567 obtained for $A(S)$ through the first approach is the common value of I and J.

The LUB-GLB Approach to Defining Definite Integral

More generally, let $y = f(x)$ denote a function that is defined and continuous at each value of an interval $[a, b]$. Consider a sequence of values $a = x_0, x_1, x_2, \ldots, x_{n-1}, x_n = b$ that partitions the interval $[a, b]$ into subintervals $[x_0, x_1], [x_1, x_2], \ldots, x_{n-1}, x_n]$. From each such sequence, we obtain a lower sum

$$\sum_{k=1}^{k=n} f(m_k)(x_k - x_{k-1}) = f(m_1)(x_1 - x_0) + \ldots + f(m_n)(x_n - x_{n-1})$$

and an upper sum

$$\sum_{k=1}^{k=n} f(M_k)(x_k - x_{k-1}) = f(M_1)(x_1 - x_0) + \ldots + f(M_n)(x_n - x_{n-1})$$

where $m_1, m_2, \ldots m_n$ are are values in $[x_0, x_1], [x_1, x_2], \ldots, [x_{n-1}, x_n]$ at which $y = f(x)$ takes on its absolute minimum value in these respective subintervals, and $M_1, M_2, \ldots, M_n$ are values in $[x_0, x_1], [x_1, x_2], \ldots, [x_{n-1}, x_n]$ at which $y = f(x)$ takes on its absolute maximum value in these respective subintervals.

Let I denote the least upper bound of the set of all lower sums:

$$I = \text{LUB of} \left\{ \sum_{k=1}^{n} f(m_k)(x_k - x_{k-1}) \right\}$$

Let J denote the greatest lower bound of the set of all upper sums:

$$J = \text{GLB of} \left\{ \sum_{k=1}^{n} f(M_k)(x_k - x_{k-1}) \right\}$$

If $I = J$, then their common value is defined as the **definite integral of $y = f(x)$ with respect to interval $[a, b]$** and is denoted, as previously noted, by

$$\int_a^b f(x)dx$$

This formulation of the definite integral concept is equivalent to those previously discussed. Its advantage is due to the fact that it allows us to side-step a rather complex limit argument. It lends itself to applications, but the approach generally employed is the previously discussed limit argument.

CHAPTER 13

Concepts with a Definite Integral Structure

Having obtained some basic results on definite integrals, let us return to concepts defined in terms of definite integrals and apply our results to them.

13.1. AREA OF A REGION

The **area of the region** S bounded above by the graph of $y = f(x)$, below by the x-axis, and on the sides by the lines $x = a$ and $x = b$ is defined by

$$A(S) = \int_a^b f(x)dx$$

This assumes that $f(x) \geq 0$ for x between a and b; if $f(x) < 0$ for x between a and b [$y = f(x)$ is a lower boundary of S and the x-axis is an upper boundary], then the area of S is defined by

$$A(S) = -\int_a^b f(x)dx$$

If S is bounded above by the graph of $y = f(x)$, below by the graph of $y = g(x)$, and on the sides by the lines $x = a$ and $x = b$, as shown in Figure 13.1, then the area of S is defined as the difference in the areas under the two curves.

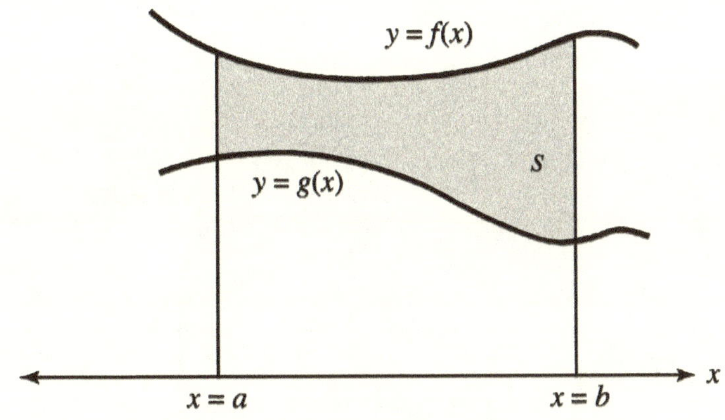

Figure 13.1

$$A(S) = \int_a^b f(x)dx - \int_a^b g(x)dx = \int_a^b [f(x) - g(x)]dx$$

There are other variations on this theme, but what we have at this point is sufficient to illustrate some basic area computations.

EXAMPLE 1

Find the area of the region S bounded by the graph of $f(x) = 16 - x^2$, the x-axis, and the lines $x = -3$ and $x = 2$.

It is a good idea to begin by drawing the graph of S. This yields the shaded region shown in Figure 13.2.

$$A(S) = \int_{-3}^{2} (16 - x^2)dx - (16x - \frac{1}{3}x^3]_{-3}^{2}$$

$$= \frac{88}{3} - (-39) = \frac{205}{3}$$

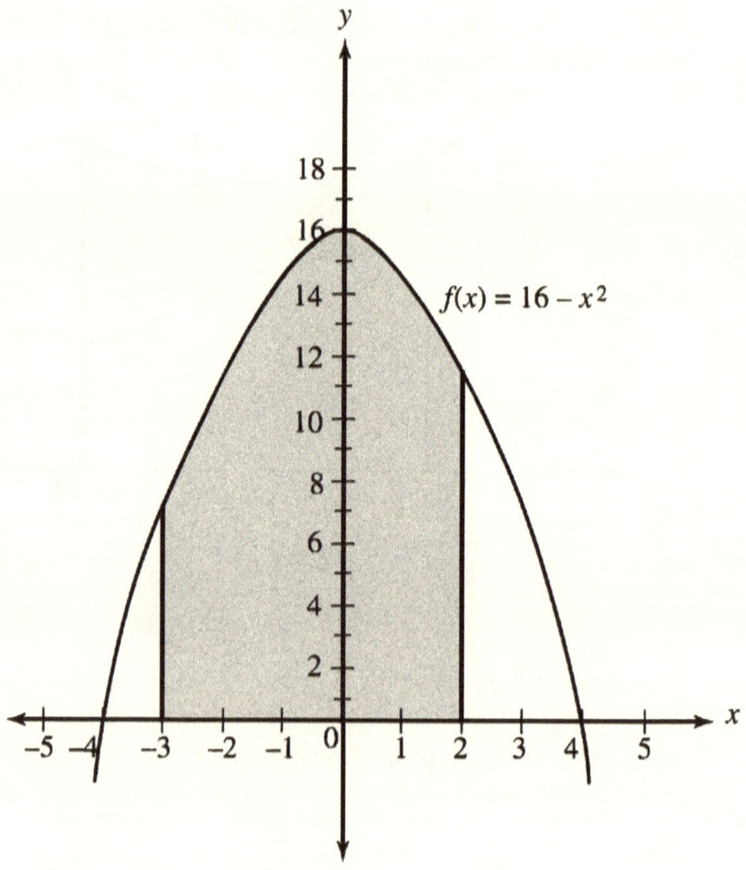

Figure 13.2

EXAMPLE 2

Find the area of the region S bounded by the graph of $y = 25 - x^2$ and $y = -2x + 17$.

Graphing S yields the shaded region shown in Figure 13.3. To determine the points of intersection of these curves, we set $25 - x^2$ equal to $-2x + 17$ and solve for x.

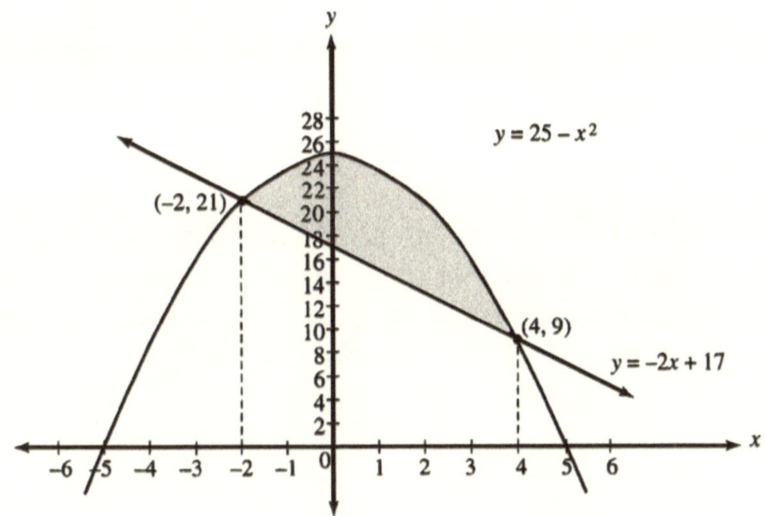

Figure 13.3

$$\begin{aligned}-2x+17&=25-x^2\\ x^2-2x-8&=0\\ (x-4)(x+2)&=0\\ (x-4)=0, &\quad x+2=0\\ x=4, &\quad x=-2\end{aligned}$$

Substituting 4 for x in either function yield y = 9. Similarly, for $x=-2$, we obtain y = 21. Thus the intersection points are $(-2,21)$ and (4, 9). Subtracting $g(x)=-2x+17$ from $f(x)=25-x^2$ yields

$$\begin{aligned}f(x)-g(x)&=25-x^2-(-2x+17)\\ &=-x^2+2x+8\end{aligned}$$

Thus the area of S is expressed by

$$A(S)=\int_{-2}^{4} f(x)-g(x))dx=\int_{-2}^{4}(-x^2+2x+8)dx$$

$$=(-\frac{1}{3}x^3+x^2+8x]_{-2}^{4}=\frac{80}{3}-\left(-\frac{28}{3}\right)=\frac{108}{3}$$

EXERCISES

In Exercises 1 through 15, find the area of the region S bounded by the graphs of the following equations. Sketch the graph of S.

1. $y=2x^2+3,\ \ x=-3,\ \ x=2,\ \ y=0$
2. $y=4x^2-1,\ \ x=-2,\ \ x=4,\ \ y=0$
3. $y=16-x^2,\ \ y=0$
4. $y=2x^2+4x+1,\ \ x=1,\ \ x=3,\ \ y=0$
5. $y=-x^2+2x+10,\ \ x=-2,\ \ x=1,\ \ y=0$
6. $y=2x^2+1,\ \ y=x+10,\ \ y=-1,\ \ x=1$
7. $y=x^3,\ \ x=-4,\ \ x=-1,\ \ y=0$
8. $y=x^3-3x-3,\ \ x=-2,\ \ x=2,\ \ y=0$
9. $y=2x^2+4,\ \ y=-x^2+16$
10. $y=x\sqrt{4-x^2},\ \ x=0,\ \ x=2,\ \ y=0$
11. $y=\sqrt{x},\ \ y=x$
12. $y=\sqrt{x},\ \ y=x^3$
13. $y=e^x,\ \ y=e^{-x},\ \ x=4$
14. $y=x(x-3)^2,\ \ y=0$
15. $y=x^2+x+1,\ \ y=3x+4$

13.2. CONSUMERS' SURPLUS

Consider the demand function $x = g(p)$, which expresses x, the amount of a commodity bought by the market per unit time, as a function of p, the unit price of the commodity. The inverse function $p = f(x)$ expresses price p as a function of quantity x. If the market price is p_1 and the corresponding demand is $x_1 = g(p_1)$, then those consumers who would be willing to pay more for the commodity than market price p_1 gain from the market price being set at p_1. The area of the region bounded below by the line $p = p_1$, on the left and right by the p-axis ($x = 0$) and the line $x = x_1$, and above by the graph of the price-quantity demand function $p = f(x)$ (see Figure 13.4) serves, under

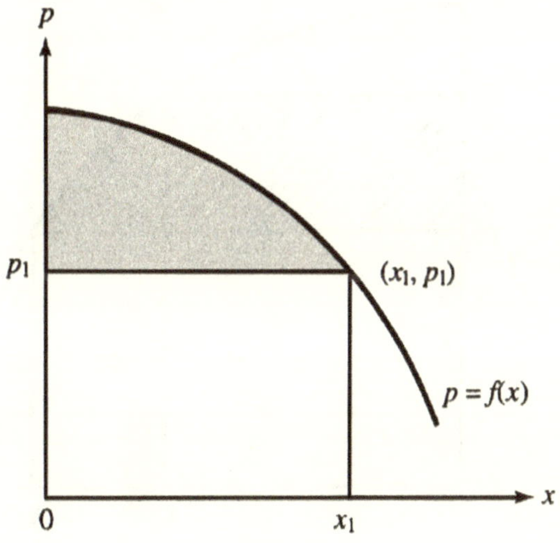

Figure 13.4

certain economic assumptions, as a numerical measure of the consumers' total gain. The area of this region, called the **consumers' surplus for price p_1**, is defined by

$$\int_0^{x_1} f(x)dx - p_1x_1$$

The product p_1x_1, of price p_1 and quantity x_1 bought at this price, expresses the amount paid by consumers for the commodity. This amount is represented geometrically by the area of the rectangle with length x_1 and height p_1 (see Figure 13.4).

If the price changes from p_1 to p_2, with a corresponding change in the quantity demanded from $x_1 = g(p_1)$ to $x_2 = f(x_2)$, then the change in consumers' surplus ΔC is the sum of the areas of regions R and S shown in Figure 13.5. If the price rises from p_1 to p_2, ΔC serves as a numerical measure of the injury suffered by consumers, and if the price falls from p_2 to p_1, ΔC serves as a numerical measure of the benefit derived by consumers.

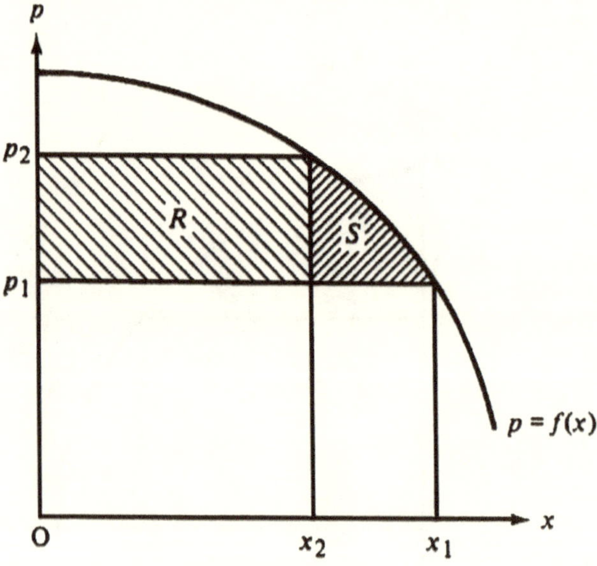

Figure 13.5

The areas of R and S are

$$A(R) = p_2x_2 - p_1x_2$$

$$A(R) = \int_{x_2}^{x_1} f(x)dx - (p_1x_1 - p_1x_2)$$

$$= \int_{x_2}^{x_1} f(x)dx - p_1x_1 + p_1x_2$$

Thus

$$A(R)+A(S)=p_2x_2-p_1x_2+\int_{x_2}^{x_1} f(x)dx-p_1x_1+p_1x_2$$
$$=\int_{x_2}^{x_1} f(x)dx-p_1x_1+p_2x_2$$
$$=\int_{x_2}^{x_1} f(x)dx-(p_1x_1-p_2x_2)$$

We thereby obtain for $\Delta C = A(R)+A(S)$

$$\Delta C=\int_{x_2}^{x_1} f(x)dx-(p_1x_1-p_2x_2)$$

$$\boxed{\Delta C=\int_{x_2}^{x_1} f(x)dx-\text{(change in expenditure)}}$$

EXAMPLE 1

The demand functions for coffee in a certain market are $p=500-x^2$ and $x=\sqrt{500-p}$, where p is the price in dollars per ton and x is the number of tons bought per week. Determine the consumers' surplus for p = 100. If p drops from \$100 to \$59 per ton, what is the change in consumers' surplus? If p rises from \$100 to \$139 per ton, what is the change in consumers' surplus?

When p = 100, x = 20. Thus the consumers' surplus for p = 100 is

$$\int_0^{20}(500-x^2)dx-100(20)=(500x-\frac{1}{3}x^3]_0^{20}-2000$$
$$=\frac{22{,}000}{3}-2000=5333.3$$

When p = 100, x = 20; when p = 59, x = 21. Thus the change in consumers' surplus ΔC for a price change from p_2 = 100 to p_1 = 59 is

$$\int_{20}^{21}(500-x^2)dx-[59(21)-100(20)]=(500x-\frac{1}{3}x^3]_{20}^{21}-(-761)$$

$$=7413-\frac{22,000}{3}+761=840.7$$

$\Delta C=840.7$ serves as a measure of the benefit derived by consumers when the price drops from \$100 to \$59 per ton.

When p = 100, x = 20; when p = 139, x = 19. Thus the change in consumers' surplus ΔC for a price change from p_1 = 100 to p_2 = 139 is

$$\int_{19}^{20}(500-x^2)dx-[100(20)-139(19)]=(500x-\frac{1}{3}x^3]_{19}^{20}-(-641)=760.7$$

$\Delta C=760.7$ serves as a measure of consumer loss when the price rises from \$100 to \$139 per ton.*

EXERCISES

1. The demand functions for a certain commodity are $p=2000-2x^2$ and $x=\sqrt{(2000-p)/2}$.

   (a) Determine the consumers' surplus for p = 200.
   (b) If p drops from 200 to 78, what is the change in consumers' surplus?
   (c) If p rises from 200 to 318, what is the change in consumers' surplus?

2. The total cost function of a coffee-producing monopolist is $c(x)=\frac{19}{2}x^2+10x+500$ dollars for an output of x tons a week. The demand functions for coffee in the market supplied are

* For further discussion of the concept of consumers' surplus, see, for example, D.S. Watson, *Price Theory and Its Uses*, 2nd ed. (Boston: Houghton-Mifflin Co., 1968), pp. 68-71, 139-140, 326-328.

$p = (50 - x)^2$ and its inverse $x = 50 - \sqrt{p}$, where p is the price of coffee in dollars per ton and x is the output in tons per week. Find the consumers' surplus for the market price determined by the given conditions. *Hint*: Determine the output for which the monopolist's profit is maximized.

13.3. PRODUCERS' SURPLUS

Consider the supply functions $x = s(p)$, which expresses x, the amount of a commodity that producers wish to sell to a market per unit time, as a function of its market price p, and the inverse function $p = h(x)$, which expresses market price as a function of quantity supplied. If the market price of the commodity is p_1 and the corresponding amount supplied to the market is $x_1 = s(p_1)$, then those producers who would be willing to supply the commodity to the market when the price is less than p_1 gain from the market price being set at p_1.

The area of the region bounded below by the price-quantity supply function $p = h(x)$, above by the line $p = p_1$, and on the left and right by the p-axis ($x = 0$) and the line $x = x_1$ (see Figure 13.6), serves, under certain economic assumptions, as a

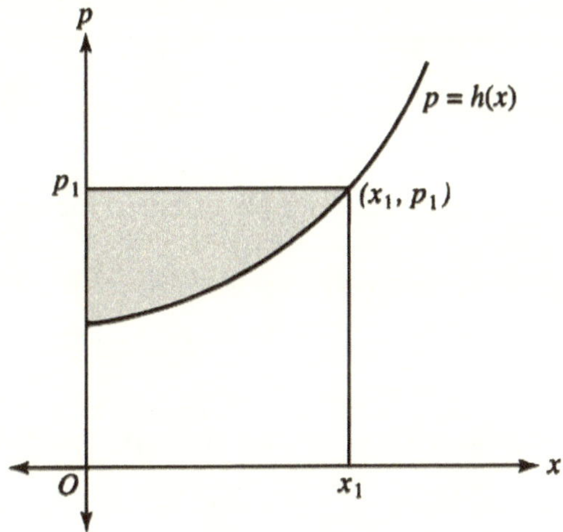

Figure 13.6

numerical measure of the total producer gain. The area of this region, called the ***producers' surplus***, is given by

$$p_1x_1 - \int_0^{x_1} h(x)dx .$$

EXAMPLE 1

The price-quantity supply function for sugar in a certain market is $p = x^2$, where x is tons of sugar supplied per month and p is the market price of sugar in dollars per ton. Determine the producers' surplus corresponding to a market price of $100 per ton.

For $p = 100$, $x = 10$. Thus the producers' surplus is

$$100(10) - \int_0^{10} x^2 dx = 1000 - \left(\frac{1}{3}x^3\right)\Big]_0^{10}$$

$$= 1000 - \frac{1000}{3}$$

$$= \frac{2000}{3}.$$

EXERCISES

1. The price-quantity supply function for coal is $p = 2x^2$.

   (a) Find the producers' surplus for $p = 200$.
   (b) Find the producers' surplus for $p = 162$.

2. The price-quantity supply function for tea is $p = \frac{1}{4}x^3$.

   (a) Find the producers' surplus for $p = 16$.
   (b) Find the producers' surplus for $p = 54$.

13.4. PRESENT VALUE OF A CASH FLOW

In preparation for a discussion of continuous cash flow, we first consider a discrete cash flow situation. An investment plan provides for payments of \$500 one year from now, \$1000 two years from now, and \$2000 three years from now. The interest rate is 6 percent per annum compounded continuously. What is the present value of this investment plan; that is, under the given interest conditions, what amount set aside today will yield \$500 in 1 year, \$1000 in 2 years, and \$2000 in 3 years?

Let us recall that, with respect to the interest rate r per annum compounded continuously, the present value of an amount A to be realized in x years is (see Section 4.2)

$$Ae^{-rx}$$

Thus the present values of the given amounts are

$$\begin{aligned} 500e^{-0.06} &= 500(0.94176) = \$470.88 \\ 1000e^{-0.12} &= 1000(0.88692) = \$886.92 \\ 2000e^{-0.18} &= 2000(0.83527) = \$1670.54 \end{aligned}$$

The total present worth of our discrete income stream is the sum \$470.88 + \$886.92 + \$1670.54 = \$3028.34. The proposed income plan is worth \$3028.34 today under the stated interest conditions. More generally, the present value of a discrete cash flow consisting of the amounts $A_1, A_2, \ldots, A_n$ is obtained by finding the present values of $A_1, A_2, \ldots, A_n$ and adding.

A **continuous cash flow** or **stream** is defined when there is given $y = f(t)$, which is defined for all real numbers in a time interval and specifies the rate of cash flow as a function of time. To say that the cash (or income, or profit) flow is $f(t)$ dollars per year means that the amount of money accumulated in a small time interval of length h is approximately equal to $f(t)h$, where t is any time value in the time interval.

Let $y = f(t)$ denote a cash-flow function that is defined on the time interval $[0, v]$ (see Figure 13.7). By introducing a sequence of values $0 = t_0, t_1, t_2, \ldots, t_{n-1}, t_n =$

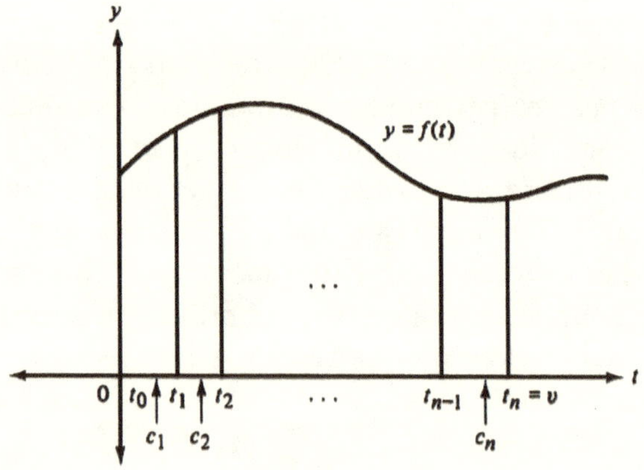

Figure 13.7

v, we partition the time interval $[0, v]$ into subintervals $[t_0, t_1], [t_1, t_2], \ldots, [t_{n-1}, t_n]$. From each of these subintervals we choose, arbitrarily, a time value: c_1 from $[t_0, t_1]$, c_2 from $[t_1, t_2], \ldots, c_n$ from $[t_{n-1}, t_n]$. Then the amount accumulated over $[t_0, t_1]$ is approximately

$$f(c_1)(t_1 - t_0)$$

With respect to the interest rate r per annum compounded continuously, the present value of this amount is

$$e^{-rc_1} f(c_1)(t_1 - t_0)$$

The amount accumulated over the second subinterval $[t_1, t_2]$ is approximately $f(c_2)(t_2 - t_1)$, with present value

$$e^{-rc_2} f(c_2)(t_2 - t_1)$$

Over the nth subinterval $[t_{n-1}, t_n]$, the amount accumulated is approximately $f(c_n)(t_n - t_{n-1})$, with present value

$$e^{-rc_n} f(c_n)(t_n - t_{n-1})$$

The present value of the discrete cash flow consisting of the n amounts $f(c_1)(t_1 - t_0), f(c_2)(t_2 - t_1), \ldots, f(c_n)(t_n - t_{n-1})$ is given by the sum of the present values of these n amounts:

$$\sum_{k=1}^{k=n} e^{-rc_k} f(c_k)(t_k - t_{k-1}) = e^{-rc_1} f(c_1)(t_1 - t_0) + \ldots + e^{-rc_n} f(c_n)(t_n - t_{n-1})$$

By approximating the continuous cash flow by a discrete structure, we have obtained a sum that serves as an approximation to what it might be reasonable to call the present value of the cash flow. To emerge with a uniquely defined value that is independent of the particular sequence of values $0 = t_0,\ t_1,\ \ldots,\ t_{n-1}, t_n = v$ used to partition $[0,\ v]$ into subintervals and the time values $c_1,\ c_2,\ \ldots,\ c_n$ chosen in these respective subintervals, we consider the behavior of the present-value sums

$$\sum_{k=1}^{k=n} e^{-rc_k} f(c_k)(t_k - t_{k-1})$$

as sequences $0 = t_0,\ t_1,\ \ldots,\ t_n = v$ with more and more values are chosen $(n \to \infty)$ in such a way that the difference between consecutive values approach zero $(t_k - t_{k-1} \to 0)$. If, under these conditions, the corresponding present-value sums approach one number Π (Π is the uppercase Greek letter pi),

$$\sum_{k=1}^{k=n} e^{-rc_k} f(c_k)(t_k - t_{k-1}) \to \Pi$$

then Π is called the **present value of the cash flow defined by $y = f(t)$ from time $t = 0$ time $t = v$ with respect to the interest rate**

***r* per annum compounded continuously**. By definition of definite integral (alternative form), we have that

$$\Pi = \int_0^v e^{-rt} f(t)dt$$

expresses the **present worth** of the envisioned continuous cash flow.

EXAMPLE 1

A certain investment is envisioned as yielding income at the constant rate of $2000 per year for 6 years. With respect to an interest rate of 8 percent per annum compounded continuously, determine the present value of this income stream.

Since $f(t) = 2000$, $r = 0.08$, and the time interval is [0, 6], we have

$$\Pi = \int_0^6 2000e^{-0.08t} dt$$

Our first task in determining the value of Π is to find an indefinite integral

$$F(t) = \int 2000e^{-0.08t} dt = 2000\int e^{-0.08t} dt$$

Let $u = -0.08t$; then $D_t u = -0.08$. To obtain $D_t u$ under the integral sign, we introduce -0.08 under the integral sign and compensate by also introducing $-1/0.08$ outside the integral sign. Thus we obtain

$$F(t) = 2000\int e^{-0.08t} dt = -\frac{2000}{0.08}\int \underbrace{(-0.08)}_{D_t u} e^{\overbrace{-0.08t}^{u}} dt$$

$$= -25,000\int e^u du = -25,000e^u$$

$$= -25,000e^{-0.08t}$$

Thus

$$\Pi = (-25{,}000e^{-0.08t}]_0^6 = -25{,}000(e^{-0.48} - e^0)$$
$$= -25{,}000(0.61878 - 1) = -25{,}000(-0.38122)$$
$$= \$9530.50$$

The present worth of a cash flow of \$2000 per year over a period of 6 years, with respect to the interest rate 8 percent, per annum compounded continuously, is \$9530.50.

EXERCISES

1. Income from a certain investment is derived at the constant rate of \$3000 per year for 5 years. With respect to the interest rate of 9 percent per annum compounded continuously, find the present value of this income stream.
2. Income from an investment is obtained at the constant rate of \$5000 per year for 8 years. With respect to the interest rate of 10 percent per annum compounded continuously, find the present value of this income stream.
3. Show that, if income from an investment is obtained at the constant rate of \$$A$ per year for x years, then, with respect to the interest rate r per annum compounded continuously, the present value of the income stream is $\frac{A}{r}(1-e^{-rx})$ dollars.
4. In developing the concept of present value of a continuous cash flow, the alternative form of the definite integral based on general sums served as the foundation. Develop the concept of present value in terms of the lower and upper sum approach to the definite integral.
5. Develop the concept of present value in terms of the LUB-GLB approach to the definite integral.

13.5. EQUIPMENT INVESTMENT ANALYSIS

An important problem in equipment-investment analysis is concerned with determining the present value to a firm of a piece of production equipment which earns revenue at the continuous rate of $R(t)$ dollars per year t years after its installation and incurs operation and maintenance expenses at the continuous rate of $E(t)$ dollars per year t years after its installation (this does not include depreciation and interest on investment). Income is derived from such a piece of equipment at the continuous rate of $f(t) = R(t) - E(t)$ dollars per year t years after its installation.

If cost after installation and salvage value considerations are ignored for the moment, then the **present value of a piece of equipment** which is assumed to have a useful life of x years is

$$\Pi = \int_0^x f(t)e^{-rt}\,dt$$

where the interest rate is r per annum compounded continuously.

The cost C of the equipment after installation is simple to deal with. Just subtract it from Π. If S is the salvage value of the equipment x years later, then

$$Se^{-rx}$$

is the present value of S, where the interest rate is r per annum compounded continuously (see Section 4.2).

Thus the **present value Π to a firm of a piece of production equipment** which earns net revenue (total revenue minus the maintenance expenses incurred) at the continuous rate of $f(t)$ dollars per year t years after its installation is given by

$$\Pi = \int_0^x f(t)e^{-rt}\,dt + Se^{-rx} - C$$

where x is the life of the equipment in years, S is its salvage value after x years, C is the cost of the equipment after installation, and r is the continuously compounded annual interest rate.*

EXAMPLE 1

Consider a firm that has acquired a machine which is assumed to have a useful life of twenty-years. It is estimated that net income is derived at the continuous rate of

$$f(t) = 2200 - 100t$$

dollars t years after installation. If the interest on investment is 10% per annum compounded continuously, the installed cost of the equipment is \$1000, and its salvage value after twenty years is \$200, find the present value of the equipment.

$$\Pi = \int_0^{20} (2200 - 100t)e^{-0.1t}dt + 200e^{-2} - 1000$$

Now

$$F(t) = -12,000e^{-0.1t} + 1000te^{-0.1t}$$

is an indefinite integral of $g(t) = (2200 - 100t)e^{-0.1t}$ (see Example 3, Section 9.3, p. 290)

$$\Pi = (-12,000e^{-0.1t} + 1000te^{-0.1t})\Big]_0^{20} + 200e^{-2} - 1000$$

* For further discussion of equipment-investment analysis, see such works as E. H. Bowman and R. B. Fetter, *Analysis for Production and Operations Management*, 3rd ed. (Homewood, Illinois: Richard P. Irwin, Inc., 1967). Chapter 10: F. and V. Lutz. *The Theory of Investment of the Firm* (Princeton, New Jersey: Princeton University Press, 1951), Chapter VIII; and G. A. D. Preinreich. "The Economic Life of Industrial Equipment," *Econometrica* (January 1940), pp. 12-44.

$$= (-12,000e^{-2} + 20,000e^{-2}) - (-12,000e^{0}) + 200e^{-2} + 1000$$

Since $e^{-2} = 0.13534$ and $e^{0} = 1$, we obtain:

$$\Pi = \$12,109.79$$

EXERCISES

1. A machine acquired by a firm is assumed to have a useful life of 10 years. It is estimated that net income is derived at the continuous rate of $f(t) = 1800 - 50t$ dollars per year t years after installation. Interest on investment is 10% per annum compounded continuously.

   (a) If installation costs and salvage value considerations are ignored, determine the present value of the machine $(e^{-1} = 0.36788)$.
   (b) If the installed cost of the machine is \$800 and its salvage value at the end of 10 years is \$100, determine the present value of the machine.

2. A piece of production equipment acquired by a company is assumed to have a useful life of 20 years. It is estimated that net income is derived at the continuous rate of $f(t) = 1000 - 20t$ dollars per year t years after installation. Interest on investment is 5% per annum compounded continuously.

   (a) If installation costs and salvage value considerations are ignored, determine the present value of the production equipment.
   (b) If the installed cost of the production equipment is \$500 and its salvage value at the end of 20 years is \$50, determine the present value of the equipment.

13.6. PRESENT VALUE OF A PERPETUAL INCOME STREAM

In Section 13.4 we saw that if income is obtained continuously over time at a rate of $f(t)$ dollars per year t years from the present ($t = 0$), where the interest rate is r per annum compounded continuously, then the **present value** Π **of the income** stream from the present ($t = 0$) to time $t = x$ years is defined by

$$\Pi = \int_0^x f(t)e^{-rt}\,dt.$$

If the income flow were to persist indefinitely, as in the case of interest from a perpetual bond or revenue from an indestructible asset such as land, then the present value Π of the income stream is defined by the improper integral

$$\Pi = \int_0^\infty f(t)e^{-rt}\,dt.$$

EXAMPLE 1

Consider a perpetual income stream flowing at the constant rate of \$1000 per year, where the interest rate is 10% per annum compounded continuously. Find the present value of this income stream.

$$\Pi = \int_0^\infty 1000e^{-0.1t}\,dt.$$

Now

$$\begin{aligned}\int_0^t 1000e^{-0.1t}\,dt &= (-10,000e^{-0.1t})\Big]_0^t \\ &= -10,000e^{-0.1t} - (-10,000e^0) \\ &= -10,000e^{-0.1t} + 10,000.\end{aligned}$$

Thus

$$\underset{t \to x}{\text{limit}} \int_0^t 1000e^{-0.1t} dt = \underset{t \to \infty}{\text{limit}}(-10,000e^{-0.1t} + 10,000) = 10,000$$

since

$$e^{-0.1t} = \frac{1}{e^{0.1t}} \to 0 \quad as\, t \to \infty .$$

Therefore

$$\Pi = \int_0^x 1000e^{-0.1t} dt = \$10,000 .$$

EXERCISES

1. Find the present value of a perpetual income stream flowing at the constant rate of \$4000 per year, where the interest rate is 10% per annum compounded continuously.
2. Find the present value of a perpetual income stream flowing at the constant rate of \$1200 per year, where the interest rate is 12% per annum compounded continuously.
3. Show that the present value of a perpetual income stream flowing at the constant rate of A dollars per year, where the interest rate is r per annum compounded continuously, is A/r dollars.

13.7. TOTAL VALUE OF A CASH FLOW

Suppose that \$500 is invested now, \$1000 is invested 1 year from now, and \$2000 is invested 2 years from now. The interest rate is 6 percent per annum compounded continuously, and the end of the investment period is 3 years from now. Let us determine the total value of this investment stream at the end of the investment period. To begin, recall that, with respect to the interest rate r per annum compounded continuously, \$$A$ invested now is worth

$$Ae^{rx} \textbf{ dollars}$$

x years from now (see Section 4.2, pp. 124-126). Thus the future value of \$500 at the end of the investment period (3 years after it is invested), \$1000 at the end of the investment period (2 years after it is invested), and \$2000 at the end of the investment period (1 year after it is invested) is given by

$$500e^{0.06(3)} = 500e^{0.18} = 500(1.1972) = \$598.60$$

$$1000e^{0.06(2)} = 1000e^{0.12} = 1000(1.1275) = \$1127.50$$

$$2000e^{0.06(1)} = 2000e^{0.06} = 2000(1.0618) = \$2123.60$$

Thus the total value of this discrete investment stream is the sum \$598.60 + \$1127.50 +\$2123.60 = \$3849.70. More generally, if $\$A_0$, $\$A_1$, $\$A_2$, . . . , $\$A_n$ are invested now and in succeeding years at the rate r per annum compounded continuously, the total value of this discrete investment stream at the end of the investment period is given by the sum

$$A_0e^{rn} + A_1e^{r(n-1)} + A_2e^{r(n-2)} + ... + A_ne^{r(n-n)} = \sum_{k=0}^{k=n} A_ke^{r(n-k)}$$

With this background in mind, let us turn our attention to a continuous cash flow. Suppose that investment is made continuously over time at a rate of $f(t)$ dollars per year t years from the present ($t = 0$), where the interest rate is r per annum compounded continuously. We further assume that the investment period is for v years, so that investment is occurring over the time interval $[0, v]$. By following in the suitably modified footsteps of the approach used to define present value of a continuous cash flow, show that the **total value of this continuous investment stream** is expressed by

$$\boxed{S = \int_0^v e^{r(v-t)} f(t)dt}$$

EXAMPLE 1

Investment is made continuously over time at the constant rate of $1000 per year at an interest rate of 10% per annum compounded continuously. Find the total value of this investment stream at the end of five years.

$$S = \int_0^5 1000e^{0.1(5-t)}dt$$

$F(t) = -10,000e^{0.1(5-t)}$ is an indefinite integral of $g(t) = -1000e^{0.1(5-t)}$.
Thus:

$$\begin{aligned} S &= (-10,000e^{0.1(5-t)})\Big]_0^5 \\ &= -10,000e^0 - 10,000e^{0.5} \\ &= 10,000(e^{0.5} - e^0) \\ &= 10,000(1.6487 - 1) \\ &= \$6487.00 \end{aligned}$$

EXERCISES

1. If money is invested continuously at the constant rate of $2000 per year at an interest rate of 10% per annum compounded continuously, find the total sum accumulated at the end of ten years.
2. If money is invested continuously at the constant rate of $3000 per year at an interest rate of 12% per annum compounded continuously, find the total sum accumulated at the end of eight years ($e^{0.96} = 2.6117$).
3. If money is invested continuously at the constant rate of $5000 per year at an interest rate of 8% per annum compounded continuously, find the total sum accumulated at the end of eleven years ($e^{0.88} = 2.4109$).

4. Show that if money is invested continuously at the constant rate of \$$A$ per year at an interest rate of r per annum compounded continuously, then the total sum accumulated at the end of x years is $\frac{A}{r}(e^{rx}-1)$ dollars.
5. If money is invested continuously at the rate $f(t) = 100 + 3t$ dollars per year at an interest rate of 10% per annum compounded continuously, find the total sum accumulated at the end of five years ($e^{0.5} = 1.6487$).
6. If money is invested continuously at the rate $f(t) = 500 - 5t$ dollars per year at an interest rate of 12% per year compounded continuously, find the total sum accumulated at the end of eight years.

13.8. WORK DONE BY A VARIABLE FORCE

If a constant force moves an object a certain distance along a straight line path, then the product of the magnitude of the force times the distance moved is called the **work done** by the force in moving the object. Thus if 20 pounds is used to move an object 10 feet, the work done in moving the object is 20(10) = 200 foot-pounds.

The work done by a constant force serves as a quantitative measure of the energy expended in moving an object a given distance. But suppose the force is not constant in magnitude. What would be an appropriate way to extend the concept of work to such a situation? This is the problem we now consider. To be specific, suppose that an object is moved six feet along a horizontal line by a variable force whose magnitude x feet from the initial position of the object is $f(x) = 2x + 3$ pounds (see Figure 13.8).

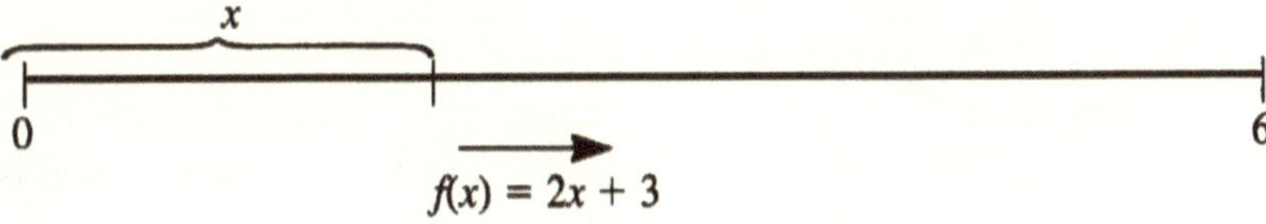

Figure 13.8

The problem, then, is to define a concept of work done by this force whose magnitude varies from point to point in moving the object six feet. However the work done by this force is defined, it should not be less than $3 \cdot 6 = 18$ foot-pounds since the minimum value of the force in the interval of motion is $f(0) = 3$ pounds and the total distance moved by the object is 6 feet. Also, the work done should not exceed $15 \cdot 6 = f(6) = 15$ pounds. Thus if we let W denote the work done by the force, we have

$$18 \leq W \leq 90$$

as the bounds for the work done.

Now suppose we regard the work done W as being done in two stages. In the first stage the object is moved, let us say, two feet, from 0 to 2, and in the second stage the object is moved four feet, from 2 to 6 (see Figure 13.9).

In the first stage the work done, call it W_1, should not be less than $3 \cdot 2 = 6$ foot-pounds (the minimum value of the force in this interval is $f(0) = 3$ pounds and

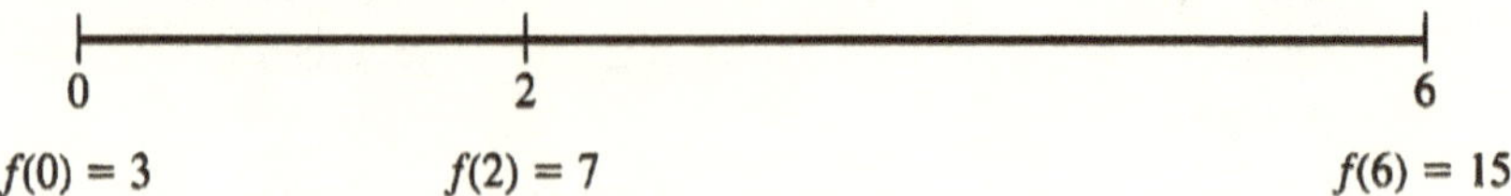

Figure 13.9

the distance moved is 2 feet). Also, W_1 should not exceed $7 \cdot 2 = 14$ foot-pounds (the maximum value of the force in this interval is $f(2) = 7$ pounds and the distance moved is 2 feet). We thus have

$$6 \leq W_1 \leq 14.$$

In the second stage the work done, call it W_2, should not be less than $7 \cdot 4 = 28$ foot-pounds (the minimum value of the force in this interval is $f(2) = 7$ pounds and the distance moved is 4 feet). Also, W_2 should not exceed $15 \cdot 4 = 60$ foot-pounds (the minimum value of the force in this interval is $f(6) = 15$ pounds and the distance moved is 4 feet). We thus have

$$28 \leq W_2 \leq 60.$$

One of the properties of the work concept for a force of constant magnitude is that if W is the work done over an interval I, and W_1, W_2, etc., represent the work done over subintervals into which I has been partitioned, then

$$W = W_1 + W_2 + \cdots.$$

That is, the work done over an interval I equals the sum of the work done over the subintervals into which I has been partitioned.

If the concept of work for a variable force which we are seeking to define is to be modeled after the concept of work for a constant force, then it should share its most important properties. We are thus led to impose the condition that the work done by a variable force over an interval I be equal to the sum of the work done over subintervals into which I has been partitioned. For the situation under consideration, this leads to the condition

$$W = W_1 + W_2.$$

Since

$$6 \leq W_1 \leq 14$$

$$28 \leq W_2 \leq 60$$

addition yields

$$6 + 28 \leq W_1 + W_2 \leq 14 + 60$$

from which we obtain

$$\boxed{34 \leq W \leq 74}$$

as the bounds for W.

Suppose we regard the work done W as being done in three stages, where, let us say, in the first stage the object is moved two feet, from 0 to 2, in the second stage it is moved three feet, from 2 to 5, and in the third stage it is moved one foot, from 5 to 6 (see Figure 13.10).

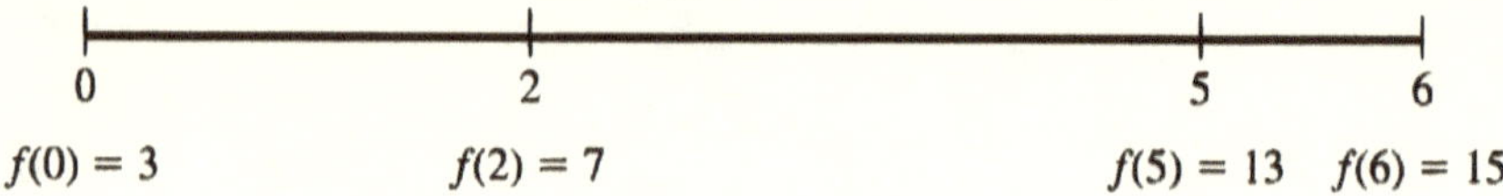

Figure 13.10

Then we obtain

$$6 \leq W_1 \leq 14$$

$$21 \leq W_2 \leq 39$$

$$13 \leq W_3 \leq 15$$

as the bounds for W_1, W_2, and W_3, the work done in each subinterval. By adding and using the condition

$$W = W_1 + W_2 + W_3$$

we obtain

$$\boxed{40 \leq W \leq 68}$$

as the bounds for W.

We thus see the same idea here in our work problem as in our area problem. A lower bound for W is obtained by partitioning the interval [0, 6] into subintervals, forming the product of the minimum value of the force function in each subinterval times the length of the subinterval, then adding. An upper bound is obtained by forming the product of the maximum value of the force function in each subinterval times the length of the subinterval, then adding.

More generally, suppose we partition the interval [0, 6] into n subintervals, $[x_0, x_1]$, $[x_1, x_2]$, . . . , $[x_{n-1}, x_n]$, by choosing a sequence of partition points $0 = x_0, x_1, x_2, \ldots, x_{n-1}, x_n = 6$ (see Figure 13.11). Then suppose we regard the work done W as

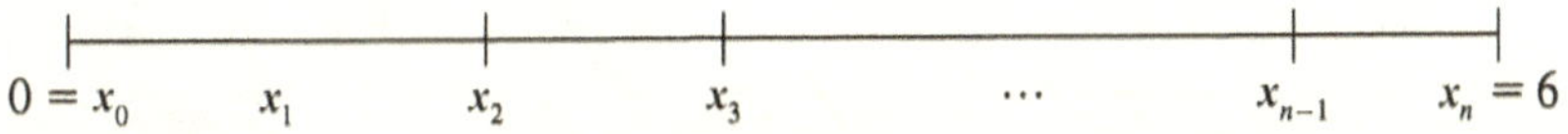

$$0 = x_0 \quad x_1 \quad x_2 \quad x_3 \quad \cdots \quad x_{n-1} \quad x_n = 6$$

Figure 13.11

being done in n stages, where in the first stage the object is moved from $0 = x_0$ to x_1, in the second stage it is moved from x_1 to x_2, in the third stage it is moved from x_2 to x_3, and so on. As bounds for $W_1, W_2, W_3, \ldots, W_n$, the work done in each of these n stages, we have the following expressions.

$$\begin{array}{c}
f(m_1)\cdot(x_1 - x_0) \le W_1 \le f(M_1)\cdot(x_1 - x_0) \\
f(m_2)\cdot(x_2 - x_1) \le W_2 \le f(M_2)\cdot(x_2 - x_1) \\
\vdots \qquad\qquad \vdots \qquad\qquad \vdots \\
f(m_n)\cdot(x_n - x_{n-1}) \le W_n \le f(M_n)\cdot(x_n - x_{n-1})
\end{array}$$

Here $m_1, m_2, \ldots m_n$ denote values in $[x_0, x_1], [x_1, x_2], \ldots, [x_{n-1}, x_n]$ at which the force function $f(x) = 2x + 3$ takes on its absolute minimum value in these respective subintervals (in our case $m_1 = x_0$, $m_2 = x_1, \ldots m_n = x_{n-1}$). $M_1, M_2, \ldots M_n$ denote values in $[x_0, x_1]$, $[x_1, x_2], \ldots, [x_{n-1}, x_n]$ at which $f(x) = 2x + 3$ takes on its absolute maximum value in these respective subintervals (in our case $M_1 = x_1$, $M_2 = x_2, \ldots, M_n = x_n$).

By adding and using the condition

$$W = W_1 + W_2 + \cdots + W_n$$

we obtain

$$\boxed{\sum_{i=1}^{i=n} f(m_i)\cdot(x_i - x_{i-1}) \le W \le \sum_{i=1}^{i=n} f(M_i)\cdot(x_i - x_{i-1})}$$

as the bounds for *W*. Each sequence of partition points $0 = x_0, x_1, x_2, \ldots, x_{n-1}, x_n = 6$ yields bounds of this sort for *W*.

We have seen that by partitioning the interval [0, 6] into more and smaller subintervals, we narrow the gap between the lower and upper bounds obtained for *W*. To close the gap entirely we proceed as discussed in Section 12.4 (p. 357). Since every lower-bound sum, $\sum f(m_i)\cdot(x_i - x_{i-1})$, is less than or equal to *W*, the least upper bound *I* of the set of all lower-bound sums is less than or equal to *W*.

$$I = LUB \text{ of } \left\{ \sum_{i=1}^{i=n} f(m_i)\cdot(x_i - x_{i-1}) \right\}$$
$$I \leq W$$

Since every upper-bound sum, $\sum f(M_i)\cdot(x_i - x_{i-1})$, is greater than or equal to *W*, the greatest lower bound of the set of all upper-bound sums is greater than or equal to *W*.

$$J = GLB \text{ of } \left\{ \sum_{i=1}^{i=n} f(M_i)\cdot(x_i - x_{i-1}) \right\}$$
$$W \leq J$$

Thus we have

$$I \leq W \leq J.$$

If $I = J$, we define *W* to be their common value.

W, we recognize, is:

$$W = \int_0^6 (2x+3)dx = (x^2 + 3x]_0^6 = 54 \text{ foot-pounds}$$

More generally, suppose a particle is moved along a line from $x = a$ to $x = b$ by a variable force whose magnitude at value *x* in [*a*, *b*] is given by the continuous function $y = f(x)$. Then the work done by the force is defined as

$$W = \int_a^b f(x)dx.$$

EXERCISES

1. A particle is moved ten feet from $x = 1$ to $x = 11$ along a line by a force whose magnitude at value x in [1, 11] is $f(x) = x^2 + 2x + 1$ pounds. Determine the work done by this force.
2. A particle is moved nine feet from $x = 0$ to $x = 9$ along a line by a force whose magnitude at value x in [0, 9] is $f(x) = \sqrt{x} + 1$ pounds. Determine the work done by this force.
3. A particle is moved four feet from $x = 3$ to $x = 7$ along a line by a force whose magnitude at value x in [3, 7] is $f(x) = 5x^4 + 8$ pounds. Determine the work done by this force.

13.9. INTEGRAL REPRESENTATIONS FOR OLD FACES

By means of the fundamental theorem of integral calculus such notions as total cost and total revenue of a firm can be expressed in terms of definite integrals of their corresponding marginal concepts. If $f(x)$ is the marginal cost function of a firm, then the total cost function $c(x)$ is an indefinite integral of $f(x)$ since, by definition, $f(x) = c'(x)$. From the fundamental theorem of integral calculus we have

$$c(x) - c(0) = \int_0^x f(x)dx.$$

Thus

$$c(x) = c(0) + \int_0^x f(x)dx. \qquad (13.1)$$

$c(0)$ represents the firm's cost when there is no output, that is, the fixed cost of the firm irrespective of output.

Equation (13.1) gives us a view of total cost in terms of fixed and marginal costs. It simply says that the total cost for output x is fixed cost plus the integral of the marginal cost function from 0 to x. Geometrically this says that the total cost for output x is the fixed cost plus the area under the marginal cost curve from 0 to x (see Figure 13.12).

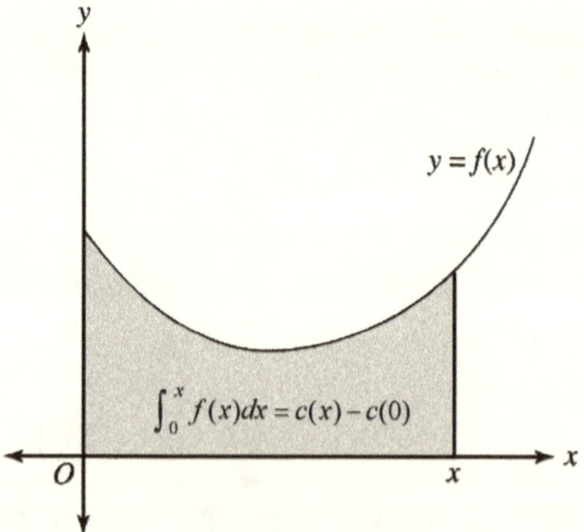

Figure 13.12

To illustrate, consider an aluminum producer whose fixed cost is \$1000 per week and whose marginal cost for an output of x tons per week is

$$f(x) = x^2 - 18x + 100\,.$$

Then his total cost for output x is

$$c(x) = 1000 + \int_0^x (x^2 - 18x + 100)dx$$

that is, $c(x)$ equals 1000 plus the area under the graph of $f(x) = x^2 - 18x + 100$ between 0 and x. Thus, for example, $c(10)$, the total cost of producing ten tons of aluminum per week, is

$$c(10) = 1000 + \int_0^{10} (x^2 - 18x + 100)dx$$

$$= 1000 + \left(\frac{1}{3}x^3 - 9x^2 + 100x\right)\Bigg]_0^{10}$$

$$= 1000 + \frac{1300}{3}$$

$$= \$1433.33$$

If $f(x)$ is the marginal revenue function of a firm, then by an analogous argument we obtain

$$R(x) = R(0) + \int_0^x f(x)dx$$

for the total revenue function $R(x)$. Since $R(0) = 0$ (no production, no revenue), we have

$$\boxed{R(x) = \int_0^x f(x)dx}$$

that is, the total revenue derived from the production and sale of x units equals the area under the marginal revenue curve between 0 and x (see Figure 13.13).

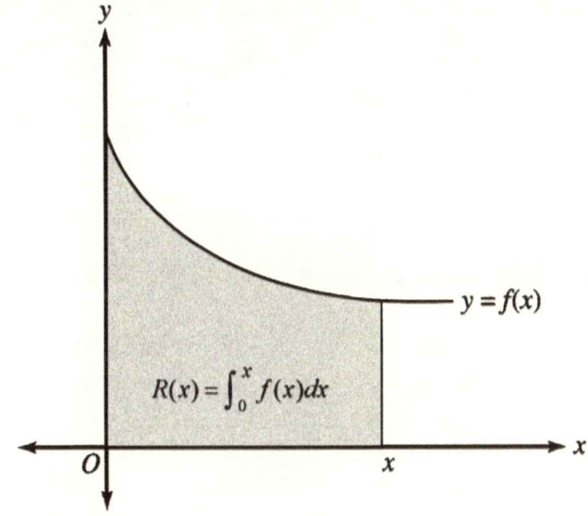

Figure 13.13

CHAPTER 14

Topics in Multivariable Calculus

14.1. MULTIVARIABLE FUNCTIONS

The idea of multivariable functions is a natural extension of the concept of function of one independent variable. For example, u is said to be a **function of two independent variables, x and y**, if for each ordered pair of numbers (x, y) there is determined, by means of some rule, exactly one value of u. If the rule of the function is denoted by f, then the value assigned to (x, y) is denoted by $f(x, y)$. The **domain of definition**, or **domain**, of the function consists of all ordered pairs of numbers (x, y) to which a value is assigned by the rule. Here, too, when the domain is not explicitly stated, it is understood to be the collection of all ordered pairs of numbers for which the rule makes sense. The function defined by

$$f(x,y)=2xy-x^2 \quad \text{or} \quad u=2xy-x^2$$

for example, is defined for all ordered pairs of numbers. For $(-1,2)$, we have

$$f(-1,2)=2(-1)2-(-1)^2=-5$$

while to $(2,-1)$ there is assigned the value

$$f(2,-1)=2(2)(-1)-(2)^2=-8$$

Taking this a step further, u is said to be a **function of three independent variables, x, y, and z**, let us say, if for each ordered triple of numbers (x, y, z) there is determined, by means of some rule, exactly one value of u. The domain of such a function is the collection of all ordered triples of numbers (x, y, z) to which a value is assigned by the rule. If, for example,

$$f(x, y, z) = 3x + yz^2 \quad \text{or} \quad u = 3x + yz^2$$

then

$$f(1,2,-1)=3(1)+2(-1)^2=5$$
$$f(-2,4,-1)=3(-2)+4(-1)^2=-2$$

The concept of function of n independent variables is defined in an analogous way. Functions of more than one independent variable are called **multivariable functions**.

EXAMPLE 1. A CAPITAL ACCUMULATION FUNCTION

The amount accumulated from an investment depends on the initial amount invested, interest conditions, and time. Specifically, the function

$$f(y, r, t)=ye^{rt}$$

describes the amount accumulated after t years when an initial amount of y dollars is invested at a rate of r per annum compounded continuously (see Section 4.2 for a discussion of continuous compounding of interest).

EXAMPLE 2. DEMAND FUNCTIONS

A demand function expresses the amount q of a commodity purchased per unit time in a market as a function of p, the price of the commodity, x, a measure of the prices of related commodities, y, a measure of the incomes of buyers, and t, a measure of the tastes of buyers. A demand function thus has the general form

$$q = f(p, x, y, t).$$

If income, the prices of related commodities, and taste are all assumed to be constant, then q becomes a function of one variable p representing the price of the commodity,

$$q = h(p)$$

One specific function type that has been used in the study of demand is the linear form

$$q = a + bp + cx + dy$$

where a, b, c, and d are constants. In his now classic study of demand functions,* Henry Schultz obtained

$$q = 3.4892 - 0.0899p + 0.0637x + 0.0187y$$

as a demand function for beef. The related commodity in this situation is pork.

Another function type that has been used in the study of demand is the nonlinear form

$$q = ap^b x^c y^d t^f$$

---

* *The Theory and Measurement of Demand* (Chicago: University of Chicago Press, 1938).

where *a, b, c, d,* and *f* are constants. In his paper "The Analysis of Market Demand,"[†] R. Stone obtained the function

$$q = 1.058 p^{0.727} x^{0.914} y^{0.136} t^{0.816}$$

in his analysis of the demand for beer in the United Kingdom. *x* represents the average retail price of all other related commodities and *t* is an index of the strength of beer which can be interpreted as an index of taste.

EXAMPLE 3. PRODUCTION FUNCTIONS

A production function expresses the output of a firm or industry per unit time as a function of the inputs utilized in the production process. If, for example, *q* is the output when amounts *x, y, z, w* of four productive factors are employed, then the production function can be expressed in function notation by

$$q = f(x, y, z, w)$$

To illustrate, suppose that

$$q = 6xy - 3x^2 - y^2$$

bushels of rye are produced per week when *x* labor-hours are employed on *y* acres of land. If 300 acres are cultivated and 500 labor-hours are employed, then the weekly output is

$$q = 6(500)(300) - 3(500)^2 - (300)^2 = 60{,}000 \text{ bushels.}$$

One widely used production function form in economic analysis, whose use was pioneered by former Senator Paul H. Douglas and his collaborators, is the Cobb-Douglas function,

$$q = ax^b y^{1-b}.$$

[†] *Journal of the Royal Statistical Socity*, vol. 108 (1945).

Here q is output, x is a measure of the quantity of labor employed, y is a measure of the quantity of capital employed, and a and b ($0 < b < 1$) are constants. Douglas and his associates derived production functions for whole economies and sectors of economies. In a 1928 paper,* Douglas and C. W. Cobb employed the production function

$$q = 1.01x^{0.75}y^{0.25}$$

in a study of manufacturing in the United States.

EXERCISES

1. For $f(x,y) = 2xy - x^2y^2$, find $f(1,3)$, $f(-1,2)$, $f(3,1)$, and $f(2,6)$.
2. For $f(x,y,z) = 2xy + y^2z$, find $f(2,3,1)$, $f(-1,2,4)$, and $f(3,2,-1)$.
3. For $f(x,y,z) = \dfrac{x^2+y^2}{z^2}$, find $f(2,3,1)$, $f(-1,2,4)$, and $f(3,2,1)$.

14.2. PARTIAL DERIVATIVES

If all but one of the independent variables of a multivariable function are held constant, then the multivariable function reduces to a function of that one variable. For example, if $y = 3$, then

$$f(x,y) = x^2y + 2x - y$$

reduces to the function of x:

$$f(x,3) = 3x^2 + 2x - 3$$

For $x = 2$, we obtain the function of y:

$$f(2, y) = 3y + 4$$

---

* "A Theory of Production," *American Economic Review*, vol. 18 (1928).

For such functions of one variable derivative questions can be posed. The derivative of

$$f(x,3) = 3x^2 + 2x - 3$$

defined by

$$\operatorname*{limit}_{h\to 0}\frac{f(x+h,3) - f(x,3)}{h}$$

is called the partial derivative of $f(x,y) = x^2y + 2x - y$ with respect to x at $(x, 3)$, and is denoted by $f_x(x, 3)$. Thus

$$f_x(x,3) = D_x(3x^2 + 2x - 3) = 6x + 2$$

The partial derivative of $f(x,y) = x^2y + 2x - y$ with respect to x at (2, 3), for example, is

$$f_x(2, 3) = 6(2) + 2 = 14$$

and its partial derivative with respect to x at $(-1,3)$ is

$$f_x(-1,3) = 6(-1) + 2 = -4$$

More generally, if $u = f(x, y)$ is a function of x and y, then the **partial derivative of $u = f(x, y)$ with respect to x at (x, y)** is defined by

$$f_x(x,y) = \operatorname*{limit}_{h\to 0}\frac{f(x+h,y) - f(x,y)}{h}$$

The partial derivative $f_x(x, y)$, with respect to x, is obtained by regarding y as being constant, thus generating a function of x, and differentiating the resulting function of x (with respect to x). Thus

For $f(x,y)=x^3y^2,\ \ f_x(x,y)=D_x(x^3y^2)=y^2D_x(x^3)=3y^2x^2$.

For $f(x,y)=y+\ln x,\ \ f_x(x,y)=D_x(y+\ln x)=0+\dfrac{1}{x}=\dfrac{1}{x}$.

If $f(x,y)=x^2y+2x-y,\ \ f(2,y)=3y+4$, and the derivative of $f(2, y)$, defined by

$$\operatorname*{limit}_{h\to 0}\frac{f(2,y+h)-f(2,y)}{h}$$

is called the partial derivative of $f(x, y)$ with respect to y at $(2, y)$, and is denoted by $f_y(2, y)$. Thus

$$f_x(x,\ y)=D_y(3y+4)=3$$

> More generally, if $u = f(x, y)$ is a function of x and y, then the **partial derivative of $u = f(x, y)$ with respect to y at (x, y)** is defined by
>
> $$f_y(x,y)=\operatorname*{limit}_{h\to 0}\frac{f(x,y+h)-f(x,y)}{h}$$

The partial derivative $f_y(x,y)$, with respect to y, is obtained by regarding x as being held constant, thus generating a function of y, and differentiating the resulting function of y. Thus

For $f(x,\ y)=x^3y^2,\ f_y(x,\ y)=D_y(x^3y^2)=x^3D_y(y^2)=2x^3y$.

For $f(x,\ y)=y+\ln x,\ f_y(x,\ y)=D_y(y+\ln x)=1+0=1$.

There are a variety of notations in general use for partial derivatives. If $u = f(x, y)$, then other notations for $f_x(x,y)$ include

$$\frac{\partial u}{\partial x},\ \frac{\partial f(x,y)}{\partial x},\ \frac{\partial f}{\partial x},\ f_x,\ u_x(x,y),\ u_x$$

For $f_y(x,y)$, we have

$$\frac{\partial u}{\partial y},\ \frac{\partial f(x,y)}{\partial y},\ \frac{\partial f}{\partial y},\ f_y,\ u_y(x,y),\ u_y$$

The notations $f_x(x,y)$, $u_x(x,y)$, $f_y(x,y)$, and $u_x(x,y)$ are most appropriate when partial derivatives at specific ordered pairs of numbers are being considered. The partial derivatives of $u = f(x,\ y)$ with respect to *x* and with respect to *y* at $(2,-1)$, say, are denoted by

$$f_x(2,-1) \text{ and } f_y(2,-1)$$

respectively. The other notations are suitable when partial-derivative functions in general are under discussion.

EXAMPLE 1

For $f(x,\ y) = 2x^4y^4 + xy$, find $f_x(2,-1)$ and $f_y(2,-1)$.

Holding *y* constant and differentiating with respect to *x* yields

$$f_x(x,y) = D_x(2x^4y^2 + xy) = 8x^3y^2 + y$$

$$f_x(2,-1) = 8(2)^3(-1)^2 + (-1) = 63$$

Holding *x* constant and differentiating with respect to *y* yields

$$f_y(x,y) = D_y(2x^4y^2 + xy) = 8x^4y + x$$

$$f_y(2,-1) = 4(2)^4(-1) + 2 = -62$$

EXAMPLE 2

For $f(x,y) = \dfrac{x^2 + 2xy}{x+y}$, find $\dfrac{\partial f(x,y)}{\partial x}$ and $\dfrac{\partial f(x,y)}{\partial y}$.

Holding y constant and differentiating with respect to x yields

$$\frac{\partial f(x,y)}{\partial x}=\frac{(x+y)D_x(x^2+2xy)-(x^2+2xy)D_x(x+y)}{(x+y)^2}$$

$$=\frac{(x+y)(2x+2y)-(x^2+2xy)1}{(x+y)^2}$$

Holding x constant and differentiating with respect to y yields

$$\frac{\partial f(x,y)}{\partial y}=\frac{(x+y)D_y(x^2+2xy)-(x^2+2xy)D_y(x+y)}{(x+y)^2}$$

$$=\frac{(x+y)(0+2x)-(x^2+2xy)(0+1)}{(x+y)^2}=\frac{x^2}{(x+y)^2}$$

EXAMPLE 3

For $f(x, y)=x\cdot\ln y$, find $f_x(x, y)$ and $f_y(x, y)$.

Holding y constant and differentiating with respect to x yields

$$f_x(x, y)=D_x(x\cdot\ln y)=(\ln y)D_x(x)=\ln y$$

Holding x constant and differentiating with respect to y yields

$$f_y(x,y)=D_y(x\cdot\ln y)=xD_y(\ln y)=\frac{x}{y}$$

Partial derivatives for functions of more than two variables are defined analogously. If, for example, u is a function of three variables

$$u=f(x, y, z)$$

then a function of one variable is obtained when the other two are held constant. If x and y are held constant, then $u = f(x, y, z)$ reduces to a function of z. The derivative of this function of z, called the **partial derivative of $u = f(x, y, z)$ with respect to z**, is defined by

$$f_z(x,y,z) = \lim_{h \to 0} \frac{f(x,y,z+h) - f(x,y,z)}{h}$$

$f_x(x, y, z)$ and $f_y(x, y, z)$ are defined in a similar way.

EXAMPLE 4

For $f(x, y, z) = xyz^2 + e^{xz}$, find $f_x(x, y, z)$, $f_y(x, y, z)$ and $f_z(x, y, z)$.

Holding y and z constant and differentiating with respect to x yields

$$\begin{aligned} f_x(x, y, z) &= D_x(xyz^2) + D_x(e^{xz}) \\ &= yz^2 D_x(x) + e^{xz} D_x(xz) \\ &= yz^2 + ze^{xz} \end{aligned}$$

Holding x and z constant and differentiating with respect to y yields

$$\begin{aligned} f_y(x, y, z) &= D_y(xyz^2) + D_y(e^{xz}) \\ &= xz^2 D_y(y) + 0 = xz^2 \end{aligned}$$

Holding x and y constant and differentiating with respect to z yields

$$\begin{aligned} f_z(x, y, z) &= D_z(xyz^2) + D_z(e^{xz}) \\ &= xy\, D_z(z^2) + e^{xz} D_z(xz) \\ &= 2xyz + xe^{xz} \end{aligned}$$

EXERCISES

Find $f_x(x, y)$, $f_y(x, y)$, $f_x(1, 3)$ and $f_y(2, 4)$ *for the following functions.*

1. $f(x,y)=x^2y$
2. $f(x,y)=3x^4y-xy$
3. $f(x,y)=4x^3y-y^2$
4. $f(x,y)=\ln xy$
5. $f(x,y)=x^2e^y$
6. $f(x,y)=\sqrt{x^2+y}$
7. $f(x,y)=\dfrac{4x^2-xy}{x^2+y^2}$
8. $f(x,y)=e^x\ln y$

Find $f_x(x, y, z)$, $f_y(x, y, z)$, *and* $f_z(x, y, z)$ *for the following functions.*

9. $f(x,y,z)=\dfrac{4x^2+y}{z}$
10. $f(x,y,z)=xe^{yz}$
11. $f(x,y,z)=x^3yz^2$
12. $f(x,y,z)=x^2\sqrt{x^2+z^2}$
13. $f(x,y,z)=x^2\ln xyz$
14. $f(x,y,z)=\dfrac{x^3y+z}{y^2+z^2}$

14.3. CONCEPTS WITH A PARTIAL DERIVATIVE STRUCTURE

The following are a few of many concepts which are defined in terms of partial derivatives.

Marginal Productivity

If $q = f(x_1, x_2, \ldots, x_n)$ is a production function which expresses output q as a function of the productive factors $x_1, x_2, \ldots, x_n$, then $\frac{\partial q}{\partial x_1}$ is defined as the **marginal productivity of factor x_1** (or **marginal product of x_1**), $\frac{\partial q}{\partial x_2}$ is defined as the **marginal productivity of factor x_2**, and so on.

To illustrate, consider the Cobb-Douglas production function,

$$q = 1.01x^{0.75}y^{0.25}$$

used in a study of manufacturing in the United States (see Example 3, Section 14.1), where x is a measure of quantity of labor and y is a measure of quantity of capital. The marginal productivity of labor is

$$\frac{\partial q}{\partial x} = 1.01y^{0.25} \cdot \frac{d(x^{0.75})}{dx} = 0.7575x^{-0.25}y^{0.25}.$$

The marginal productivity of capital is

$$\frac{\partial q}{\partial y} = 1.01x^{0.75} \cdot \frac{d(y^{0.25})}{dy} = 0.2525x^{0.75}y^{-0.75}.$$

$\frac{\partial q}{\partial x}$ expresses the **instantaneous rate of change of output q with respect to labor**, assuming that capital is not changing; $\frac{\partial q}{\partial y}$ expresses the **instantaneous rate of change of output** q with respect to capital, assuming that labor is not changing.

Price Elasticity of Demand*

Suppose that the demand functions for n related commodities $A_1, A_2, \ldots, A_n$ in terms of their respective prices $p_1, p_2, \ldots, p_n$ and income y are

$$q_1 = f_1(p_1, p_2, \ldots, p_n, y)$$
$$q_2 = f_2(p_1, p_2, \ldots, p_n, y)$$
$$\vdots$$

* It might be useful to review price elasticity of demand in terms of one independent variable at this point. See Section 6.8, p. 205.

.

$$q_n = f_n(p_1, p_2, \ldots, p_n, y).$$

Then we have the following definitions.

$$\eta_{11} = -\frac{p_1}{q_1} \cdot \frac{\partial q_1}{\partial p_1}$$

is called the ***partial elasticity of demand for commodity A_1 with respect to its price p_1***.

$$\eta_{22} = -\frac{p_2}{q_2} \cdot \frac{\partial q_2}{\partial p_2}$$

is called the ***partial elasticity of demand for commodity A_2 with respect to its price p_2***, and so on. These concepts are direct extensions of ordinary demand elasticity in which demand is regarded as a function of only the price of the good concerned.

Consider the demand function obtained by Henry Schultz for beef (see Example 2, Section 14.1.)

$$q_1 = 3.4892 - 0.0899p_1 + 0.0637p_2 + 0.0187y$$

where p_1 is the price of beef, p_2 is the price of pork, and y is a measure of income. We obtain

$$\eta_{11} = -\frac{p_1}{q_1} \cdot \frac{\partial q_1}{\partial p_1} = -\frac{p_1}{q_1}(-0.0899) = \frac{(0.0899)p_1}{q_1}$$

as the partial elasticity of demand for beef with respect to its price.

η_{11} corresponds to point elasticity of demand for the price of the commodity itself, η, discussed in Section 6.8 (p. 205). In both cases the underlying assumption is that the prices of related commodities are not changing.

Price Cross-Elasticity of Demand

Again, suppose that the demand functions for n related commodities $A_1, A_2, \ldots, A_n$ in terms of their respective prices $p_1, p_2, \ldots, p_n$ and income y are

$$
\begin{aligned}
q_1 &= f_1(p_1, p_2, \ldots, p_n, y) \\
q_2 &= f_2(p_1, p_2, \ldots, p_n, y) \\
&\vdots \\
q_n &= f_n(p_1, p_2, \ldots, p_n, y).
\end{aligned}
$$

Partial elasticities of demand for one commodity with respect to the price of another commodity are defined as follows.

$\eta_{12} = \frac{p_2}{q_1} \cdot \frac{\partial q_1}{\partial p_2}$	The partial elasticity of demand of A_1 with respect to the price of A_2.
$\eta_{13} = \frac{p_3}{q_1} \cdot \frac{\partial q_1}{\partial p_3}$	The partial elasticity of demand of A_1 with respect to the price of A_3.
$\eta_{21} = \frac{p_1}{q_2} \cdot \frac{\partial q_2}{\partial p_1}$	The partial elasticity of demand of A_2 with respect to the price of A_1.

More generally, the ***partial elasticity of demand of good A_i with respect to the price of good A_j*** is defined by

$$\eta_{ij} = \frac{p_j}{q_i} \cdot \frac{\partial q_i}{\partial p_j}$$

From Henry Schultz's demand function for beef,

$$q_1 = 3.4892 - 0.0899 p_1 + 0.0637 p_2 + 0.0187 y$$

we obtain

$$\eta_{12} = \frac{p_2}{q_1} \cdot \frac{\partial q_1}{\partial p_2} = \frac{p_2}{q_1}(0.0637) = \frac{0.0637 p_2}{q_1}$$

as the partial elasticity of demand of beef with respect to the price of pork.

Price cross-elasticities of demand for a commodity are quantitative measures of changes in demand for the commodity to changes in price of related commodities.

Income Elasticity of Demand

Once again, suppose that the demand functions for n related quantities $A_1, A_2, \ldots, A_n$ in terms of their respective prices p_1, $p_2, \ldots, p_n$ and income y are

$$\begin{aligned} q_1 &= f_1(p_1, p_2, \ldots, p_n, y) \\ q_2 &= f_2(p_1, p_2, \ldots, p_n, y) \\ &\vdots \\ q_n &= f_n(p_1, p_2, \ldots, p_n, y). \end{aligned}$$

Then the ***income elasticity of demand for commodity A_1*** is defined as

$$\eta_{1y} = \frac{y}{q_1} \cdot \frac{\partial q_1}{\partial y}.$$

The ***income elasticity of demand for commodity A_2*** is defined as

$$\eta_{2y} = \frac{y}{q_2} \cdot \frac{\partial q_2}{\partial y}$$

and so on.

From Henry Schultz's demand function for beef,

$$q_1 = 3.4892 - 0.0899 p_1 + 0.0637 p_2 + 0.0187 y$$

we obtain

$$\eta_{1y} = \frac{y}{q_1} \cdot \frac{\partial q_1}{\partial y} = \frac{y}{q_1} \cdot (0.0187) = \frac{0.0187 y}{q_1}$$

as the income elasticity of demand for beef.

Income elasticities of demand for related commodities are quantitative measures of changes in demand for the commodities to changes in consumer income.*

EXERCISES

1. The demand functions for beef and pork in Bell City are

$$q_1 = 5 - p_1 + 1.2 p_2 ,$$
$$q_2 = 6 + 2 p_1 - 1.1 p_2 ,$$

where q_1 and q_2 express the amount demanded (in millions of pounds per week) of beef and pork, respective, and p_1 and p_2 are the unit prices (in dollars per pound) of beef and pork, respectively. For p_1 = 2.00 and p_2 = 1.50 find:

   (a) The partial elasticity of demand of beef with respect to its price. How is this result interpreted?
   (b) The partial elasticity of demand of beef with respect to the price of pork. How is this result interpreted?

2. The production function that expresses the output of steel is

* For further discussion of these concepts, see, for example, C. E. Ferguson, *Microeconomic Theory*, rev. ed. (Homewood, Illinois: Richard D. Irwin, Inc., 1969), Chapter 4; R. G. D. Allen, *Mathematical Analysis for Economists* (New York: Macmillan Company, 1938), Section 12.6.

$$f(x, y) = 60x^{2/3}y^{1/3},$$

where x is in units of labor and y is in units of capital. Find the marginal productivities with respect to labor and capital.

3. The demand functions for beef, pork, and lamb in Ralph City are

$$q_1 = 5 - p_1 + 1.2p_2 + 2p_3$$
$$q_2 = 6 + 2p_1 - 1.1p_2 + 1.5p_3$$
$$q_3 = 4 + p_1 + 0.9p_2 - 2p_3,$$

where q_1, q_2, and q_3 express the amount demanded (in millions of pounds per week) of beef, pork, and lamb, respectively, and p_1, p_2, and p_3 are their respective unit prices (in dollars per pound). For $p_1 = 2.00$, $p_2 = 1.50$ and $p_3 = 2.50$ find and state an interpretation for:

(a) The partial elasticity of demand for beef with respect to its price.
(b) The partial elasticity of demand for beef with respect to the price of pork.
(c) The partial elasticity of demand for beef with respect to the price of lamb.
(d) The partial elasticity of demand for pork with respect to the price of lamb.

14.4. HIGHER ORDER PARTIAL DERIVATIVES

Analogous to higher order derivatives of functions of one variable are higher-order partial derivatives. For example, the second partial derivatives of $u = f(x, y)$ are defined as follows:

$$\frac{\partial}{\partial x}\left(\frac{\partial u}{\partial x}\right) = \frac{\partial^2 u}{\partial x^2} = \frac{\partial^2 f}{\partial x^2} = u_{xx}(x, y) = f_{xx}(x, y)$$

$$\frac{\partial}{\partial y}\left(\frac{\partial u}{\partial y}\right)=\frac{\partial^2 u}{\partial y^2}=\frac{\partial^2 f}{\partial y^2}=u_{yy}(x,y)=f_{yy}(x,y)$$

$$\frac{\partial}{\partial x}\left(\frac{\partial u}{\partial y}\right)=\frac{\partial^2 u}{\partial x\partial y}=\frac{\partial^2 f}{\partial x\partial y}=u_{yx}(x,y)=f_{yx}(x,y)$$

$$\frac{\partial}{\partial y}\left(\frac{\partial u}{\partial x}\right)=\frac{\partial^2 u}{\partial y\partial x}=\frac{\partial^2 f}{\partial y\partial x}=u_{xy}(x,y)=f_{xy}(x,y)$$

It is worthy of note that the subscripts are reversed when taking higher-order partial derivatives in one notation, but not in the other. For example,

$$\frac{\partial}{\partial x}\left(\frac{\partial u}{\partial y}\right)=\frac{\partial^2 u}{\partial x\partial y}=u_{yx}(x,y)$$

EXAMPLE 5

For $f(x, y)=x \cdot \ln y$, find $f_{yy}(x, y)$, $f_{yx}(x, y)$, and $f_{xy}(x, y)$.

From Sec. 14.2, Example 3, we have

$$f_y(x,y)=\frac{x}{y},\quad f_x(x,y)=\ln y$$

Thus

$$f_{yy}(x,y)=D_y\left(\frac{x}{y}\right)=xD_y\left(\frac{1}{y}\right)=-\frac{x}{y^2}$$

$$f_{yx}(x,y)=D_x\left(\frac{x}{y}\right)=\frac{1}{y}D_x(x)=\frac{1}{y}$$

$$f_{xy}(x,y)=D_y(\ln y)=\frac{1}{y}$$

Observe that $f_{xy}(x, y) = f_{yx}(x, y) = 1/y$

More generally, it can be shown that if the partial derivatives of $u = f(x, y)$ satisfy suitable conditions, then

$$f_{xy}(x, y) = f_{yx}(x, y)$$

Third and still higher-order partial derivatives are defined analogously.

EXERCISES

1. Find all second partial derivatives of $f(x, y) = 4x^2y + 2y$ and $f_{xx}(-2,1)$, $f_{yy}(-1,3)$, and $f_{xy}(3, 2)$.
2. Find all second partial derivatives of $f(x,y) = xy^2 - 3x^2y$ and $f_{xx}(-1,4)$, $f_{yy}(-3,1)$, and $f_{xy}(3, 4)$.
3. If $f(x,y) = x^2 - 4xy^2$, find $4xf_{xx}(x, y) + f_{yy}(x, y)$.
4. Find all second partial derivatives of $f(x, y, z) = 3xy^2z^3$.

14.5. OPTIMIZATION PROBLEMS

Definition of absolute and local maximum and minimum values are analogous to those given for one-variable functions. A function $u = f(x, y)$ is said to have an **absolute minimum value at (a, b)** if $f(a, b) \leq f(x, y)$ for all (x, y) in the domain of definition of $u = f(x, y)$. The value $f(a, b)$ itself is called the **absolute minimum value**. $f(a, b)$ is said to be a **local minimum value** if $f(a, b) \leq f(x, y)$ for all (x, y) in some circle centered at (a, b) which are also in the domain of $u = f(x, y)$. By inspection it is clear that the function

$$f(x, y) = x^2 + y^2 + 5$$

has an absolute minimum value at (0, 0), and that this minimum value is $f(0, 0) = 5$. For any ordered pair (x, y) other than (0, 0),

x^2+y^2 is positive, and x^2+y^2+5 exceeds 5. The definitions of absolute and local maximum value are obtained by replacing ≤ by ≥ in the preceding definitions. The maximum and minimum values of a multivariable function, both local and absolute, are called its **extreme values**.

Analogous to critical values for one-variable functions are critical points for multivariable functions, points in the domain of definition of a multivariable function that may give rise to extreme values. Critical points (a, b) for a function of two variables $u=f(x, y)$ are found where boundary curves or lines of its domain of definition intersect and at interior points of the domain for which both partial derivatives $f_x(a, b)$ and $f_y(a, b)$ are zero. We shall restrict our attention to this latter situation. To obtain such critical points of $u=f(x, y)$, determine the partial derivatives $f_x(x, y)$ and $f_y(x, y)$, set them equal to zero, and solve for x and y. To illustrate, let us return to

$$f(x, y)=x^2+y^2+5$$

The first partial derivatives are

$$f_x(x, y)=2x, \qquad f_y(x, y)=2y$$

Setting these partial derivatives equal to zero yields

$$2x=0, \qquad 2y=0$$

$$x=0, \qquad y=0$$

Thus (0, 0) is a critical point, which comes as no surprise. As we have observed, it is clear from the nature of $f(x, y)=x^2+y^2+5$ that $f(0, 0)=5$ is an absolute minimum value.

More generally, the following theorem states a condition that enables us to distinguish critical points which yield extreme values from those which do not.

Theorem. Let $u = f(x, y)$ denote a function that is defined and "well behaved" in some circular region centered at the critical point (a, b) $[f_x(a, b) = 0, \quad f_y(a, b) = 0]$. Consider

$$A = [f_{xx}(a,b)][f_{yy}(a,b)] - [f_{yx}(a,b)]^2$$

1. If $A > 0$ and $f_{xx}(a, b) > 0$, then $f(a, b)$ is a local minimum value.
2. If $A > 0$ and $f_{xx}(a, b) < 0$, then $f(a, b)$ is a local maximum value.
3. If $A < 0$, then $f(a, b)$ is not an extreme value.
4. If $A = 0$, then the test is inconclusive and no conclusion can be drawn.

By $u = f(x, y)$ is "well behaved" in some circular region centered at (a, b), we mean that certain technical conditions, which we shall not describe here, must be satisfied for the theorem to be applicable. These conditions are satisfied by all function considered in this section.

To illustrate, we again return to

$$f(x, y) = x^2 + y^2 + 5$$

for which $(0, 0)$ is a critical point, and $f(0, 0) = 5$ is the absolute maximum value. The first partial derivatives are

$$f_x(x, y) = 2x, \qquad f_y(x, y) = 2y$$

from which we obtain the following second partial derivatives:

$$f_{xx}(x, y) = 2, \qquad f_{yy}(x, y) = 2, \qquad f_{yx}(x, y) = 0$$

At the critical point $(0, 0)$, we have

$$f_{xx}(0, 0) = 2, \qquad f_{yy}(0, 0) = 2, \qquad f_{yx}(0, 0) = 0$$

Thus

$$A = [f_{xx}(0,0)][f_{yy}(0,0)] - [f_{yx}(0,0)]^2$$
$$= 2(2) - (0)^2 = 4$$

Since $A = 4$ is positive and $f_{xx}(0, 0) = 2$ is positive, it follows from part 1 of our theorem that $f(0, 0) = 5$ is a local minimum value. The nature of the function allows us to state the stronger result that $f(0, 0) = 5$ is an absolute minimum value.

EXAMPLE 1

Determine the extreme values of $f(x, y) = 2x^3 + x^2y^2 - 24x + 5$.

To obtain the critical points of this function, we determine $f_x(x, y)$ and $f_y(x, y)$, set them equal to zero, and solve for x and y.

$$f_x(x, y) = 6x^2 + 2xy^2 - 24, \quad f_y(x, y) = 2x^2y$$

Setting these partial derivatives equal to zero and solving for x and y yields

$$6x^2 + 2xy^2 - 24 = 0 \qquad (14.1)$$
$$2x^2y = 0 \qquad (14.2)$$

From Equation (14.2) we obtain $x = 0$ or $y = 0$. Substituting $x = 0$ into Equation (14.1) yields $-24 = 0$. Thus $x = 0$ is not possible. Substituting $y = 0$ into Equation (14.1) yields

$$6x^2 - 24 = 0, \quad x^2 = 4, \quad x = -2 \text{ and } 2$$

Thus $(-2, 0)$ and $(2, 0)$ are critical points. To determine whether or not $f(-2, 0)$ and $f(2, 0)$ are extreme values, we must find A for $(-2, 0)$ and $(2, 0)$. The second partial derivatives of $f(x, y)$ are

$$f_{xx}(x, y) = 12x + 2y^2, \qquad f_{yy}(x, y) = 2x^2, \qquad f_{yx}(x, y) = 4xy$$

For $(-2,0)$, we have

$$f_{xx}(-2,0) = -24, \quad f_{yy}(-2,0) = 8, \quad f_{yx}(-2,0) = 0$$

Thus

$$\begin{aligned} A &= [f_{xx}(-2,0)][f_{yy}(-2,0)] - [f_{yx}(-2,0)]^2 \\ &= (-24)8 - (0)^2 = -192 \end{aligned}$$

Since $A = -192$ is negative, by part 3 of our theorem, $f(-2,0)$ is not an extreme value.

For (2, 0) we have

$$f_x(2,\ 0) = 24, \quad f_{yy}(2,0) = 8, \quad f_{yx}(2,\ 0) = 0$$

Thus

$$\begin{aligned} A &= [f_{xx}(2,0)][f_{yy}(2,0)] - [f_{yx}(2,0)]^2 \\ &= 24(8) - (0)^2 = 192 \end{aligned}$$

Since A = 192 is positive and $f_{xx}(2,\ 0) = 24$ is positive, it follows from part 1 of our theorem that $f(2,0) = -27$ is a local minimum value.

The Method of Lagrange Multipliers

Many situations of interest lead to the problem of maximizing or minimizing a multivariable function whose variables are required to satisfy a prescribed equation. The method of Lagrange multipliers, whose fundamentals we take up here, is a powerful general technique for solving such problems.

To illustrate, consider the following area maximization problem. A rectangular playground with a perimeter of 3000 feet is to be constructed. What are the dimensions of the region with largest area? If we let x and y denote the dimensions of the

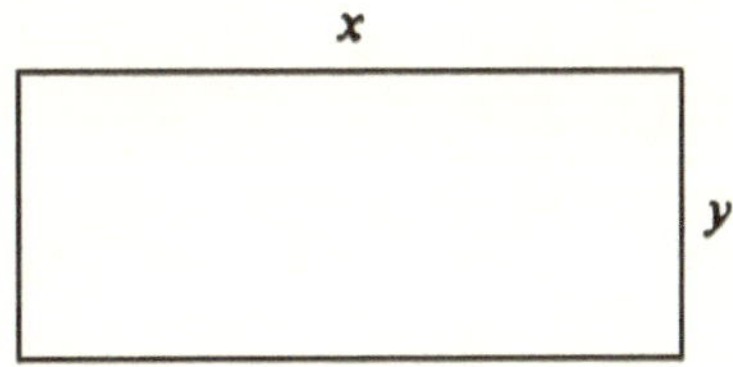

Figure 14.1

playground to be constructed (see Figure 14.1), then the area function to be maximized is

$$A(x, y) = xy$$

subject to the perimeter constraint $2x + 2y = 3000$, or, equivalently,

$$x + y = 1500$$

One approach is to reduce the problem to an equivalent optimization problem involving a function of one variable y using the constraint equation to express one variable in terms of the other. Solving $x + y = 1500$ for y in terms of x yields

$$y = 1500 - x$$

Substituting 1500 - x for y in our expression for area yields the area function

$$F(x) = 1500x - x^2$$

from which we obtain $x = 750$ as yielding the maximum are value. From $y = 1500 - x$, we have $y = 1500 - 750 = 750$.

The success of this strategy presupposes that the constraint equation can be solved for one variable in terms of the other, and that it is not too difficult to do so.

Apart from its generality, a major advantage of the Lagrange multiplier method lies in its freedom from this restriction. Step 1 in applying the Lagrange multiplier method is to express the constraint equation in the form $g(x, y) = 0$. For our perimeter constraint, we have

$$g(x, y) = x + y - 1500 = 0$$

We next introduce a parameter L, called a **Lagrange multiplier**, and define a three-variable function $H(x, y, L)$ of x, y and L in terms of the given function $A(x, y)$, the constraint function $g(x, y)$, and L as follows:

$$\boxed{H(x, y, L) = A(x, y) - L \cdot g(x, y)}$$

For our area problem, we have

$$\begin{aligned} H(x, y, L) &= xy - L(x + y - 1500) \\ &= xy - Lx - Ly + 1500L \end{aligned}$$

The function $H(x, y, L)$ is called the **Lagrangian function**, and the method of Lagrange multipliers reduces the problem of obtaining the extreme values of $A(x, y)$ subject to the constraint $g(x, y) = 0$ to the problem of determining the extreme values of the Lagrangian function without constraints. Thus our next step is to apply the techniques discussed earlier in this section to the Lagrangian function $H(x, y, L)$; that is, determine the partial derivatives $H_x(x, y, L)$, $H_y(x, y, L)$, and $H_L(x, y, L)$, set them equal to zero, and solve for x and y. For our area problem, we have

$$\begin{aligned} H_x(x, y, L) &= y - L \\ H_y(x, y, L) &= x - L \\ H_L(x, y, L) &= -x - y + 1500 \end{aligned}$$

Setting these partial derivatives equal to zero yields

$$y - L = 0 \qquad (14.3)$$

$$x - L = 0 \qquad (14.4)$$

$$-x - y + 1500 = 0 \qquad (14.5)$$

To solve these equations for x and y, we first solve the first two equations for L in terms of x and y, respectively. This yields

$$y = L, \qquad x = L$$

Thus

$$y = x$$

Substituting x for y in Equation (14.5) yields

$$-2x + 1500 = 0, \quad x = 750$$

By substituting $x = 750$ into Equation (14.5), we obtain $y = 750$. Does (750, 750) yield a maximum value for our area function? From our earlier solution of this problem we know that the answer is yes, but looking at the situation from the point of view of the Lagrange multiplier technique, a word of caution is in order.

The method of Lagrange multipliers does not by itself distinguish critical values that yield extreme values from those which do not, nor does it distinguish maximum values from minimum values. Other methods, not taken up here, must be used to make such distinctions. However, for many problems the lack of such methods is not a severe drawback, since the nature of the problem itself allows one to draw conclusions about the kind of critical point or extreme value.

Let us observe that in taking the partial derivative of the Lagrangian function

$$H(x, y, L) = xy - Lx - Ly + 1500L$$

with respect to L and setting it equal to zero to obtain Equation (14.5),

$$-x - y + 1500 = 0$$

we merely reproduced the constraint equation

$$g(x, y) = x + y - 1500 = 0$$

Such is the case in general, as we now show. Consider the Lagrangian function

$$H(x, y, L) = A(x, y) - L \cdot g(x, y)$$

Differentiating with respect to L yields

$$H_L(x, y, L) = 0 - g(x, y) = -g(x, y)$$

Setting $H_L(x, y, L)$ equal to zero gives us

$$-g(x, y) = 0$$

Multiplying by -1 yields

$$g(x, y) = 0$$

which is our constraint equation. Thus we need not differentiate the Lagrangian function with respect to L and set the result equal to zero to obtain the final partial derivative equation; it suffices to reproduce the constraint equation.

EXAMPLE 2

The following problem, which we examine from the point of view of the Lagrange multiplier technique is presented and analyzed from another point of view in Section 7.1, Example 6 (p. 229). The director of the studies-abroad program at Ecap University is planning to have 5000 copies of a poster describing his forthcoming program in Rome and Athens printed. The type plate containing the printed material is to be 96 square inches, the top and bottom margins are each to be 1 inch, and the side margins are each to be $1\frac{1}{2}$ inches. An all-out effort to advertise the program is being made, and the poster is to be printed on high-quality paper.

Of interest is the question of what the dimensions of the type plate should be so as to use the least amount of paper.

Let x and y denote the length and width, respectively, of the type plate (see Figure 14.2). The length of the poster is $x + 2$ inches and its width is $y + 3$ inches.

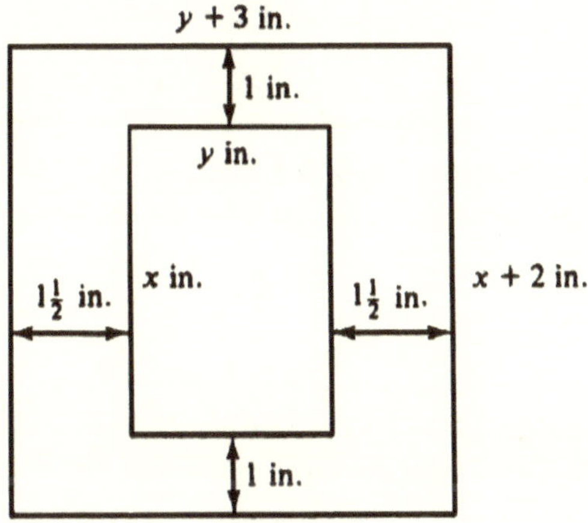

Figure 14.2

Thus the area function to be minimized is

$$\begin{aligned} A(x, y) &= (x + 2)(y + 3) \\ &= xy + 2y + 3x + 6 \end{aligned}$$

subject to the condition that the area of the type plate is 96 square inches. This area condition is expressed by the constraint

$$g(x, y) = xy - 96 = 0$$

The Lagrangian function in this situation is defined by

$$\begin{aligned} H(x, y, L) &= A(x, y) - L \cdot g(x, y) \\ &= xy + 2y + 3x + 6 - L(xy - 96) \\ &= xy + 2y + 3x + 6 - Lxy + 96L \end{aligned}$$

Taking the partial derivatives of $H(x, y, L)$ with respect to x and y, setting them equal to zero, and restating the constraint equation yields

$$y+3-Ly=0 \tag{14.6}$$

$$x+2-Lx=0 \tag{14.7}$$

$$xy-96=0 \tag{14.8}$$

Solving Equation (14.6) for L yields

$$L=\frac{y+3}{y} \tag{14.8}$$

Solving Equation (14.7) for x yields

$$x-xL=-2, \quad x(1-L)=-2$$

$$x=\frac{-2}{1-L} \tag{14.9}$$

Substituting $L = (y + 3)/y$, stated in Equation (14.8), into Equation (14.9 and simplifying yields

$$x=\frac{-2}{1-\frac{y+3}{y}}=\frac{-2}{\frac{y}{y}-\frac{(y+3)}{y}}=\frac{-2}{\frac{y-y-3}{y}}$$

$$=\frac{-2}{\frac{-3}{y}}=-2\cdot\left(\frac{-y}{3}\right)=\frac{2}{3}y$$

Substituting $\frac{2}{3}y$ for x in Equation (14.8), yields

$$\frac{2}{3}y^2 - 96 = 0, \quad y^2 = 144, \quad y = 12$$

Substituting 12 for y in $xy - 96 = 0$ yields $x = 8$. Thus we obtain $x = 8$, $y = 12$ as the dimensions of the poster of least area that satisfies the given conditions. This finding agrees with the solution obtained in Section 7.1, Example 6 (p. 229).

EXAMPLE 3

The profit function of the Richmond Company, a producer of coffee and tea, is

$$P(x, y) = 50x + 30y - x^2 - y^2$$

where x is output of coffee in tons per week, y is output of tea in tons per week, and $P(x, y)$ is profit in dollars per week. If total output is 40 tons per week, what output levels maximize profit?

The problem then is to maximize $P(x, y)$ subject to the constraint

$$g(x, y) = x + y - 40 = 0$$

The Lagrangian function is

$$H(x, y, L) = P(x, y) - L \cdot g(x, y)$$

$$= 50x + 30y - x^2 - y^2 - Lx - Ly + 40L$$

Taking the partial derivatives of the Langrangian function with respect to x and y yields

$$H_x(x, y, L) = 50 - 2x - L$$

$$H_y(x, y, L) = 30 - 2y - L$$

Setting these partial derivatives equal to zero and appending the constraint equation gives us

$$50-2x-L=0 \quad (14.10)$$
$$30-2y-L=0 \quad (14.11)$$
$$x+y-40=0 \quad (14.12)$$

By solving Equations (14.10) and (14.11) for L, we obtain

$$L=50-2x, \quad L=30-2y$$

Setting these expressions for L equal to each other and solving for y in terms of x yields

$$30-2y=50-2x$$
$$-2y=20-2x$$
$$y=x-10$$

Substituting $x-10$ for y in Equation (14.12) and solving for x gives us

$$x+x-10-40=0$$
$$2x=50$$
$$x=25$$

Substituting 25 for x in $x+y-40=0$ and solving for y yields y = 15. Thus we have

$$x=25, \quad y=15$$

In the absence of techniques to test for extreme values, it can be argued that x = 25, y = 15 yields a maximum value for $P(x,y)=50x+30y-x^2-y^2$, since profit functions, by their nature, have a unique extreme value that is a maximum value.

EXERCISES

Examine the following functions for extreme values.

1. $f(x,y) = -x^2 + y^2 + 6$
2. $f(x,y) = x^2 - xy + 2y^2 + 3$
3. $f(x,y) = x^2 + 6y^2 - 8x$
4. $f(x,y) = x^3 - 3x - y^2 + 10$
5. $f(x,y) = 2x^2 - 3y^2$
6. $f(x,y) = 3x^2 - xy + 2y^2$
7. $f(x,y) = x^2 + 4xy + 2y^2 - 8y + 4x$
8. $f(x,y) = x^3 - 2xy^2 + x^2$
9. Find the minimum value of $f(x, y) = x^2 + y^2$ subject to the constraint $x + y = 4$.
10. Find the minimum value of $f(x, y) = x^2 + 4y + 12$ subject to the constraint $x + y = 6$.
11. Find the maximum value of $f(x,y) = 12x - x^2 + 4y - y^2$ subject to the constraint $x + y = 8$.
12. Find the maximum value of $f(x,y) = 8x - 2x^2 + 8y - y^2$ subject to $x + y = 9$.
13. Find the minimum value of $f(x,y) = x^2 + y^2 - 2xy$ subject to $x + 2y = 12$.
14. Find the minimum value of $f(x, y) = x^2 + y^2$ subject to $xy = 4$.
15. The profit function of the Jensen Coal Company is $P(x,y) = 400x + 200y - x^2 - y^2$, where x is output of bituminous coal in tons per day, y is output of anthracite coal in tons per day, and $P(x, y)$ is profit in dollars per day. If the total output is 300 tons per day, for what output levels is profit maximized?

SELF-TESTS FOR CHAPTERS 1-14

Allow 90 or so minutes for each self-test. Go over each one before proceeding to the next.

Self-Test 1

1. $c(x) = \sqrt{1500x + 5000}$ is the cost function of a firm. Find the marginal cost function.

2. A producer has revenue and cost functions $R(x) = 5000x - 3x^2$, $c(x) = x^2 + 16x + 2000$ where output is in tons per week. Find the optimal output level.
3. With respect to the conditions described in (2) what would the optimal output level be if a lump-sum tax of \$1500 per week were imposed?
4. With respect to the conditions described in (2) what would the optimal output level be if an output tax of \$416 per ton were imposed?
5. With respect to the conditions described in (2) suppose that an output tax of t dollars per ton were imposed. State the function that describes tax revenue in terms of t.
6. With respect to (5), find t for which the tax revenue function is maximized.
7. For $f(x) = e^{3x^2+2}$, find $f'(x)$.
8. State the definition of $f'(2)$ for $y = f(x)$.
9. Find the tangent line to the graph of $f(x) = x^2 + 2x + 1$ at $P(2,9)$.
10. Find $\int (x^3 + x - 2)dx$
11. Find $\int x^3(x^4 + 1)^5 dx$
12. Find the derivative of the inverse of $y = 2x^3 + x$.
13. Find the critical values of $f(x) = \frac{1}{3}x^3 + \frac{3}{2}x^2 - 10x - 12$.
14. For $f(x) = x^3 \cdot \ln x$ find $f'(x)$.
15. Find $f''(2)$ for $f(x) = \ln x$.
16. Find, if it exists, $\lim_{x \to 6} \frac{2x^2 - 72}{x - 6}$. If this limit does not exist or is ∞ or $-\infty$, so state. Explain the basis for your conclusion.

Self-Test 2

In questions (1) through (3) find, if it exists, the limit required. If the limit does not exist or is infinity or -∞, so state. Explain the basis for your conclusion.

1. $\lim\limits_{x \to 3} f(x)$, where $f(x) = \begin{cases} 3x+5, \text{ for } x<3 \\ 8, \text{ for } x=3 \\ 4x-1, \text{ for } x>3 \end{cases}$

2. $\lim\limits_{x \to 2} \dfrac{12}{(x-2)^2}$

3. $\lim\limits_{x \to \infty} \dfrac{5x^3 - 3x^2 + 1}{7x^3 + 2x - 4}$

4. Is the function $f(x)$ defined in (1) continuous at 3? Answer YES or NO and explain your answer in terms of the *definition* of continuity.

In questions (5) through (7) find the derivatives of the functions given

5. $y = 2e^{4x^2+3}$
6. $y = \ln(5x^3 - 3x)$
7. $y = (3x^2 + 3)e^x$
8. Determine and simplify $g(h) = \dfrac{f(x+h) - f(x)}{h}$, where $f(x) = 4x^2 - 2x + 1$.
9. Find, if it exists, $\lim\limits_{h \to 0} g(h)$ from (8). If this limit exists, what is it called?
10. Find $f''(x)$ for $f(x) = 3x^2 + \dfrac{1}{x}$.
11. Define $f'(5)$ for $y = f(x)$.
12. Define $\int_a^b f(x)dx$.
13. Find the critical values of $f(x) = \dfrac{1}{3}x^3 + \dfrac{1}{2}x^2 - 20x$.
14. Find and identify all extreme values of the function given in (13).
15. The revenue and cost functions of a firm are

$$R(x) = 590x - 2x^2, \quad c(x) = \frac{1}{2}x^2 + 10x + 500$$

where x is output in tons per week. Find the optimal output level.

16. Four the situation described in (15) suppose that an output tax of t dollars per ton were imposed. State the function which describes tax revenue in terms of t.
17. For the situation described in (16) find the output tax t for which tax revenue is maximized.
18. Define e.
19. Find the tangent line to the graph of $y = 2x^2 + 2x$ where $x = 3$.
20. For what x is $f(x) = \dfrac{x-1}{3x+1}$ not defined? Explain.

In questions (21) through (23) determine the integrals stated.

21. $\int (2x^2 + x)dx$
22. $\int \dfrac{6xdx}{(3x^2+1)^4}$
23. $\int_0^2 (x^2 - 3x + 1)dx$
24. Find the area of the region bounded by $y = 9 - x^2$ and the x-axis.
25. Income from an investment is obtained at the constant rate of \$6000 per year for 8 years. With respect to an interest rate of 10% per annum compounded continuously find the present value of this income stream. Values involving e may be left as such.

Self-Test 3

In (1) through (13) state whether you agree or disagree with the statement made. Explain the basis for your conclusion.

1. The domain of $f(x) = 1/(x-3)$ is all real numbers except $x = 0$.
2. $f(x) = \dfrac{x-1}{x-1}$ and $g(x) = 1$ are equal functions.

3. Even if $f(3) = 0$, $\lim_{x \to 3} 1/f(x)$ could exist.
4. If $f(x)$ gets closer and closer to L as x gets closer and closer to a, then $f(a) = L$.
5. It is possible for $f(a)$ to exist and $\lim_{x \to a} f(x)$ to exist but $f(x)$ not be continuous at $x = a$.
6. If a function $y = f(x)$ is continuous at a it must have a derivative at a.
7. If a function $y = f(x)$ has a derivative at a, then it must be continuous at a.
8. If $f(a)$ is not defined, then $\lim_{x \to a} f(x)$ does not exist.
9. If $\lim_{x \to a} f(x)$ does not exist, then $f(a)$ is not defined.
10. If $f'(a) = 0$ and $f''(a) = 0$, then $y = f(x)$ has a maximum value at a.
11. If $F(x)$ is an indefinite integral of $y = f(x)$, then $\int_a^b f(x)dx = F(b) - F(a)$.
12. If $\lim_{x \to a} f(x) = \lim_{x \to a} g(x) = 0$, then $\lim_{x \to a} \frac{f(x)}{g(x)}$ does not exist.
13. If $h(x) = f(x) + g(x)$ and $\lim_{x \to a} h(x) = L$, then the components $f(x)$ and $g(x)$ must also have limits as $x \to a$.
14. If a car rental service charges $20 per day plus 10 cents for each mile over 100 (i.e., the first 100 miles are free), find an expression for the total daily cost as a function of x, the total number of miles driven.
15. Find $f_x(x, y, z)$ for $f(x, y, z) = 2(y^3 z)\, e^{x^2 y}$
16. The demand functions for coffee and tea in Matt City are

$$q_1 = 4 - p_1 + p_2$$
$$q_2 = 5 + 2p_1 - 2p_2$$

where q_1 and q_2 express the amount demanded (in thousands of pounds per week) of coffee and tea, respectively, and p_1 and

p_2 are the unit prices (in dollars per pound) of coffee and tea, respectively.

Consider the price levels $p_1 = 2.00$ and $p_2 = 1.50$.

(a) Find the partial elasticity of demand of coffee with respect to the price of coffee.
(b) How is the value obtained in part (a) interpreted?
(c) Find the partial elasticity of demand of coffee with respect to the price of tea.
(d) How is the value obtained in part (c) interpreted?

17. Find the critical points and extreme values of

$$f(x,y) = 2x^3 + 3x^2 - 12x + y^2 - y + 2.$$

Use the second derivative test to justify your results.

18. Suppose x units for labor and y units of capital can produce

$$f(x,y) = 20x^{4/5}y^{1/5}$$

units of a certain product. Each unit of labor costs 32 dollars, each unit of capital costs 100 dollars, and 10,000 dollars is available to spend on production. How many units of labor and how many units of capital should be used so as to maximize production?

Self-Test 4

1. Money is invested at the continuous rate of $f(t) = 8000 + 20t$ dollars per year at time t at an interest rate of 15% per annum compounded continuously. Find the total sum accumulated at the end of 5 years.
2. Find the present value of a perpetual income stream flowing at the constant rate of \$1500 per year at time t, where the interest rate is 15% per annum compounded continuously.

3. A piece of production equipment acquired by a firm is assumed to have a useful life of 10 years. It is further assumed that net income is derived at the continuous rate of

$$f(t) = 5000 - 20t$$

dollars per year t years after installation. The underlying interest rate is 20% per annum compounded continuously. The cost of the machinery after installation is $1000 and the estimated salvage value at the end of 10 years is $200. Find the present value of this piece of production equipment.
4. $F(x,y) = 60\, x^{3/4}\, y^{1/4}$ units of a certain product can be produced by employing x units of labor and y units of capital. Each unit of labor costs $100 whereas each unit of capital costs $200. The amount available to spend on production is $30,000. Find the number of units of labor and capital that should be utilized to maximize production.
5. Determine the extreme values of $f(x,y) = 2x^4 + y^2 - x^2 + 2y$. Employ the second derivative test to justify your results.
6. State the function that describes the amount that must be initially invested at a rate of 9 percent per annum compounded continuously if $4000 is to be available at the end of x years. How much must be initially invested if $4000 is to be available at the end of 3 years?
7. Find the extreme values of $f(x) = x^3 + \frac{11}{2}x^2 - 4x + 5$.
8. Find the extreme values of $f(x) = xe^{-x}$.
9. Sketch the graph of $f(x) = x^3 + 6x$.
10. Sketch the graph of $f(x) = xe^{-x}$.
11. Define e.

Self-Test 5

1. A chain of dairy-product outlets has 30 stores in a certain region, each doing an average of $10,000 worth of business per day. Studies indicate that, if new stores are opened in the region, the amount of business done by each store will drop

by \$200 per day for each new store that is opened. How many new stores should be opened if the total average amount of business done per day is to be maximized?

2. An information pamphlet describing the programs offered by the history department of Ecap University is to be printed. The type plate containing the printed matter is to be 40 square inches, the top margin is to be $\frac{3}{4}$ inch, the bottom margin is to be $\frac{1}{2}$ inch, and the side margins are each to be 1 inch. What should the dimensions of the type plate be so that the area of the pamphlet is minimized?

Find the following integrals.

3. $\int_0^7 \sqrt{9+x}\,dx$ 4. $\int_{-1}^3 2x(x^2+1)^3 dx$ 5. $\int_0^4 100e^{0.05x} dx$

6. $\int_0^3 150e^{0.08x} dx$ 7. $\int_0^4 100e^{0.06(4-t)} dt$

Find the area of the region S bounded by the graphs of the following equations. Sketch the graph of S.

8. $y = x^{1/3}, x = 1, x = 8, y = 0$

9. $y = \frac{1}{2}x^2, \quad y = x + 4$

10. $y = x - 2, \quad y = 2x - x^2$

11. A certain investment is envisioned as yielding income at the constant rate of \$2000 per year for 4 years. With respect to the interest rate of 9 percent per annum compounded continuously, determine the present value of this continuous investment stream.

12. If money is invested at the constant rate of \$1000 per year at the interest rate of 7 percent per annum compounded continuously, find the total value of this continuous investment stream at the end of 6 years.

13. The tangent line to a curve at a point (x, y) has slope $6x^2 - 1$. Find the curve with this slope that passes through the point $(1, 4)$.
14. Remains of organic origin obtained in an unearthed settlement are found to contain 50 percent of the original amount of carbon-14. Find the date of the remains.
15. The marginal cost function of a chocolate producer is $f(x) = \frac{1}{5}x + 10$, where x is output in tons per week. The chocolate producer incurs a fixed cost of \$1500 per week. Determine the total cost function.

Find $f_x(x, y)$, $f_y(x, y)$, $f_{xx}(x, y)$, $f_{yy}(x, y)$, and $f_{xy}(x, y)$ *for the following functions.*

16. $f(x, y) = 3x^4y - y^2x$
17. $f(x, y) = 5x^4y - xy^2$
18. $f(x, y) = \sqrt{3x^2y + y^3}$
19. $f(x, y) = xe^y$
20. Find the extreme values of $f(x, y) = x^3 + x^2y^2 - 27x + 4$.
21. Find the minimum value of $f(x, y) = x^2 + 2y^2$ subject to the constraint $x + 2y = 6$.
22. Find the maximum value of $f(x, y) = 12x - x^2 + 8y - 2y^2$ subject to the constraint $x + y = 8$.

APPENDIX
ON TABLES

Table 1 COMPOUND INTEREST

$(1 + i)^n$

n	1%	2%	3%	4%	5%
1	1.01000	0.02000	1.03000	1.04000	1.05000
2	1.02010	1.04040	1.06090	1.08160	1.10250
3	1.03030	1.06121	1.09273	1.12486	1.15762
4	1.04060	1.08243	1.12551	1.16986	1.21551
5	1.05101	1.10408	1.15927	1.21665	1.27628
6	1.06152	1.12616	1.19405	1.26532	1.34010
7	1.07214	1.14869	1.22987	1.31593	1.40710
8	1.08286	1.17166	1.26677	1.36857	1.47746
9	1.09369	1.19509	1.30477	1.42331	1.55133
10	1.10462	1.21899	1.34392	1.48024	1.62889
11	1.11567	1.24337	1.38423	1.53945	1.71034
12	1.12683	1.26824	1.42576	1.60103	1.79586
13	1.13809	1.29361	1.46853	1.66507	1.88565
14	1.14947	1.31948	1.51259	1.73168	1.97993
15	1.16097	1.34587	1.55797	1.80094	2.07893
16	1.17258	1.37279	1.60471	1.87298	2.18287
17	1.18430	1.40024	1.65285	1.94790	2.29202
18	1.19615	1.42825	1.70243	2.02582	2.40662
19	1.20811	1.45681	1.75351	2.10685	2.52695
20	1.22019	1.48595	1.80611	2.19112	2.65330
21	1.23239	1.51567	1.86029	2.27877	2.78596
22	1.24472	1.54598	1.91610	2.36992	2.92526
23	1.25716	1.57690	1.97359	2.46472	3.07152
24	1.26973	1.60844	2.03279	2.56330	3.22510
25	1.28243	1.64061	2.09378	2.66584	3.38635
26	1.29526	1.67342	2.15659	2.77247	3.55567
27	1.30821	1.70689	2.22129	2.88337	3.73346
28	1.32129	1.74102	2.28793	2.99870	3.92013
29	1.33450	1.77584	2.35657	3.11865	4.11614
30	1.34785	1.81136	2.42726	3.24340	4.32194
31	1.36133	1.84759	2.50008	3.37313	4.53804
32	1.37494	1.88454	2.57508	3.50806	4.76494

33	1.38869	1.92223	2.65234	3.64838	5.00319
34	1.40258	1.96068	2.73191	3.79432	5.25335
35	1.41660	1.99989	2.81386	3.94609	5.51602
36	1.43077	2.03989	2.89828	4.10393	5.7182
37	1.44508	2.08069	2.98523	4.26809	6.08141
38	1.45953	2.12230	3.07478	4.43881	6.38548
39	1.47412	2.16474	3.16703	4.61637	6.70475
40	1.48886	2.20804	3.26204	4.80102	7.03999

Table 2 PRESENT VALUE

$(1+i)^{-n}$

n	1%	2%	3%	4%	5%
1	.99010	.98039	.97087	.96154	.95238
2	.98030	.96117	.94260	.92456	.90703
3	.94059	.94232	.91514	.88900	.86384
4	.96098	.92385	.88849	.85480	.82270
5	.95147	.90573	.86261	.82193	.78353
6	.94205	.88797	.83748	.79031	.74622
7	.93272	.87056	.83109	.75992	.71068
8	.92348	.85349	.78941	.73069	.67684
9	.91434	.83676	.76642	.70259	.64461
10	.90529	.82035	.74409	.67556	.61391
11	.89632	.80426	.72242	.64958	.58468
12	.88745	.78849	.70138	.62460	.55684
13	.87866	.77303	.68095	.60057	.53032
14	.86996	.75788	.66112	.57748	.50507
15	.86135	.74301	.64186	.55526	.48102
16	.85282	.72845	.62317	.53391	.45811
17	.84438	.71416	.60502	.51337	.43630
18	.83602	.70016	.58739	.49363	.41552
19	.82774	.68643	.57029	.47464	.39573
20	.81954	.67297	.55368	.45639	.37689
21	.81143	.65978	.53755	.43883	.35894
22	.80340	.64684	.52189	.42196	.34185
23	.79544	.63416	.50669	.40573	.32557
24	.78757	.62172	.49193	.39012	.31007
25	.77977	.60953	.47761	.37512	.29530
26	.77205	.59758	.46369	.36069	.28124
27	.76440	.58586	.45019	.34682	.26785
28	.75684	.57437	.43708	.33348	.25509
29	.74934	.56311	.42435	.32065	.24295
30	.74192	.55207	.41199	.30832	.23138
31	.73458	.54125	.39999	.29646	.22036
32	.72730	.53063	.38834	.28506	.20987

33	.72010	.52023	.37703	.27409	.19987
34	.71297	.51003	.36604	.26355	.19035
35	.70591	.50003	.35538	.25342	.18129
36	.69892	.49022	.34503	.24367	.17266
37	.69200	.48061	.33498	.23430	.16444
38	.68515	.47119	.32523	.22529	.15661
39	.67837	.46195	.31575	.21662	.14915
40	.67165	.45289	.30656	.20829	.14205

Table 3 VALUES OF e^x AND e^{-x}

x	e^x	e^{-x}	x	e^x	e^{-x}
0.00	1.00000	1.00000			
0.01	1.01005	.99004	1.60	4.95302	.20189
0.02	1.02020	.98019	1.70	5.47394	.18268
0.03	1.03045	.97044	1.80	6.04964	.16529
0.04	1.04081	.96078	1.90	6.68589	.14956
0.05	1.05127	.95122	2.00	7.38905	.13533
0.06	1.06183	.94176			
0.07	1.07250	.93239	2.10	8.16616	.12245
0.08	1.08328	.92311	2.20	9.02500	.11080
0.09	1.09417	.91393	2.30	9.97417	.10025
0.10	1.10517	.90483	2.40	11.02316	.09071
			2.50	12.18248	.08208
0.11	1.11628	.89583	2.60	13.46372	.07427
0.12	1.12750	.88692	2.70	14.87971	.06720
0.13	1.13883	.87810	2.80	16.44463	.06081
0.14	1.15027	.86936	2.90	18.17412	.05502
0.15	1.16183	.86071	3.00	20.08551	0.4978
0.16	1.17351	.85214			
0.17	1.18530	.84366	3.50	33.11545	.03020
0.18	1.19722	.83527	4.00	54.95815	.01832
0.19	1.20925	.82696	4.50	90.01713	.01111
			5.00	148.41316	.00674
0.20	1.22140	.81873	5.50	224.69193	.00409
0.30	1.34985	.74081			
0.40	1.49182	.67032	6.00	403.42879	.00248
0.50	1.64872	.60653	6.50	665.14163	.00150
0.60	1.82211	.54881			
0.70	2.01375	.49658	7.00	1096.63316	.00091
0.80	2.22554	.44932	7.50	1808.04241	.00055
0.90	2.45960	.40656			
1.00	2.71828	.36787	8.00	2980.95799	.00034
			8.50	4914.76884	.00020
1.10	3.00416	.33287			
1.20	3.32011	.30119	9.00	8130.08393	.00012
1.30	3.66929	.27253	9.50	13359.72683	.00007
1.40	4.05519	.24659			
1.50	4.48168	.22313	10.00	22026.46579	.00005

ANSWERS TO ODD-NUMBERED EXERCISES AND SELF-TESTS

Section 1.2 (p. 12)

1. -6
3. 6
5. -1
7. 10
9. 1
11. 1
13. $15a$
15. $5wz$
17. $2ab$
19. $10xyz$
21. $-4x-6y-8z$
23. $-2x-9$
25. $2n-8m$
27. $3xy+a$
29. $(2(x+2y)$
31. $x(y+z)$
33. $4xy(z+2w)$
35. $2x(1+2x)$
37. $a^2(y+z+w)$
39. $k^2(4w-3+a)$
41. $3a(bc+2b+c)$
43. $x(2y+3z+1)$
45. yes, since $(-x+4)+(x-4)=0$
47. no, since $-2(2)\neq 1$
49. 1 is the multiplicative inverse of 1 since $1\cdot 1=1$.
51. $\frac{1}{6},-1,4,1/\sqrt{2}$, and $1/\pi$, respectively

Section 1.2 (p. 18)

53. -2

55. -7 57. 5

59. 1

61. 10 63. -5

65. 10

67. -3 69. -9

71. 1/6

73. $-1/8$ 75. 0

77. $-6a+8b-6c$

79. $-3a-4b+2c$

81. $-8x-5y$

83. $7x+10y+7z$

85. $-7xy$

87. $-5ab+2a$

89. $x(y-4z-5)$

91. $mn(4-5n+3p)$

93. $\frac{1}{2}a(1-4b)$ or $a(\frac{1}{2}-2b)$

95. $\frac{1}{3}xy(1+18z)$ or $xy(\frac{1}{3}+6z)$

Section 1.3 (p. 23)

1. incorrect

3. incorrect

5. incorrect

7. incorrect

9. $\frac{1}{3}$

11. $\frac{3}{11}$

13. $\frac{3ac}{4}, b \neq 0$

21. $\frac{2a+b}{a+b}$

23. $-1, x \neq 3$

25. $\frac{2x+1}{2x}$

27. 2, where $x \neq 2$

29. 1, where $x \neq -1$

31. 4, where $x-4y \neq 0$

33. 3/2, where $x \neq 5$

15. $4, a \neq -b$

17. $\dfrac{2+(x+y)}{x+y}$

19. $\dfrac{3(x+y)}{2}, x \neq y$

35. $1/x^2$, where $4m + 3n \neq 0$

37. c/e, where $4d - 1 \neq 0$

39. $ab/3$, where $3 - c \neq 0$

Section 1.3 (p. 29)

41. not equal

43. not equal

45. equal

53. $4/b$, where $a \neq 0$

55. $\dfrac{3yx}{(x+1)(6y-1)}$

57. $12/x$, where $y \neq 0$

59. $1/2x$, where $3x - 2 \neq 0$

61. $-\dfrac{3}{4}$, where $x \neq 2$

63. $-8/n$, where $n \neq 3$

65. $\dfrac{2}{3}$

47. $\dfrac{4}{7}$

49. $\dfrac{4}{5}$

51. $\dfrac{3(8y-1)}{4y}$, where $x \neq 0$

67. $-\dfrac{12}{5}$

69. $-\dfrac{1}{8}$

71. $2a^2/3$, where $a \neq 0, b \neq 0$

73. $2/b$, where $a \neq 0$

77. $-\dfrac{1}{2}$, where $x \neq 2$,

79. $-1/a$, where $a \neq 3$

Section 1.4 (p. 32)

1. x^3+8x^2+4x-4
3. $-8x^2h+5xh+h^2-1$
5. $4x^2y-3xy-3y+7$
7. $2x^2-x-5$
9. $-5x^2h-3xh+2x+1$

Section 1.4 (p. 34)

11. $3x^2+12x$
13. $-x^3+x^2-x$
15. $-3x^4+12x^3-21x$
17. $6x^3h^2+8xh^2-2xh$
19. x^2+x-2
21. x^2-h^2
23. $2x^2-x-3$
25. $6x^2+x-2$
27. $x^3-3hx^2+h^2x+h^3$
29. $x^4+x^3-6x^2+4x-8$
31. $x^5+2x^4-3x^3-4x^2+5x+2$
33. $9x^4+6x^3-8x^2+21x-14$
35. $16x^4+12x^3+12x^2+37x+21$
37. $10x^5-30x^4+5x^3-15x^2$
39. $x^3-5x^2-18x+72$
41. $x^5+5x^4h+10x^3h^2+10x^2h^3+5xh^4+h^5$

Section 1.4 (p. 40)

43. $x(5+3y+z)$
45. $2a(3x-2ay)$
47. $xy(2+x-y)$
49. $(x+6)(x-1)$
51. $(x+11)(x+1)$
53. $(t-5)(t-1)$
55. $(x-7)(x-3)$
57. does not factor
59. $3x(x-7)(x-2)$
61. $t(t-10)(t-6)$
67. $2(x-8), x\neq 1$
69. $3x(x-2), x\neq 7$

63. $4x(x-4)(x+2)$

65. $\dfrac{x^2+9}{x+3}$

71. $2(x+2), x \neq 3$

Section 1.4 (p. 42)

73. $x=5, x=-5$

75. $x=0, x=-4$

77. $x=1, x=8$

79. $x=0, x=2, x=7$

81. $x=-3/2, x=2/3$

83. $x=-5, x=11$

85. $x=\dfrac{1+\sqrt{21}}{2}$

Section 1.5 (p. 47)

1. -32

3. $-\dfrac{1}{8}$

5. $\dfrac{3}{16}$

7. $-\dfrac{9}{8}$

9. 1000

11. $\dfrac{1}{324}$

13. 1

15. $-\dfrac{25}{64}$

21. 1

23. 1

25. $6x^2y^2$

27. $25/3x^8$

29. $2+(1/x)+(1/x^3)$

31. $\dfrac{1}{8a^3b^6}$

33. $\dfrac{a^3}{8b^6}$

35. $\dfrac{4x^4}{3y^5}$

17. $4/xy^2$

19. x^6

Section 1.5 (p. 50)

37. 8

39. –4

41. –2

43. 3

45. 5

47. 4

49. 6

51. $\frac{25}{2}$

Section 1.5 (p. 52)

53. $4\sqrt{3}$

55. $3\sqrt{6}$

57. $3\sqrt{14}$

59. $15\sqrt{7}$

61. $-\sqrt{7}$

63. $2\sqrt[4]{3}$

65. $2a\sqrt{b}$

67. $2x\sqrt{xy}$

69. $2\sqrt{x^2+4y^2}$

Section 1.5 (p. 55)

71. 5

73. 5

75. $\frac{1}{6}$

77. $\frac{1}{2}$

85. 36

87. 32

89. $3x^2$

91. x

79. 9

81. 27

83. $\dfrac{1}{16}$

93. $1/x^5$

95. x^3

97. $x^{47/30}$

Section 1.6 (p. 62)

1. $y = 2x + 2$

3. $y = -x + 3$

5. $y = -2x - 1$

7. $3y = x + 2$

9. $x = 1$

Section 2.1 (p. 66)

1. $f(1) = -4,\ f(3) = 2,\ f(-1) = -2,\ f(-2) = 2,\ f(0) = -4$

3. $f(1) = -4,\ f(-3) = 29,\ f(\frac{1}{2}) = \frac{17}{4},\ f(3) = 17,\ f(10) = 276$

5. $f(0) = f(1) = f(-1) = f(10) = f(-2) = -1$

7. $f(1) = 1,\ f(2) = 4,\ f(3) = 7,\ f(-1) = -4.$

Section 2.2 (p. 75)

1.

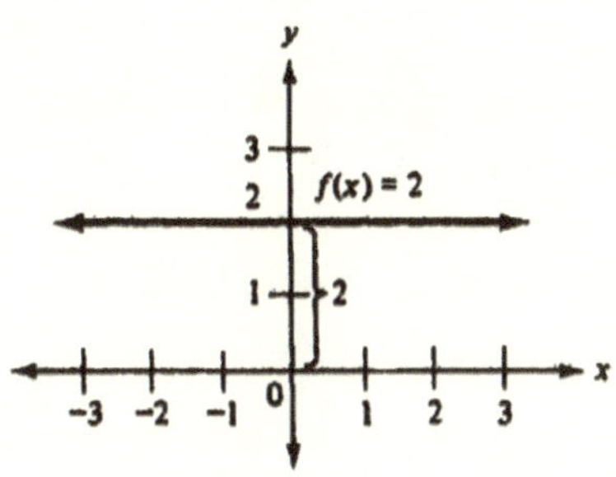

7.

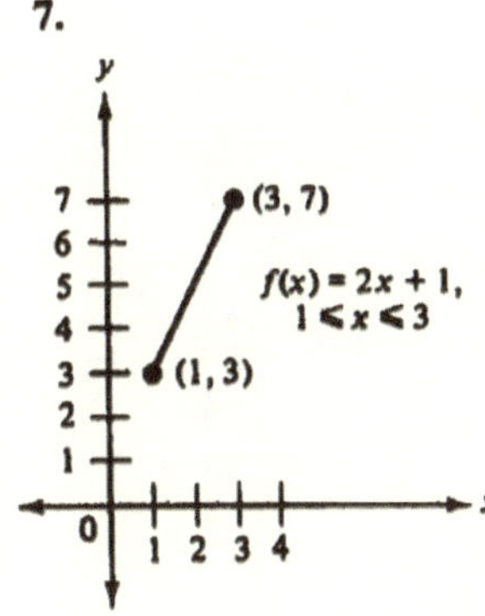

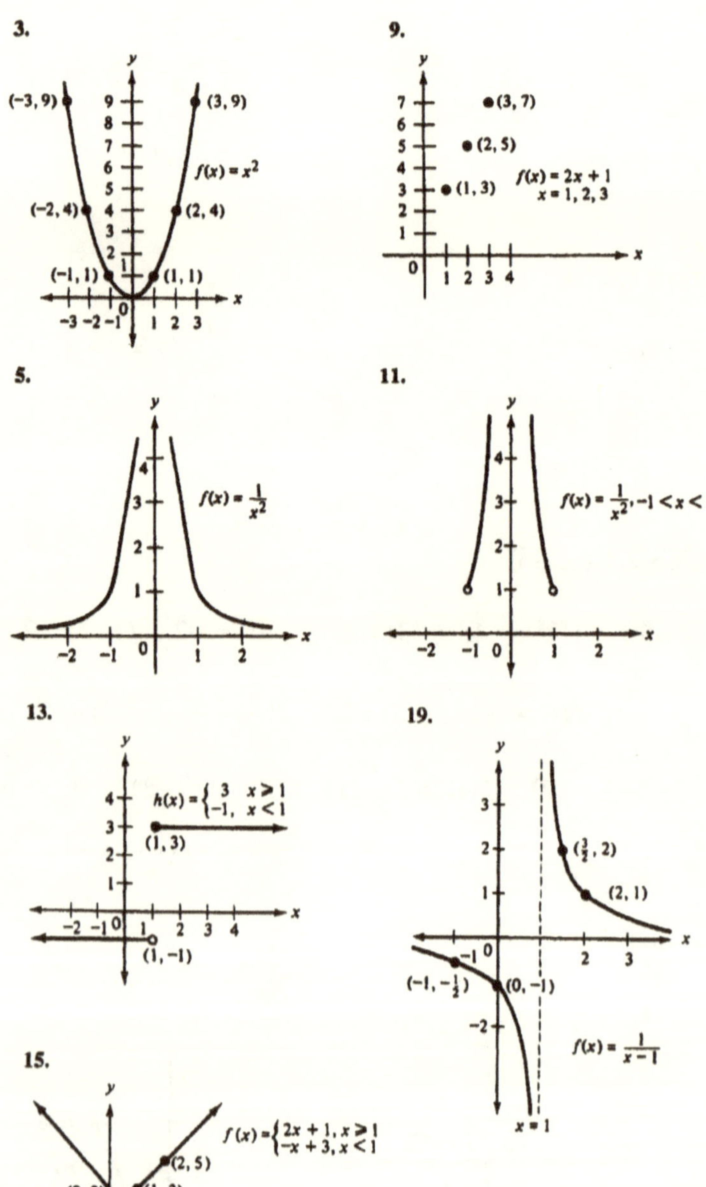
3.
(−3, 9)
(3, 9)
f(x) = x²
(−2, 4)
(2, 4)
(−1, 1)
(1, 1)
9.
(3, 7)
(2, 5)
(1, 3)
f(x) = 2x + 1
x = 1, 2, 3
5.
f(x) = 1/x²
11.
f(x) = 1/x², −1 < x < 1
13.
h(x) = { 3 x ≥ 1; −1, x < 1
(1, 3)
(1, −1)
19.
(3/2, 2)
(2, 1)
(−1, −1/2)
(0, −1)
f(x) = 1/(x − 1)
x = 1
15.
f(x) = { 2x + 1, x ≥ 1; −x + 3, x < 1
(2, 5)
(0, 3)
(1, 3)
(1, 2)

17.

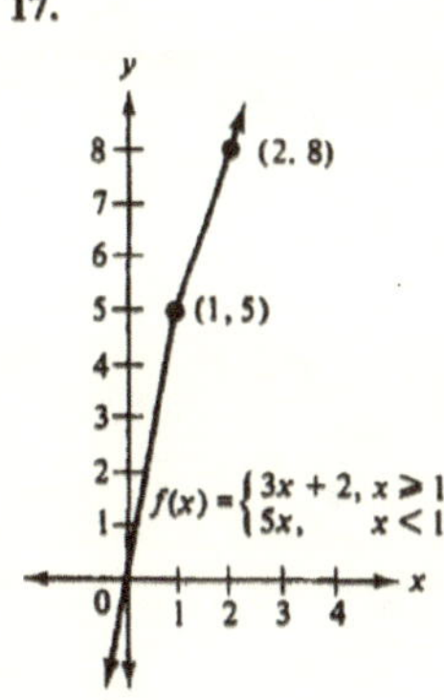

21.

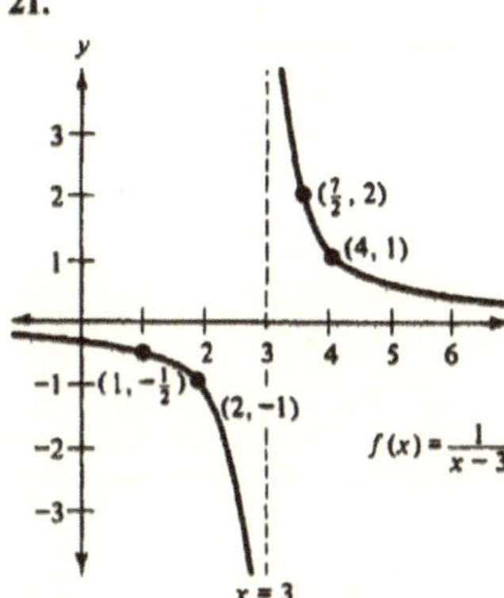

23.

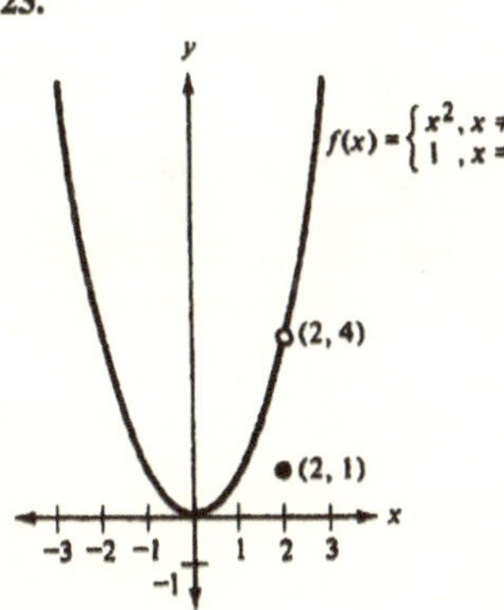

25.

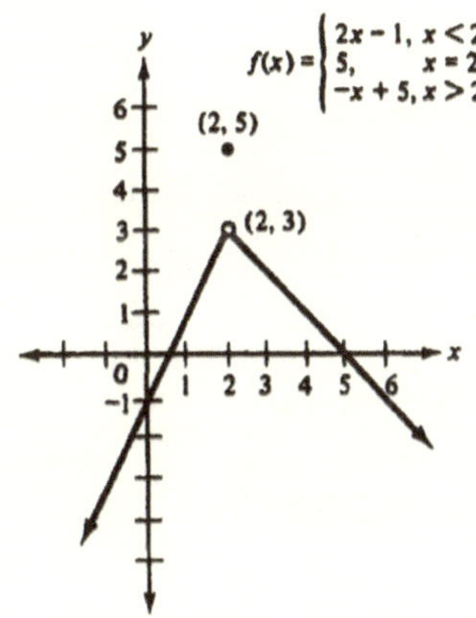

Section 2.3 (p. 83)

1. $C(x)=\begin{cases}8, & \text{for } 0\le x\le 50\\ 8+0.09(x-50), & \text{for } x>50\text{, where } x \text{ is the number of message units accumulated.}\end{cases}$

3. $I(x)=\begin{cases}1.50x, & \text{for } 0\le x\le 5000\\ 7500+2(x-5000), & \text{for } x>5000\text{, where } x \text{ is the number of books sold}\end{cases}$

5. $T(x)=\begin{cases}0, & \text{for } 0\le x\le 5000\\ 20+0.01(x-5000), & \text{for } 5000<x\le 25{,}000\\ 220+0.03(x-25{,}000), & \text{for } x>25{,}000\text{, where } x \text{ is taxable income in dollars}\end{cases}$

7. $V(t) = 2000 - 200t$, for $0 \le t \le 10$;

   $V(3) = 1400$; $V(t) = C - Crt$, for $0 \le t \le 1/r$

Section 3.1 (p. 93)

1. 7
3. 16
5. 8
7. 6
9. $\text{limit}_{x \to 0} f(x)$ does not exist; as x approaches 0, $f(x)$ clusters around -1 and 1.
11. 6
13. 2
15. $\text{limit}_{x \to 2} f(x)$ does not exist; as x approaches 2, $f(x)$ clusters around 1 and 3.
17. $\text{limit}_{x \to 2} f(x)$ does not exist; as x approaches 2, $f(x)$ increases without bound and decreases without bound.
19. $\underset{x \to 1}{\text{limit}}\, f(x)$ does not exist.

Section 3.2 (p. 101)

1. 80
3. 14
5. 108
7. 17
9. 198
11. $\frac{9}{8}$
13. 7
15. -4
17. 364
19. 6
21. 5
23. 3

Section 3.3 (p. 106)

1. Yes; $\text{limit}_{x \to 2} f(x) = 7$ and $f(2) = 7$.
3. Yes; $\text{limit}_{x \to 2} f(x) = 5$ and $f(2) = 5$.
5. Yes; $\text{limit}_{x \to 0} f(x) = 1$ and $f(0) = 1$.

7. Yes; $\text{limit}_{x \to 1} f(x) = 5$ and $f(1) = 5$.
9. No; $\text{limit}_{x \to 0} f(x)$ does not exist.

Section 3.4 (p. 114)

1. $\text{limit}_{x \to \infty} f(x) = 5$
3. $\text{limit}_{x \to 0} f(x) = \infty$
5. $\text{limit}_{x \to 2} f(x) = \infty$
7. $\text{limit}_{x \to \infty} f(x) = 4$
9. $\text{limit}_{x \to \infty} f(x) = e^4$
11. $\text{limit}_{x \to \infty} f(x)$ does not exist
13. $\text{limit}_{x \to -2} f(x)$ does not exist
15. $\text{limit}_{x \to -\infty} f(x) = \frac{2}{5}$
17. $\text{limit}_{x \to \infty} f(x) = 2e$
19. $\text{limit}_{x \to 5} f(x) = \infty$
21. $1 - e^{-2k}$

Section 4.1 (p. 122)

1. (a) 10% per annum (b) (i) 1 year, (ii) \$105, (iii) \$605
3. (a) \$4105.71, (b) \$4118.37
5. \$4973.76

Section 4.1 (p. 124)

7. 12.55%

Section 4.2 (p. 129)

1. (a) 6.183%, (b) 8.328%, (c) 10.517%, (d) 12.750%
3. \$4912.38, \$4444.86, \$3639.18, \$2979.48

Self-Tests for Chapters 1-4

Self-Test 1 (p. 130)

1. (a) No; the cancellation of $(x-2)$ from the numerator and denominator of $y = \frac{(x-2(x+3)}{(x-2)(x-1)}$ is permissible for $x \neq 2$. It is

absurd to carry out this cancellation for $x \neq 0$ and then substitute 2 for x. As to 0/0, it is not defined, cannot be defined, and no amount of hocus-pocus will define it. (b) No, (c) No; it's correct that 0 times 0 is 0, but it's irrelevant to the issue. (d) No; compound nonsense.

2. & 3. $\lim_{x \to 4} f(x) = 17$; as x approaches 4, $f(x)$ approaches 17.
4. No; $\lim_{x \to 4} f(x) = 17 \neq f(4) = 12$.
5. $e = \lim_{n \to \infty} \left(1 + \frac{1}{n}\right)^n$
6. 11
7. As x approaches 3, $f(x)$ approaches 11.
8. 17
9. $y = 3x - 1$
10. 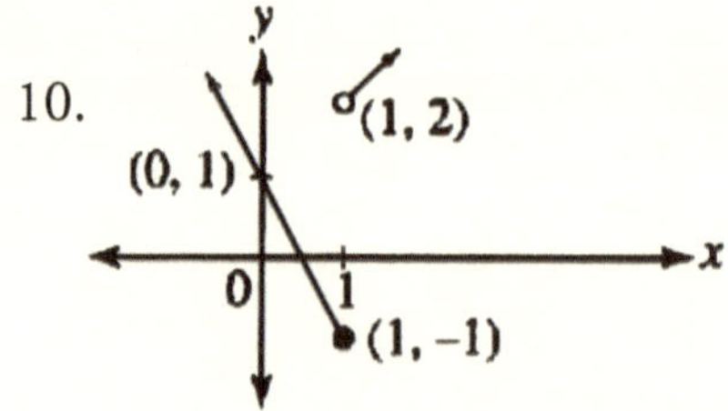

11. 7/3
12. $g(n) = 2h + 13,\ h \neq 0$
13. 13
14. & 15. ∞; as x approaches 3, $f(x)$ increases without bound.
16. all real numbers except 1 and -2.
17. $15x^2 + 17x - 4$
18. $e^{0.15} - 1$

Self-Test 2 (p. 133)

1. $\lim_{x \to 5} f(x)$ does not exist; as x approaches 5, $f(x)$ clusters about two values, 22 and 18.
2. $\lim_{x \to 1} f(x) = -2$, as x approaches -1, $f(x)$ approaches -2.
3. $\lim_{x \to 6} \frac{x^2 - 36}{x - 6} = \lim_{x \to 6}(x + 6) = 12$

4. ∞; as x approaches 2, $\dfrac{3}{(x-2)^2}$ increases without bound.

5. 4/9; as x increases without bound, $\dfrac{4x^3+2x-1}{9x^3-2x^2+2}$ approaches 4/9.

6. $e^{0.08}-1$

7. $T(x)=\begin{cases} 0, & \text{for } 0 \le x \le 10{,}000 \\ 200+0.15(x-10{,}000), & \text{for } x>10{,}000, \text{ where } x \text{ is taxable income} \end{cases}$

8. $g(h) = 2h + 12,\ h \neq 0$ 9. 12

10. −4 and 1/2; for these values the denominator of $f(x)$ is 0 and division by 0 is not defined.

11. $e = \lim_{n\to\infty}\left(1+\dfrac{1}{n}\right)^n$

12. $\lim_{x\to 3} h(x)$ does not exist; as x approaches 3, $h(x)$ both increases without bound (for $x > 3$) and decreases without bound (for $x < 3$).

13. $7000e^{-0.45}$ 14. 55/7

15. ∞; as x approaches 4, $\dfrac{3}{(x-4)^2}$ increases without bound.

16. The point of view is that of interest accumulation occurring at each time point, as opposed to a time interval. It is captured in quantitative terms by examining the behavior of basic interest relationships defined on a time interval as we force the length of the time interval to go to zero.

17.

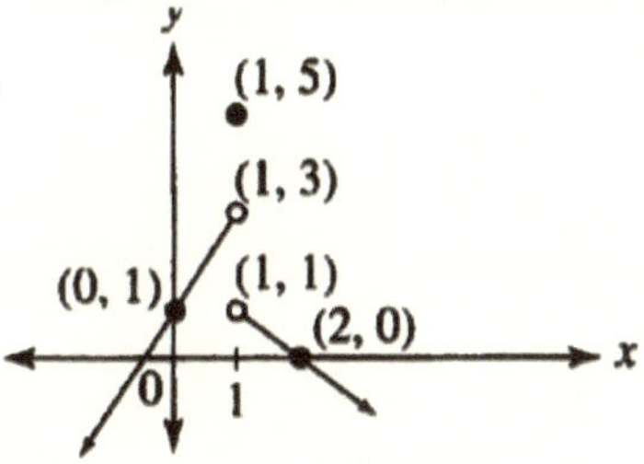

18. No; $\lim_{x\to 1} f(x)$ does not exist.

Self-Test 3 (p. 135)

1. 9/8
2. This is a vertical line; slope is not defined for vertical lines.

3. $4\sqrt{4a^2 - b^2}$
4. & 5. $\lim_{x\to 0} f(x)$ does not exist; as x approaches 0, $f(x)$ increases without bound (for $x > 0$) and decreases without bound (for $x < 0$).
6. No; since $\lim_{x\to 0} f(x)$ does not exist.
7. & 8. $\lim_{x\to\infty} f(x) = 0$; as x increases without bound, $f(x)$ approaches 0.
9. e^{-3} 10. $5000e^{0.45}$
11. 4; as x approaches 1, $\dfrac{x^2+2x-3}{x-1} = x+3, x \neq 3$, approaches 4.
12. Does not exist; as x approaches 6, $\dfrac{5}{x-6}$ increases without bound (for $x > 6$) and decreases without bound (for $x < 6$).
13. 5/3; as x increases without bound, $\dfrac{5x^3+2x^2-1}{3x^3-x+2}$ approaches 5.3.
14. 5; as x approaches 1, $\dfrac{5x-5}{x-1} = 5$ remains constant at 5.
15. 3; as x increases without bound $\frac{4}{x} \to 0, \frac{8}{x^2} \to 0$, and $3 + \frac{4}{x} - \frac{8}{x^2}$ approaches 3.
16. 23/21
17. 12; as x approaches 3, $f(x)$ approaches 12.
18. ∞; as x approaches 0, $\dfrac{1}{x^2}$ increases without bound.

Section 5.1 (p. 160)

1. 3
3. 7
5. 8x
7. 0
9. The tangent line to the graph of $f(x) = x^2 + 50$ at (3, 59) has slope 6. If $f(x) = x^2 + 50$ is the time-distance function of an object in motion, then the instantaneous velocity of the object at time 3 is 6. If $f(x) = x^2 + 50$ is the revenue function of a firm,

then the marginal revenue for an output of 3 units is 6. If $f(x) = x^2 + 50$ is the cost function of a firm, then the marginal cost for an output of 3 units is 6.

Section 5.2 (p. 162)

1. (a) No, $y = f(x)$ is not differentiable at A and B, because the corners at $(A, f(A))$ and $(B, f(B))$ are too sharp to support a tangent line at these points. $y = f(x)$ is not differentiable at C and D because the graph has breaks at these values. (b) No, $y = f(x)$ is not continuous at C and D because the graph has breaks at these values. (c) Yes, A and B; although the corners at $(A, f(A))$ and $(B, f(B))$ are too sharp to support a tangent line at these points and we do not have differentiability, the graph is in one piece at these points and we do have continuity. (d) No, if $y = f(x)$ is differentiable at a value, it must also be continuous at the value, differentiability implies a graph that is in one piece, which is the essence of continuity.

Section 6.1 (p. 166)

1. $f'(x)=0, \quad f'(2)=0$
3. $f'(x)=12x^{11}, \quad f'(-1)=-12$
5. $f'(x)=\dfrac{1}{3(\sqrt[3]{x})^2}, \quad f'(8)=\dfrac{1}{12}$
7. $f'(x)=-\dfrac{3}{x^4}, \quad f'(-2)=-\dfrac{3}{16}$
9. $f'(x)=8x^7, \quad f'(2)=8(2)^7$
11. $f'(x)=0, \quad f'(3)=0$

Section 6.2 (p. 171)

1. $35x^6$

25. $(5x^2+3)4x^3+10x^5$

3. $-45x^{14}$

5. $2x^{-3/4}$

7. $-9x^{-4}$

9. $-20x^{-5}$

11. $6x+2$

13. $20x^4-3-x^{-2}$

15. $\frac{1}{5}x-5$

17. $\frac{3}{2}x-10+4x^{-2}$

19. $(4x^2+2)(12x^3+2)+(3x^4+2x)8x$

21. $(x^3-3x)(2x-4)+(x^2-4x)(3x^2-3)$

23. $5x^2(-4+2x^5)+10x(2-4x+\frac{1}{3}x^6$

27. $\frac{2}{(1-x)^2}$

29. $\frac{6000x}{(500x^2+10)^2}$

31. $\frac{-3x^2+10x-3}{(1-x^2)^2}$

33. $\frac{(5-7x)(6x^5+\frac{1}{2}x^{-1/2}+7(x^6+\sqrt{x})}{(5-7x)^2}$

Section 6.3 (p. 178)

1. $D_x y=7(x^2+2)^6 D_x(x^2+2)=14x(x^2+2)^6$

3. $D_x y=5(4x^5-3x+1)^4 D_x(4x^5-3x+1)$

$=5(20x^4-3)(4x^5-3x+1)^4$

5. $D_x y=\frac{1}{3}(4x^3+2x-10)^{-1/2} D_x(4x^3+2x-10)$

$=(6x^2+1)(4x^3+2x-10)^{-1/2}$

7. $$D_x y = \frac{1}{4}(4x^6 - 7x^4 + 10)^{-3/4} D_x(4x^6 - 7x^4 + 10)$$
$$= (6x^5 - 7x^3)(4x^6 - 7x^4 + 10)^{-3/4}$$

9. We first use the product theorem to obtain

$$D_x y = x^2 D_x(8x^6 + 4x^2 - 1)^3 + (8x^6 + 4x^2 - 1)^3 D_x(x^2)$$
$$= x^2 \cdot 3(8x^6 + 4x^2 - 1)^2 D_x(8x^6 + 4x^2 - 1) + (8x^6 + 4x^2 - 1)^3 2x$$
$$= 3x^2(8x^6 + 4x^2 - 1)^2(48x^5 + 8x) + 2x(8x^6 + 4x^2 - 1)^3$$

11. We first use the quotient theorem to break the function into components and then differentiate the components.

$$D_x y = \frac{4x^2 D_x(4x^3 - 2x)^3 - (4x^3 - 2x)^3 D_x(4x^2)}{16x^4}$$
$$= \frac{4x^2 3(4x^3 - 2x)^2 D_x(4x^3 - 2x) - (4x^3 - 2x)^3 8x}{16x^4}$$
$$= \frac{12x^2(4x^3 - 2x)^2(12x^2 - 2) - 8x(4x^3 - 2x)^3}{16x^4}$$

13. We first use the product theorem to break the function into components and then differentiate the components.

$$D_x y = (5x^3 - 3)^4 D_x(3x^2 - 1) + (3x^2 - 1)D_x(5x^3 - 3)^4$$
$$= (5x^3 - 3)^4 6x + (3x^2 - 1)^4(5x^3 - 3)^3 D_x(5x^3 - 3)$$
$$= 6x(5x^3 - 3) + 60x^2(3x^2 - 1)(5x^3 - 3)^3$$

15. $$\frac{4(x^2 + 1)^{1/2} - x(4x + 3)(x^2 + 1)^{-1/2}}{x^2 + 1}$$

Section 6.4 (p. 181)

1. $\dfrac{-2y}{x}$
3. $\dfrac{-2xy}{x^2-1}$
5. $\dfrac{1-2xy}{x^2}$
7. $-x/y$
9. $\dfrac{-2x-y}{x}$
11. $\dfrac{3}{y}$

Section 6.5 (p. 183)

1. $x=y-10$
3. $y=\dfrac{1}{3}x^{1/3}$
5. $x=10^y$
7. $y=-\dfrac{1}{5}x+6$
9. $D_yx=\dfrac{1}{4}$
11. $D_yx=\dfrac{1}{3x^2}$
13. $D_xy=\dfrac{1}{3(y+4)^2}$

Section 6.6 (p. 191)

3. $2x+e^x$
5. $x^2e^x+2xe^x$
7. $12x^3-5e^x$
9. $4xe^x+4e^x$
11. $6x+2+xe^x+e^x$
13. $4xe^x+4e^x+4x^3$
15. $\dfrac{3xe^x+9e^x}{(x+4)^2}$
17. $\dfrac{2xe^x-(2x+5)(xe^x+e^x)}{x^2e^{2x}}$
19. $24e^{6x}$
21. $\dfrac{4xe^x-10e^x-2}{(2x-3)^2}$
23. $-x^2e^{-x}+2xe^{-x}$
25. $-2xe^{-x^2}$

Section 6.7 (p. 193)

1. $\log_6 36 = 2$
3. $\log_{25} 5 = \frac{1}{2}$
5. $\log_{27} 3 = \frac{1}{3}$
7. $\log_5 \frac{1}{125} = -3$
9. $\log_9 27 = \frac{3}{2}$
11. $\log_{81} \frac{1}{27} = -\frac{3}{4}$
13. $2^4 = 16$
15. $5^3 = 125$
17. $10^{-1} = 0.1$
19. $7^3 = 343$
21. $10^{-2} = 0.01$
23. 3
25. 1
27. $-\frac{1}{3}$

Section 6.7 (p. 200)

31. $\frac{[(3x+1)/x]-3(\ln x)}{(3x+1)^2}$
33. $\frac{3e^x}{x} + 3e^x \ln x$
35. $\frac{3}{3x+10}$
37. $\frac{1}{2x\sqrt{\ln x}}$
39. $\frac{(2x+1)[1+(1/x)]-2(x+\ln x)}{(2x+1)^2}$
41. $\frac{2(\ln x)}{x}$
43. $\frac{3(\ln x)^2 e^x - 6e^x(\ln x)(1/x)}{(\ln x)^4}$

Section 6.8 (p. 209)

1. $3x + 2y = 8$
3. 132
5. −500
7. $-x + y = 1$

9. $8x\ 5y = 26$
11. $100e^{-kt}$
13. $\eta = -\frac{p}{x}\cdot\frac{dx}{p} = -\frac{p}{x}(-3000e - p/30)$; for $p = 30, \eta = 1$.
15. $R(x) = \frac{800x}{x+10} - 4x; R'(x) = \frac{8000}{(x+10)^2} - 4$; for $p = 36, x = 10$ and $R'(10) = 16$.
17. $c'(x) = \frac{100x^3 + 32,500x^2 + 100,000x}{(50x + 10,000)^2}$; $c'(10) = \frac{87}{2505}$
19. $f'(1) = 1/3; x - 3y = -2$

Section 6.9 (p. 213)

1. $f'(x) = 6x - 2,\ f'(-2) = -14;\ f''(x) = 6,\ f''(-2) = 6;\ f'''(x) = 0,$ $f'''(-2) = 0$

3. $f'(x) = 20x^3 + 6x^2 - 2,\ f'(-2) = 138;\ f''(x) = 60x^2 + 12x,\ f''(-2) = 216;$ $f'''(x) = 120x + 12,\ f'''(-2) = -228$

5. $f'(x) = 1/x,\ f''(x) = (-1/x^2),\ f'''(x) = 2/x^3$. Since $f(x)$ = 1n x is defined for positive real numbers, as are its derivatives; $f'(-2), f''(-2), f'''(-2)$ are not defined, despite the almost overwhelming temptation to substitute -2 for x.

Section 7.1 (p. 224)

1. $f(-2) = 1$ is a minimum value.
3. $f(1) = -1$ is a minimum value.
5. $f(0)$ is not an extreme value; $f(1) = -1$ is a minimum value.
7. $f(2) = -\frac{7}{3}$ is a minimum value; $f(-1) = \frac{13}{6}$ is a maximum value.
9. $f(\frac{1}{3}) = \frac{148}{27}$ is a minimum value; $f(-3) = 24$ is a maximum value.

Section 7.1 (p. 232)

11. $f(-1)=-2$ is a minimum value
13. $f(0) = 0$ is a maximum value; $f(2)=-16$ is a minimum value.
15. $f(1) = 2$ is a maximum value; $f(5) = 10$ is a minimum value.
17. $f(0) = 1$ is a maximum value; $f(\frac{1}{2})=\frac{43}{48}$ is a minimum value; $f(-1)=\frac{1}{3}$ is a minimum value.
19. $f(-1)=-1/e$ is a minimum value.
21. $f(e) = 1/e$ is a maximum value.
23. The Thomas income function is $I(x)=(20,000-200x)(60+x)=-200x^2+800x+1,200,000$, where the domain of definition consists of all non-negative integers through 99.

    (a) Not directly since $I(x)$ is not differentiable.
    (b) Consider the function $I_1(x)$ with the same rule but defined on the interval $0 \leq x \leq 99$.
    (c) $I_1(x)$ is differentiable and the methods developed in Section 7.1 are applicable to it; $x = 20$ yields the optimal value for $I_1(x)$. Since 20 is in the domain of definition of $I(x)$ it yields the optimal value of $I(x)$ as well.
    (d) If *I* were satisfied that the income function is realistic.

25. (a) 9%, (b) $r/2$
27. 125 by 125
29. 5
31. 6 and 6

Section 7.2 (p. 243)

1. (a) 50 tons per day; \$4000, (b) 50 tons per day; \$2500, (c) 50 tons per day; \$3200, (d) 40 tons per day; \$2200, (e) 50 tons per day; \$2400, (f) \$100 per ton, (g) 25 tons per day; \$250, (h) 50 tons per day; \$1500
3. (a) 76 tons per day; \$21,904, (b) 76 tons per day; \$15,333, (c) \$304 per ton, (d) 38 tons per day; \$4576, (e) 76 tons per day; \$10,352

Section 7.3 (p. 249)

1. 6.25 years

Section 8.1 (p. 258)

1. Decreasing for $x \leq 0$, increasing for $x \geq 0$, concave upward for all x, no points of inflection.

3. Increasing for $x \leq 0$, decreasing for $x \geq 0$, concave downward for all x, no points of inflection.

5. Decreasing for $x < 0$, increasing for $x > 0$, concave downward for $x < 0$ and $x > 0$, no points of inflection.

7. Increasing for $x \leq -1$, decreasing for $-1 \leq x \leq 2$, increasing for $x \geq 2$, concave downward for $x < \frac{1}{2}$, concave upward for $x > \frac{1}{2}$; $(\frac{1}{2}, \frac{1}{12})$ is an inflection point.

9. Increasing for $x \leq -2$, decreasing for $-2 \leq x \leq 0$ and $0 < x \leq 2$, increasing for $x \geq 2$, concave downward for $x < 0$, concave upward for $x > 0$, no inflection points.

11. Decreasing for $x \leq -1$, increasing for $x \geq -1$, concave downward for $x < -2$, concave upward for $x > -2$; $(-2, -2/e^2)$ is a point of inflection.

Section 8.2 (p. 268)

1.

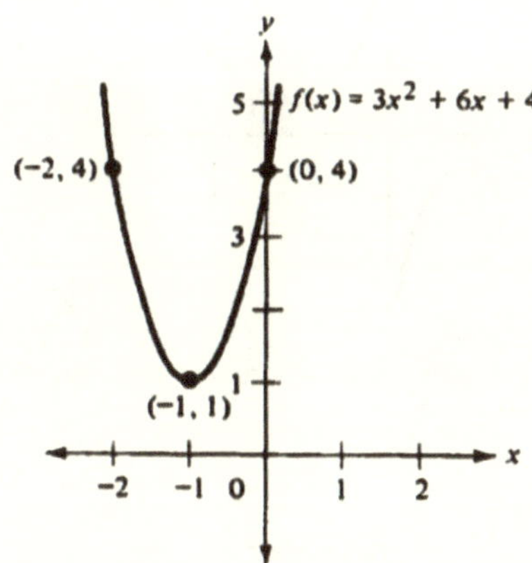

3.

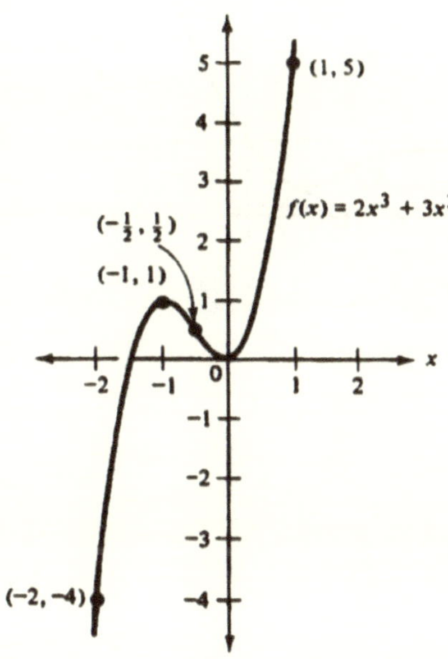

5.

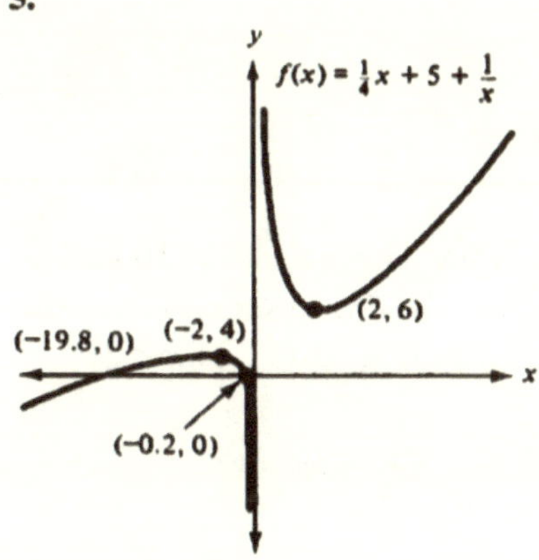

9.

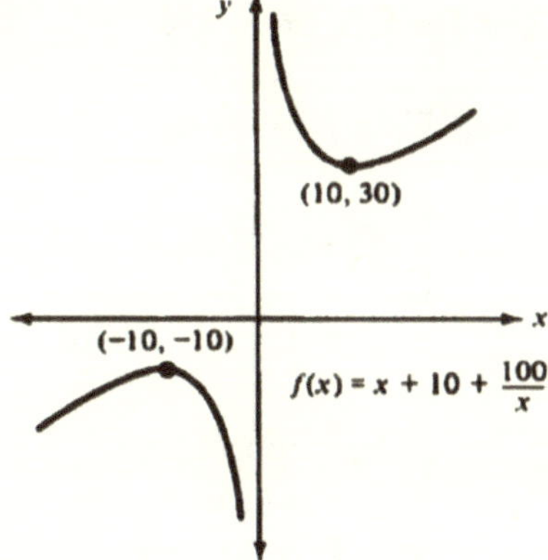

7.

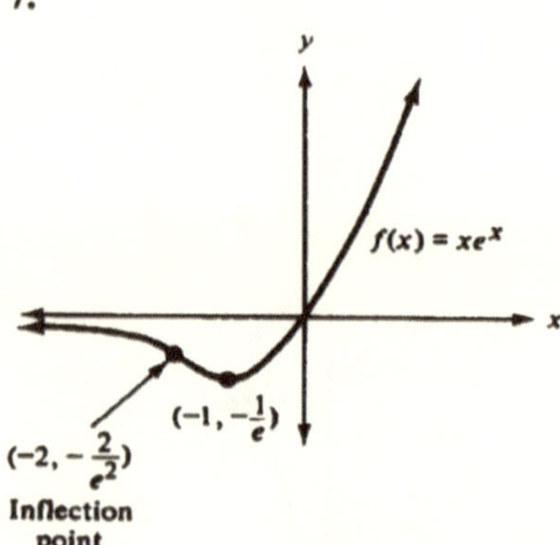

11.

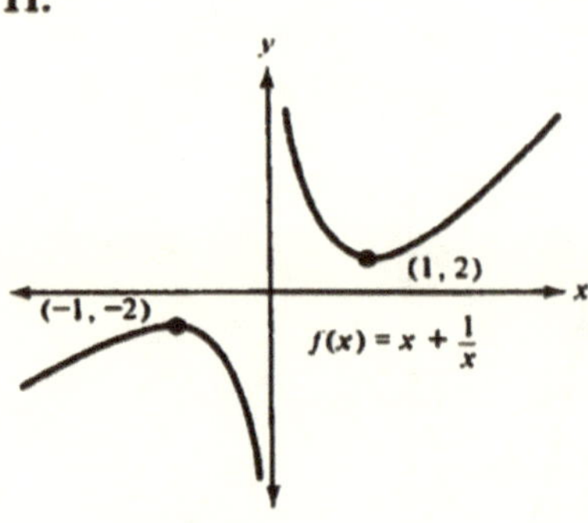

13.

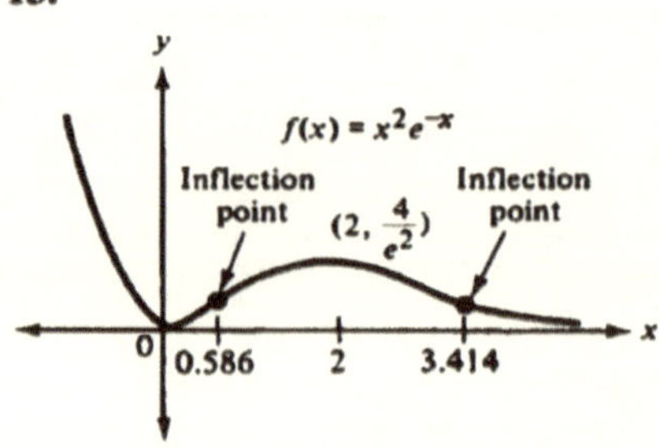

Self-Tests for Chapters 1-8

Self-Test 1 (p. 269)

1. Does not exist; as x approaches 2, $f(x)$ clusters about 10 and 8.
2. Does not exist; as x approaches 2, $f(x)$ increases without bound (for $x > 2$) and decreases without bound (for $x < 2$).
3. e^7
4. ∞; as x approaches 3, $f(x)$ increases without bound.
5. 2; as x approaches 3, $f(x)$ is constant at 2.
6. No, since $\operatorname{limit}_{x\to 2} f(x)$ does not exist.
7. $f'(x) = \operatorname{limit}_{h\to 0} \dfrac{f(x+h)-f(x)}{h}$, provided that this limit exists.
8. $f'(3) = \operatorname{limit}_{h\to 0} (2h+15) = 15$

9. No, since it is not continuous at 2 (see 6), $f(x)$ cannot have a derivative at 2.

10. $2x^4e^x + 8x^3e^x$

11. $\dfrac{12x^2+3}{4x^3+3x}$

12. $(-10x+15)e^{-x2+3x}$

13. $\dfrac{10x+2}{2\sqrt{5x^2+2x+4}}$

14. $y=12x-14$

15. $-1/4$

16. 0, 1/3, 2

17. 205

18. 205

19. 125

20. $T(t)=t(205-\frac{1}{6}t),\ \ t>0$

21. 615

Self-Test 2 (p. 270)

1. 2/27

2. $\dfrac{6x+4}{3x^2+4x}$

3. $3 + x + 2x\ln x$

4. $x = (y/2)^{1/3}$

5. 1150

6. 1150

7. 1100

8. $T(t)=1150t-\frac{1}{7}t^2,\ \ t>0$

9. 4025

10. $-1/12$

11. $-1, 2, 1/3$

12. $f'(3)=\lim_{h\to 0}\dfrac{f(3+h)-f(3)}{h}$, provided that this limit exists.

13. $f'(0)=\lim_{h\to 0} g(h)$, where $g(h)=\begin{cases}1, \text{ for } h>0\\ -1, \text{ for } h<0\end{cases}$.

This limit does not exist since as h approaches 0 $g(h)$ clusters around 1 (for $h > 0$) and -1 (for $h < 0$).

14. Yes. $f(0) = 0$ and $\lim_{x\to 0} f(x)=0$.

15. $f''(-1)$ does not exist since $f(x) = \ln x$ is not defined at -1.

16. $1/6x$

17. $\dfrac{(3x^2+x)(12x^2-2)-(4x^3-2x)(6x+1)}{(3x^2+x)^2}$

18. $5e^{5x+2}$

19. $\dfrac{3x^2+1}{\sqrt{2x^3+2x-1}}$

20. $\lim\limits_{n\to\infty}\left(1+\dfrac{1}{n}\right)^n$

21. 

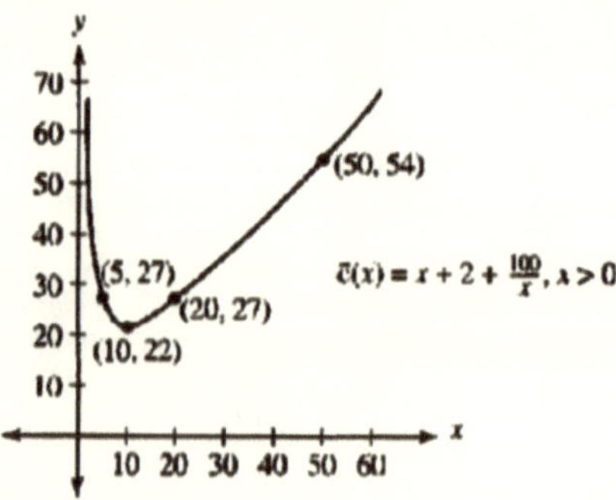

Self-Test 3 (p. 272)

1. 13; as x approaches 3, $f(x)$ approaches 13.
2. 9; as x approaches 2, $f(x)$ approaches 9.
3. 3/4; as x increases without bound, $f(x)$ approaches 3/4.
4. 6; as x approaches 3, $f(x)$ remains constant at 6.
5. ∞; as x approaches 2, $f(x)$ increases without bound.
6. Does not exist; as x approaches 1, $f(x)$ increases without bound (for $x > 1$) and decreases without bound (for $x < 1$).
7. 6; as x increases without bound, $f(x)$ approaches 6.
8. $f(3) = 10$ does not equal $\lim\limits_{x\to 3} f(x) = 13$.
9. No; since $f(x)$ is not continuous at 3, it does not have a derivative at 3.

10. 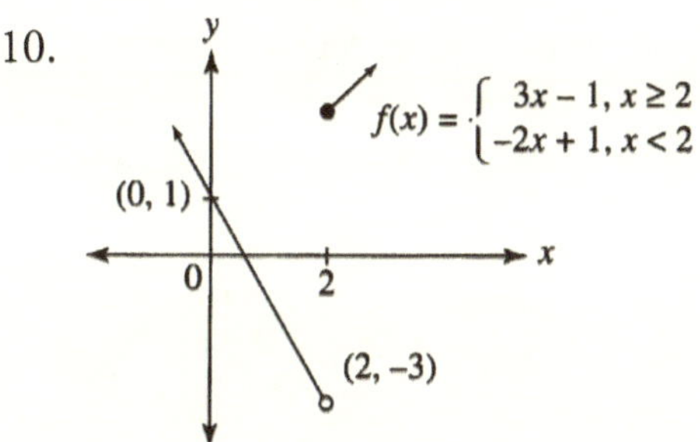

11. All real numbers except $-4/3$. $f(-4/3)$ is not defined because division by 0 is not defined.
12. $e^{0.006} - 1 = 6.183\%$
13. An amount of money invested at 6% per annum compounded continuously will earn the same interest in a year as the rate 6.183% per annum.
14. If $y = f(x)$ is a time-distance function which describes the motion of an object, then the instantaneous velocity of the object at time x is defined by:

$$f'(x) = \operatorname*{limit}_{h \to 0} \frac{f(x+h) - f(x)}{h} \qquad (1)$$

If $y = f(x)$ is an output-cost function of a firm producing a commodity, then (1) defines the marginal cost of the commodity for output level x.

Structurally speaking, there is no difference. Instantaneous velocity and marginal cost are derivatives of the underlying functions that emerge. The difference is in the settings from which these functions emerge and what they represent.

This perspective is expressed in Figure 5.8 of Chapter 5 (p. 160).

15. $T(x) = \begin{cases} 0, \text{for } 0 \leq x \leq 10{,}000 \\ 200 + 0.05(x - 10{,}000), \text{for } x > 10{,}000, \text{where } x \text{ is taxable income.} \end{cases}$
16. The cost function to be minimized is $C(x) = 80{,}000 - 4000x \cdot$ $+5000\sqrt{x^2 + 25}$, where $0 \leq x \leq 20$. To minimize cost P should be 20/3 miles from C.
17. No; see (13) and (14) of Self-Test 2.
18. 1/12

19. $f(0) = 2$ is a local maximum value; $f(1) = 1$ is a local minimum value.
20. It's applicable to 1 because $f''(a) \neq 0$; it's not applicable to 0 because $f''(0) = 0$.

Section 9.1 (p. 280)

1. $\frac{1}{6}x^6 + C$
3. $\frac{1}{13}x^{13} + C$
5. $-\frac{1}{3}x^{-3} + C$
7. $\frac{2}{5}x^{5/2} + C$
9. $\frac{4}{3}x^{3/4} + C$
11. $\frac{25}{6}x^{6/5} + C$
13. $\frac{5}{3}x^3 - \frac{3}{2}x^2 + x + C$
15. $\frac{3}{5}x^5 - 2x^{3/2} - x + C$
17. $-3x^{-1} - \frac{4}{3}x^3 + C$
19. $\frac{5}{8}x^8 - \frac{3}{4}x^4 + 4x + C$
21. $2x^3 - 3e^x + C$
23. $\frac{1}{2}x^2 + \ln x + C$

Section 9.2 (p. 287)

1. $\frac{1}{5}(4x^2 + 4)^5 + C$
3. $\frac{1}{9}(3x + 8)^9 + C$
5. $\frac{1}{60}(3x^2 - 1)^{10} + C$
7. $\frac{4}{3}(x^2 + 4)^{3/2} + C$
9. $\frac{1}{4}e^{4x-1} + C$
11. $\frac{1}{2}e^{x^2} + C$
13. $-\frac{1}{9}(4x^3 + 1)^{-3} + C$
15. $\ln(x^5 + 1) + C$
17. $\ln f(t) + C$
19. $-\frac{1}{3}e^{-3x+1} + C$

Section 9.3 (p. 290)

1. $-xe^{-x} - e^{-x} + C$
3. $-13,000e^{-0.1t} + 500te^{-0.1t} + C$
5. $-\frac{1}{2}xe^{-2x} - \frac{1}{4}e^{-2x} + C$
7. $-4687.5e^{-0.08t} + 125te^{-0.08t} + C$
9. $\frac{1}{2}x^2 \ln x - \frac{1}{4}x^2 + C$

Section 10.2 (p. 301)

1. (a) and (b) Mathematics is precise in the sense that its methods yield valid conclusions with respect to the assumptions made. Sometimes different assumptions are made about a situation under study, and these assumptions lead to different conclusions (solutions), each of which is valid (inescapable consequence) with respect to its assumptions. Different valid conclusions are not in conflict insofar as validity is concerned because they come from different assumptions. They are in conflict insofar as truth is concerned.

   (c) No; although both values are mathematically correct (that is, valid with respect to their respective assumptions), we are not assured that the larger value is more realistic than the smaller one, and that the true maximum profit is the predicted 230,000 value.

   (d) The assumptions of the M1 model were highly unrealistic.

Section 11.1 (p. 308)

1. $f(x) = \frac{5}{2}x^2 + 4x - 5$
3. $y = -2x^2 + 2x + \ln 5 + 5$
5. $f(x) = 10e^{-3x}$
7. $f(x) = \frac{1}{20}x^2 + 5x + 1000$
9. $s(t) = 2t^2 + 5t$

Section 11.3 (p. 319)

1. 4225 years
3. 1846 years

Section 11.4 (p. 324)

1. No; both models are portraits of growth, with neither one being viewed as a "perfect" description of the phenomenon. Both portraits are useful in the appropriate place, which depends on the aspect of the growth phenomenon under study, the accuracy required, and the degree to which each model satisfies this degree of accuracy.
3. (a) $N(t) = (3.06)^9 e^{0.02(t-1961)}$ (c) 200,000 billion. The Earth's surface area is approximately 1,860,000 billion square feet, 80% of which is covered by water. According to this predicted value by the year 2510 there would be 9.3 square feet per person, assuming that living on water is tenable.
4. (a) $N(t) = 2e^{0.4t}$
5. Advances in medical technology which have resulted in sharply increased life spans; very high levels of legal and illegal immigration.

Section 12.1 (p. 343)

1. (a) $\sum_{k=1}^{k=5} 3^k$, (b) $\sum_{k=1}^{k=4} x_k 2^{k+1}$,

   (c) $\sum_{k=1}^{k=4} (x_k)^2 y_{k-1}$ (d) $\sum_{k=2}^{k=5} k(x_{k-1} + x_{k-2})$
3. (a) -6, (b) 1, (c) -29, (d) 107

5.

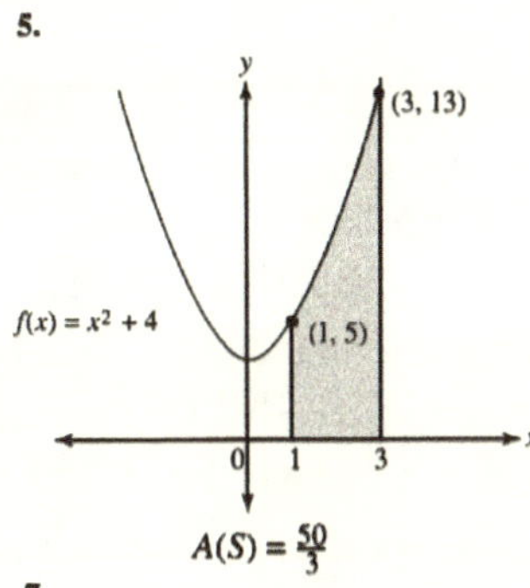

$A(S) = \frac{50}{3}$

9.

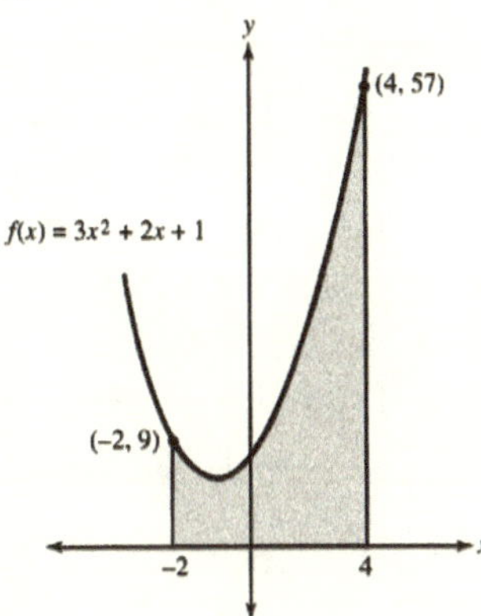

$A(S) = 90$

7.

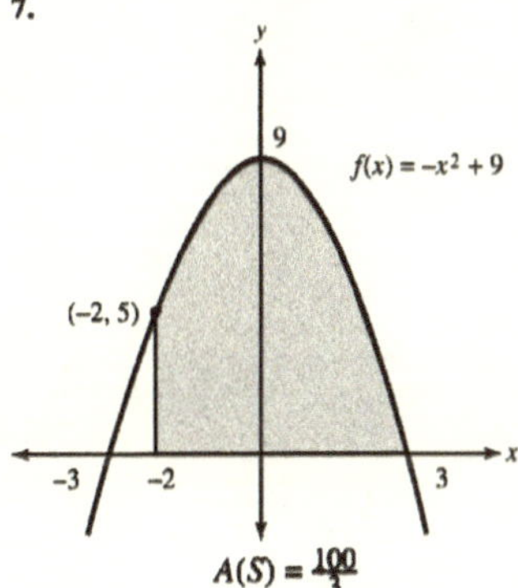

$A(S) = \frac{100}{3}$

Section 12.2 (p. 353)

1. 42
3. $\frac{140}{3}$
5. $\frac{184}{3}$
7. $\ln 3$
9. 402
11. −3,029,184
13. $20(e^2 - 1)$
15. 1
17. $2(\ln 2)$
19. $(\ln 5)/2$
21. $5000(e^{0.18} - 1)$
23. $25{,}000(e^{0.4} - 1)$

Section 12.3 (p. 356)

1. 1/2
3. Does not exist
5. 1/2
7. Does not exist

Section 12.4 (p. 358)

1. No; 4.99 is not the least upper bound of the set S of numbers between 2 and 5 since it is not even an upper bound of this set. 4.991, for example, is in S and $4.991 > 4.99$. LUB of $S = 5$.

   $1.99 \neq$ GLB of S since there are lower bounds of S which exceed 1.99, 1.991, for example. GLB of $S = 2$.
3. The GLB of this set is 1; 1 is a lower bound of this set and no number larger than 1 is a lower bound.

Section 13.1 (p. 366)

1. $\frac{115}{3}$
3. $\frac{256}{3}$
5. 24
7. $\frac{255}{4}$
9. 32
11. $\frac{1}{6}$
13. $e^4 + e^{-4} - 2$
15. $\frac{32}{3}$

Section 13.2 (p. 370)

2. \$4333.3
3. \$27,534

Section 13.3 (p. 372)

1. (a) 4000/3 (b) 972

Section 13.4 (p. 377)

1. 12,079
3. $\int_0^x Ae^{-rt}dt = (-\frac{A}{r}e^{-rt}]_0^x = \frac{A}{r}(1-e^{-rx})$

Section 13.5 (p. 380)

1. (a) $\Pi = \int_0^{10}(1800-50t)e^{-0.01t}dt = (-13,000e^{-0.1t} + 500te^{-0.1t}]_0^{10}$
   $= 10,057$

   (b) $\Pi = 10,057 + 100e^{-1} - 800 = \$9294$

Section 13.6 (p. 382)

1. \$40,000
3. $\Pi = \int_0^\infty Ae^{-rt}dt = \underset{u\to\infty}{\text{limit}}\left[\frac{A}{r}\left(1-\frac{1}{e^{ru}}\right)\right] = \frac{A}{r}$

Section 13.7 (p. 384)

1. \$34,365.60
3. \$88,181.25
5. \$693.31

Section 13.8 (p. 391)

1. 1720/3 ft. pounds
3. 16,596 ft. pounds

Section 14.1 (p. 398)

1. $f(1,3) = -3,\ f(-1,2) = -8,\ f(3,1) = -3,\ f(2,6) = -120$
2. $f(2,3,1) = 13,\ f(-1,2,4) = 5/16,\ f(3,2,1) = 13$

Section 14.2 (p. 404)

1. $f_x(x,y) = 2xy,\ f_x(1,3) = 6,\ f_y(x,y) = x^2,\ f_y(2,4) = 4$
3. $f_x(x,y) = 12x^2y,\ f_x(1,3) = 36,\ f_y(x,y) = 4x^3 - 2y,\ f_y(2,4) = 24$
5. $f_x(x,y) = 2xe^y,\ f_x(1,3) = 2e^3,\ f_y(x,y) = x^2e^y,\ f_y(2,4) = 4e^4$
7. $f_x(x,y) = \dfrac{(x^2+y^2)(8x-y)-(4x^2-xy)2x}{(x^2+y^2)^2},\ f_x(1,3) = \frac{12}{25},$

   $f_y(x,y) = \dfrac{(x^2+y^2)(-x)-(4x^2-xy)2y}{(x^2+y^2)^2},\ f_y(2,4) = -\frac{13}{50}$
9. $f_x(x,y,z) = \dfrac{8x}{z},\ f_y(x,y,z) = \dfrac{1}{z},\ f_z(x,y,z) = -\dfrac{(4x^2+y)}{z^2}$
11. $f_x(x,y,z) = 3x^2yz^2,\ f_y(x,y,z) = x^3z^2,\ f_z(x,y,z) = 2x^3yz$
13. $f_x(x,y,z) = x + 2x(\ln xyz),\ f_y(x,y,z) = \dfrac{x^2}{y},\ f_z(x,y,z) = \dfrac{x^2}{z}$

Section 14.3 (p. 409)

1. (a) $\eta_{11} = -\frac{p_1}{q_1} \cdot \frac{\partial q_1}{\partial p_1} = -\frac{2}{4.8}(-1) = 0.42$. For $p_1 = 2.00$ and $p_2 = 1.50$ a small percentage change in the price of beef will result in an approximate .42% change in the demand for beef, assuming that the price of pork remains at \$1.50 per pound.

   (b) $\eta_{12} = \frac{p_2}{q_1} \cdot \frac{\partial q_1}{\partial p_2} = \frac{1.5}{4.8} \cdot (1.2) = 0.375$. For $p_1 = 2.00$ and $p_2 = 1.50$ a small percentage change in the price of pork will result in an approximate .375% change in the demand for beef, assuming that the price of beef remains at \$2.00 per pound.

3. (a) 0.204 (b) 0.184

Section 14.4 (p. 412)

1. $f_{xx}(x, y) = 8y, \; f_{xx}(-2, 1) = 8,$

   $f_{yy}(x, y) = 0, \; f_{yy}(-1, 3) = 0,$

   $f_{xy}(x, y) = 8x, \; f_{xy}(3, 2) = 24$

3. 0

Section 14.5 (p. 425)

1. No extreme values; (0, 0) is a critical point.
3. $f(4,0) = -16$ is a minimum value.
5. No extreme values; (0, 0) is a critical point.
7. No extreme values; $(6, -4)$ is a critical point.
9. $f(2, 2) = 8$ is a minimum value.
11. $f(6, 2) = 40$ is a maximum value.
13. $f(4, 4) = 0$ is a minimum value.
15. $x = 200, y = 100$

Self-Tests for Chapters 1-14

Self-Test 1 (p. 425)

1. $\dfrac{750}{\sqrt{1500x+5000}}$
2. 623
3. 623
4. 571
5. $T(t)=623t-\frac{1}{8}t^2,\ t>0$
6. 2492
7. $6xe^{3x^2+2}$
8. $\lim_{h\to 0}\dfrac{f(2+h)-f(2)}{h}$, provided that this limit exists.
9. $y=6x-3$
10. $\frac{1}{4}x^4+\frac{1}{2}x^2-2x+C$
11. $\frac{1}{24}(x^4+1)^6+C$
12. $\dfrac{1}{6x^2+1}$
13. −5 and 2
14. $x^2+3x^2\ln x$
15. $-1/4$
16. 24; as x approaches 6, $\dfrac{2x^2-72}{x-6}=$ $2x+12, x\neq 6$, approaches 24.

Self-Test 2 (p. 426)

1. Does not exist; as x approaches 3, $f(x)$ clusters around 14 and 11.
2. ∞; as x approaches 2, $f(x)$ increases without bound.
3. 5/7; as x increases without bound, $f(x)$ approaches 5/7.
4. No; $\lim_{x\to 3} f(x)$ does not exist.
5. $16xe^{4x^2+3}$
6. $\dfrac{15x^2-3}{5x^3-3x}$
7. $(3x^2+3)e^x+6xe^x$
8. $8x+4h-2, h\neq 0$
9. $8x-2$. This is $f'(x)$, the derivative of $f(x)$ at x.
10. $6+2/x^3$

11. $\underset{h\to 0}{\text{limit}}\dfrac{f(5+h)-f(5)}{h}$, provided that this limit exists.
12. See Sections 12.2 and 12.4.
13. −5 and 4
14. $f(-5)$ is a local max. value; $f(4)$ is a local min. value
15. 116
16. $T(t)=116t-\frac{1}{5}t^2,\ t>0$
17. 290
18. $\underset{n\to\infty}{\text{limit}}\left(1+\dfrac{1}{n}\right)^n$
19. $y=14x-18$
20. $-1/3$; the denominator of $f(x)$ is 0 for this value and division by 0 is not defined
21. $\frac{2}{3}x^3+\frac{1}{2}x^2+2$
22. $-\frac{1}{3}(3x^2+1)^{-3}+C$
25. $(-60{,}000e^{-0.10t}\Big]_0^8$
24. 36

Self-Test 3 (p. 428)

1. Disagree; $f(0)=-1/3$. $f(3)$ is not defined because division by 0 is not defined.
2. Disagree; $f(1)$ is not defined whereas $g(1) = 1$.
3. Agree; the limit behavior of $1/f(x)$ as x approaches 3 does not depend on the nature of $f(3)$.
4. Disagree; if $f(x)$ gets closer and closer to L as x gets closer and closer to a, then $\underset{x\to a}{\text{limit}}\, f(x)=L$. L may or may not be $f(a)$; $f(a)$ may or may not be defined.
5. Yes; if $\underset{x\to a}{\text{limit}}\, f(x)=L$, but $L \neq f(a)$, then $f(x)$ is not continuous at a.
6. Disagree. $f(x)=\begin{cases} x+1, x\geq 0 \\ -x+1, x<0 \end{cases}$ is continuous at 0 but does not have a derivative at 0. See Section 3.3. Example 4 (p. 105) and Section 5.1, Example 2 (p. 139).
7. Agree; see Section 5.2, p. 161.
8. Disagree; the limit behavior of $f(x)$ as x approaches a has nothing to do with whether $f(a)$ is defined or not.

9. Disagree; see answer to (8).
10. Disagree; if $f'(a) = f''(a) = 0$, then the second derivative test is inconclusive. $f(a)$ may be a maximum value, or a minimum value, or neither. See Section 7.1.
11. Not necessarily. The fact that $f(x)$ has an indefinite integral $F(x)$ does not imply that $\int_a^b f(x)dx$ exists. If we impose the condition that $f(x)$ is continuous on $[a, b]$, then the existence of $\int_a^b f(x)dx$ is guaranteed and the cited result holds.
12. Disagree; we have encountered a number of examples that make clear that this is not the case. If $f(x) = 2x^2 - 72$ and $g(x) = x - 6$, then $\lim_{x \to 6} f(x) = \lim_{x \to 6} g(x) = 0$, but $\lim_{x \to 6} \frac{f(x)}{g(x)} = 24$. See Self-Test 2, problem 16.
13. Disagree; consider $f(x) = \begin{cases} 2, \text{ for } x \le 3 \\ -2, \text{ for } x > 3 \end{cases}$ and $g(x) = \begin{cases} -2, \text{ for } x \le 3 \\ 2, \text{ for } x > 3 \end{cases}$. Then $h(x) = 0$ for all x and $\lim_{x \to 3} h(x) = 0$; at the same time $\lim_{x \to 3} f(x)$ and $\lim_{x \to 3} g(x)$ do not exist.
14. $C(x) = \begin{cases} 20, \text{ for } 0 \le 100 \\ 20 + 0.10(x - 100), \text{ for } x > 100, \end{cases}$ where x is the number of miles driven per day.
15. $4xy^3ze^{x^2y}$
16. (a) 0.4 (b) For coffee at \$2.00 per pound and tea at \$1.50 per pound a small percentage change in the price of coffee will result in an approximate 0.4% change in the demand for coffee, assuming that the price of tea remains at \$1.50 per pound. (c) 0.6 (d) For coffee at \$2.00 per pound and tea at \$1.50 per pound a small percentage change in the price of tea will result in an approximate 0.6% change in the demand for coffee, assuming that the price of coffee remains at \$2.00 per pound.
17. $(1, \frac{1}{2})$ and $(-2, \frac{1}{2})$ are critical points. $f(1, \frac{1}{2})$ is a local minimum value; $f(-2, \frac{1}{2})$ is not an extreme value.
18. $x = 250, y = 20$

Self-Test 4 (p. 430)

1. $S = \int_0^5 (8000 + 20t)e^{0.15(5-t)}dt = 59,930$
2. $\Pi = \lim_{x\to\infty} \int_0^x 1500e^{-0.15t}dt = 10,000$
3. $\Pi = \int_0^{10} (5000 - 20t)e^{-0.2t}dt + 200^{e-2} - 1000 = \$20,238$
4. $x = 225, y = 37.5$
5. $f(-\frac{1}{2},-1)$ and $f(\frac{1}{2},-1)$ are local minimum values; $f(0,-1)$ is not an extreme value.
6. $b(x) = 4000e^{-0.09x}$, $b(3) = 3054$
7. $f(1/3) = 4.31$ is a local minimum value; $f(-4) = 13$ is a local maximum value.
8. $f(1) = 1/e$ is a maximum value.

10. $f(x)$ is increasing for $x \leq 1$ and decreasing for $x \geq 1$. $(2, 2/e^2)$ is a point of inflection. The graph of $f(x) = xe^{-x}$ is concave downward for $x < 2$ and concave upward for $x > 0$. The graph is shown in Figure 1.

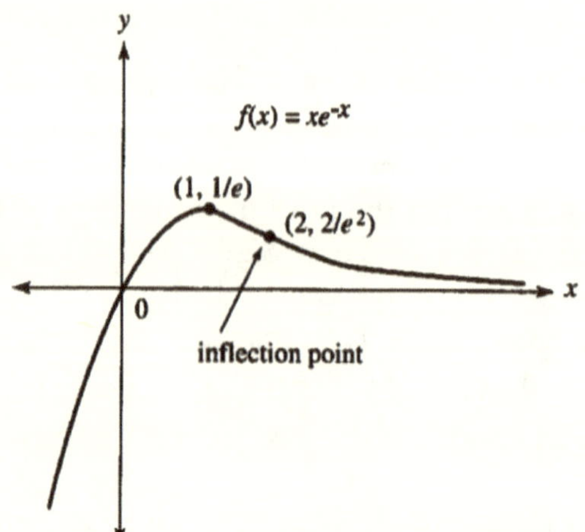

Figure 1

11. $\underset{n \to \infty}{\text{limit}} \left(1 + \frac{1}{n}\right)^n$

Self-Test 5 (p. 431)

1. $I_1(x) = (10,000 - 200x)(30 + x), x > 0; 10$
2. length 5 in; width 8 in.
3. 74/3
4. 2496
5. 442.8
6. $1875(e^{0.24} - 1)$
7. $1666.7(e^{0.24} - 1)$
8. $A(S)$ = 45/4; see Figure 2.

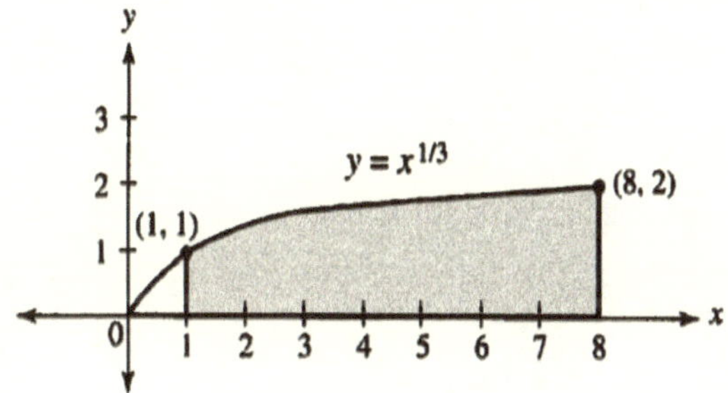

Figure 2

9. $A(S)$ = 18; see Figure 3.
10. $A(S)$ = 27/6; see Figure 4.

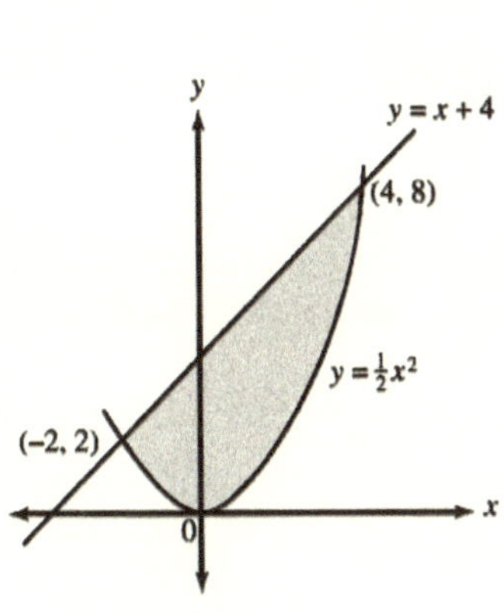

Figure 3

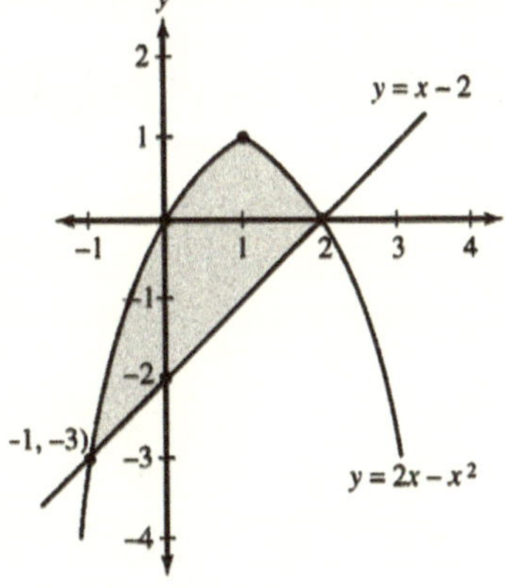

Figure 4

11. $\dfrac{2000}{0.09}(1-e^{-0.36})$
12. $\dfrac{1000}{0.07}(e^{0.42}-1)=\$7457$
13. $y=2x^3-x+3$
14. 5733
15. $c(x)=\dfrac{1}{10}x^2+10x+1500$
16. (a) $12x^3y-y^2$, (b) $3x^4-2yx$
    (c) $12x^3-2y$, (d) $36x^2y$, (e) $-2x$
17. (a) $20x^3y-y^2$, (b) $5x^4-2xy$, (c) $60x^2y$, (d) $-2x$, (e) $20x^3-2y$
18. $f_x(x,y)=3xy(3x^2y+y^3)^{-1/2}$

    $f_x(x,y)=\dfrac{1}{2}(3x^2y+y^3)^{-1/2}(3x^2+3y^2)$

    $f_{xy}(x,y)=-\dfrac{3xy}{2}(3x^2y+y^3)^{-3/2}(3x^2+3y^2)+(3x^2y+y^3)^{-1/2}$

    $f_{xx}(x,y)=-\dfrac{3xy}{2}(3x^2y+y^3)^{-1/2}6xy+3y(3x^2y+y^3)^{-1/2}$

    $f_{yy}(x,y)=(3x^2y+y^3)^{-1/2}3y-\dfrac{1}{4}(3x^2+3y^2)(3x^2y+y^3)^{-3/2}(3x^2+3y^2)$
19. *(a)* e^y, *(b)* xe^y, *(c)* 0, *(d)* xe^y, *(e)* e^y
20. $f(3,0)=-50$ is a minimum value; $f(-3,0)$ is not an extreme value.
21. $x=y=2$
22. $f(6,2)$ = 44 is a minimum value

INDEX

E

F

G

H

I

Companion to Calculus

William J. Adams

Pace University

CONTENTS

PREFACE

The objective of *Companion to Calculus* is to help you to optimize the use of *Fundamentals of Calculus with Applications* for studying calculus. It provides detailed solutions to, discussions of, and answers to the even numbered exercises.

Needless to say, the benefits of this supplement will only be reaped if it is used wisely. Wise use entails giving the exercises your best effort and only turning to this *Companion* when you are stuck or as a check when you have completed an exercise. Work as much as possible on your own. Reading worked out solutions first and then trying the problems is counterproductive in that it leads to a false perspective on the difficulties involved in the exercises. The only way to appreciate and come to grips with these difficulties is to experience them first hand.

W. J. A.

ANSWERS, SOLUTIONS, AND DISCUSSION OF EXERCISES

Chapter 1. Miscellaneous Preliminaries

Section 1.2 (p. 12)

2. 1
4. −6
6. −5
8. −2
10. 0
12. 6
14. 9xy
16. $9x^2$
18. 7ab
20. 6y + 9a + 12z
22. $-4w+(-2)v+(-3)z$
24. $(-2)a+(-4)b+(-6)c$
26. $-x+(-2)y+1$
28. $x^2+(-1)y$
30. 9(n + m)
32. 3a(b + 2c)
34. $3x^2(y + 2z)$
36. 2a(b + 2c + 4d)
38. 2mn(p + 2v + 5w)
40. m(3xy + 2y + 5x)
42. $2w^2(2v + x + 4)$
44. 3rs(t + 2 + 4k)
46. No; since their sum is not 0.
48. 0 is the additive inverse of 0 because 0 + 0 = 0.
50. $-6, 1, -1.4, -\sqrt{2}, -\pi$, respectively.

Section 1.2 (p. 18)

52. −2	54. −7	56. −7
58. 8	60. 3	62. 11
64. 2	66. −9	68. −1/2
70. −8	72. −1/8	74. not defined
76. −1/16	78. $-9a+6b+9c$	80. $a-2b-4c$
82. $-x+2y+3z$	84. ab	86. 0
88. $-xy+6y$	90. $b(4a^2-3x-1)$	92. $2y^2(x-2+3x)$

94. $gh(1-3h)$

96. $xy(3xy^2-2y+1)$

Section 1.3 (p. 23)

2. No
4. Yes
6. No
8. No
10. $\frac{1}{3}$
12. $\frac{2x}{3y}$
14. $\frac{3x+1}{x+1}$
16. $\frac{3}{x+3}$
18. $\frac{3a}{4x}, x \neq y$
20. $2, a \neq -2b$
22. $\frac{x+y+1}{y+1}$
24. $1, x \neq 5$
26. $2, x \neq -5$
28. $-2, a \neq 5$
30. $\frac{2+3b}{2(2+3ab)}, a \neq 0$
32. $3 + 4z, x \neq 0, y \neq 0$
34. $-x, y \neq -\frac{1}{3}$
36. $1/3, yz \neq 3ys$
38. $\frac{5}{x}, x \neq y$
40. $x \neq 3;\ x \neq 5;\ x \neq 0;\ x \neq -5;\ x \neq 2$

Section 1.3 (p. 29)

42. No
44. Yes
46. No
48. $\frac{2}{7}$
50. $\frac{10a}{9c}, b \neq 0$
52. $\frac{x^2}{z}, y \neq 0$
54. $\frac{x}{3}$
56. $\frac{x}{x+1}$
58. $-\frac{1}{3}, x \neq 5$
60. $2a, a \neq 1$
62. $\frac{2a}{b^2}, a \neq -\frac{1}{4}$
64. $3, x \neq 0, v \neq 3$
66. $-\frac{1}{4}$
68. 6
70. $-\frac{1}{9}$
72. $\frac{3(y-1)}{y}, x \neq 0, y \neq 1$
74. $\frac{(x-1)^2}{6y}, x \neq 1$
76. $\frac{a}{a+b}, a \neq b$
78. $4, a \neq -4$

Section 1.4 (p. 32)

2. $2x^4 + x^3 - x^2 + 9$
4. $2x^3 - x^3h - 7x^2h - xh^2 + 4x$
6. $2x^2 - 4x + 6$
8. $2x^2h + 7xh - h + 7$
10. $x^2h^2 + 3x^2h + 5xh + 4$

Section 1.4 (p. 34)

12. $4x^3 + 2x$
14. $6x^3 - 6x^2 + 8x$
16. $-6x^4 - 8x^3 + 18x^2$
18. $15x^3h - 9x^2h^2 + 12xh$
20. $x^2 - 9$
22. $x^2 - 2hx + h^2$
24. $5x^2 - 14x - 3$
26. $10x^2 - 26x + 12$
28. $x^3 + 2x^2 + x - 4$
30. $2hx^3 + 4hx^2 - 3x + 2h^2x^2 + 4h^2x - 3h$
32. $8x^2 - 14x - 30$
34. $2hx^4 + 6hx^3 - 8x^2 + h^2x^2 + 3h^2x - 4h$
36. $4x^3 + 12x^2 - 40x$
38. $x^4 - 4hx^3 + 6h^2x^2 - 4h^3x + h^4$
40. $6x^3 + 38x^2 + 66x + 18$
42. $x^5 - 5hx^4 + 10h^2x^3 - 10h^3x^2 + 5h^4x - h^5$

Section 1.4 (p. 40)

44. $a(3a + 2b + 1)$
46. $3b(1 + 2b + 4b^2)$
48. $ab(2 - 3ab + a^2b)$
50. $(x + 3)(x + 1)$
52. $(x - 3)(x + 3)$
54. cannot be factored
56. $2(x - 8)(x - 1)$
58. $2y(y - 4)(y - 2)$
60. $(x - 12)(x - 6)$
62. $(x - 8)^2$
64. $x + 3, x \neq 3$
66. $t - 5, t \neq 1$
68. $x - 2, x \neq -2$
70. $x - 8, x \neq 8$

Section 1.4 (p. 42)

72. -6 and 1
74. -3 and -1
76. -11 and -1
78. 6 and 12

80. $\frac{1}{2}$ and $\frac{5}{2}$

82. $\frac{-7 \pm \sqrt{41}}{2}$

84. $\frac{8 \pm \sqrt{56}}{4}$

86. $\frac{2 \pm \sqrt{32}}{2}$

Section 1.5 (p. 47)

2. $\frac{1}{9}$

4. 1

6. $\frac{1}{9}$

8. $\frac{8}{81}$

10. $\frac{(-2)^{-4}}{5^{-2}} = \frac{5^2}{(-2)^4} = \frac{25}{16}$

12. $[(-2)^{-3}]^{-2} = (-2)^6 = 64$

14. $\frac{-1}{432}$

16. $\frac{3y}{x^2}$

18. $\frac{3b^2}{a}$

20. $\frac{y^2}{x^2}$

22. $(1 + a)^4$

24. $(m + 2n)^2$

26. $\frac{1}{2x^3y}$

28. $6x^6 + 8$

30. $\frac{2y^6}{3z}$

32. $\frac{1}{16a^4b^6}$

34. $72x^7y$

36. $\frac{x^4}{y^8}$

Section 1.5 (p. 50)

38. -3

40. 7

42. 2

44. -5

46. -9

48. 125

50. $\frac{1}{3}$

52. No; $\sqrt{(-2)^2}$ expresses the principal square root of $(-2)^2 = 4$, which is (positive) 2.

Section 1.5 (p. 52)

54. $6\sqrt{2}$ 56. $10\sqrt{3}$ 58. $20\sqrt{2}$

60. $3\sqrt{5}$ 62. $5\sqrt[3]{2}$ 64. $3\sqrt[3]{4}$

66. $3a^2\sqrt{c}$ 68. $\sqrt{x^2-4}$ 70. $x+1, x \geq -1$

Section 1.5 (p. 55)

72. 7 74. $\frac{1}{8}$ 76. $\frac{1}{1000}$

78. $\frac{1}{6}$ 80. $8^{-4/3} = \frac{1}{8^{4/3}} = \frac{1}{(\sqrt[3]{8})^4} = \frac{1}{2^4} = \frac{1}{16}$

82. $\frac{1}{4}$ 84. $\frac{1}{\sqrt{2}}$ 86. $32^{1/6}$

88. 3 90. x^{10} 92. $x^{1/2}$

94. $(x^{-1/4}y^{1/2})^4 = (x^{-1/4})^4(y^{1/2})^4 = x^{-1}y^2 = \frac{y^2}{x}$

96. $\frac{(\sqrt[3]{x})^2}{\sqrt[3]{x^5}} = \frac{(x^{1/3})^2}{(x^5)^{1/3}} = \frac{x^{2/3}}{x^{5/3}} = \frac{1}{x}$

Section 1.6 (p. 62)

2. $-3x+y=6$ 4. $x + 2y = 2$ 6. $-x+2y=5$

8. $y=1$ 10. $-2x+y=1$

Chapter 2. Functions

Section 2.1 (p. 66)

2. $f(0) = 1, f(1) = 3, f(2) = 15, f(3) = 43, f(-1)=3, f(-2)=3$

4. $x \geq -1$; the square root of x +1 is not defined within the real-number system for x < 1. f(–2) is not defined; $f(-1) = 0$, f(3) = 2, $f(2) = \sqrt{3}$.
6. All real numbers except 2; 2 cannot be substituted for x because this leads to division by zero, which is not defined.
8. $f(1) = -1$, f(2) = 1, $f\left(\frac{1}{2}\right) = -1$, $f(0) = -1$, $f(-1) = -1$

Section 2.2 (p. 75)

2.

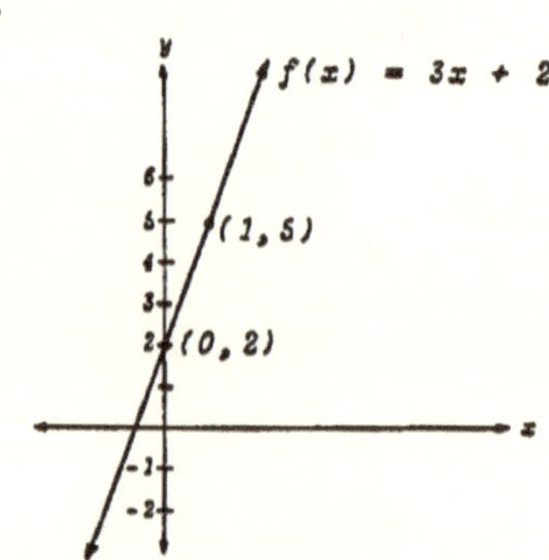

Figure 2.1

4.

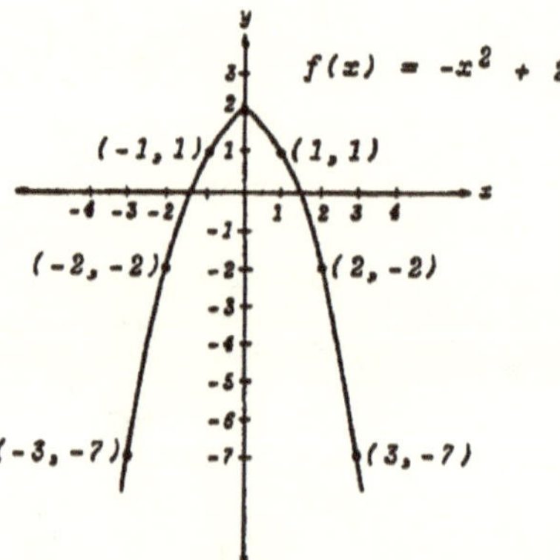

Figure 2.2

6.

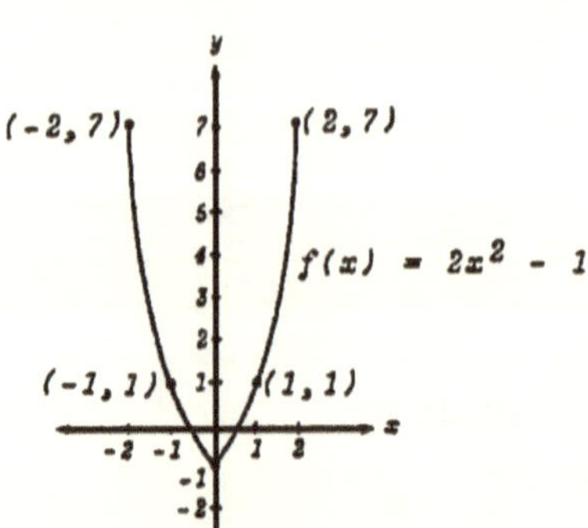

Figure 2.3

8.

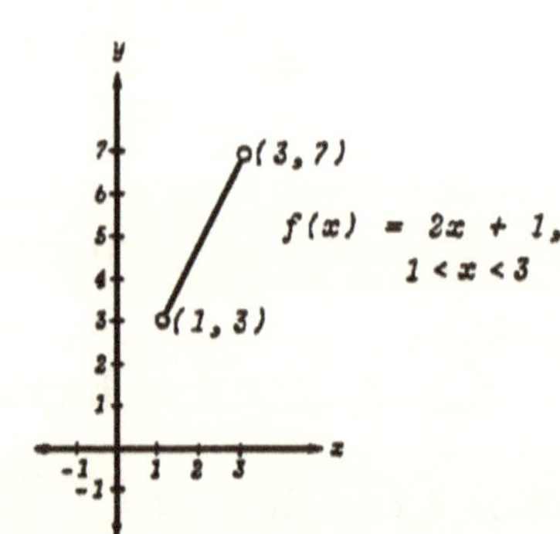

Figure 2.4

10.

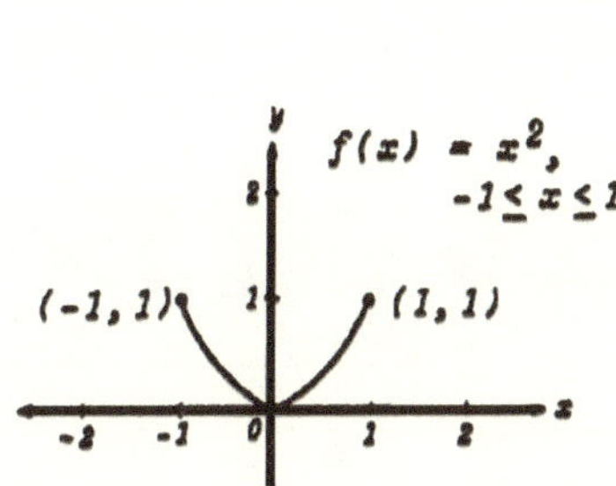

Figure 2.5

12.

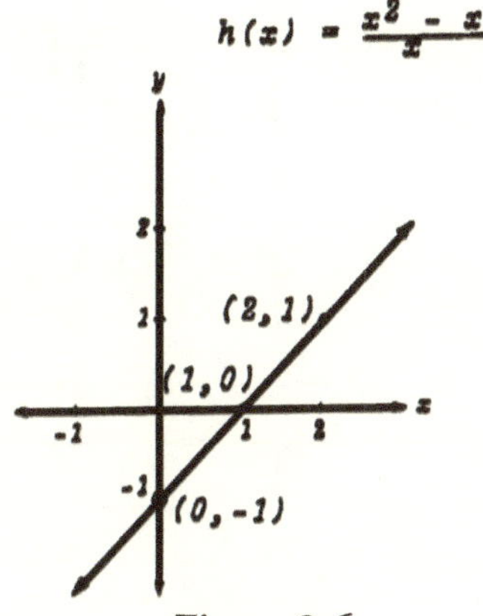

Figure 2.6

14.

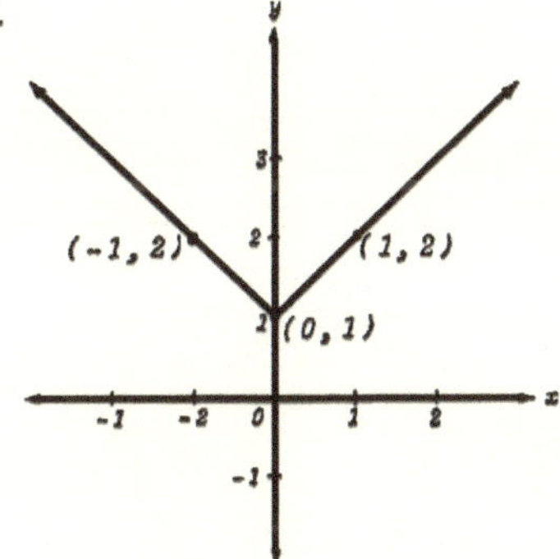

Figure 2.7

16.

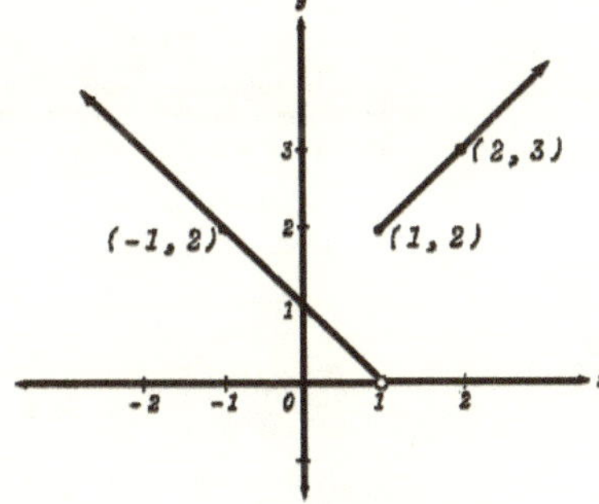

Figure 2.8

18.

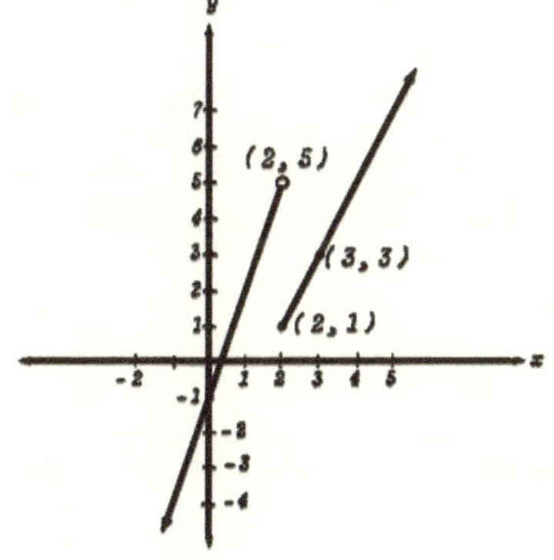

Figure 2.9

20.

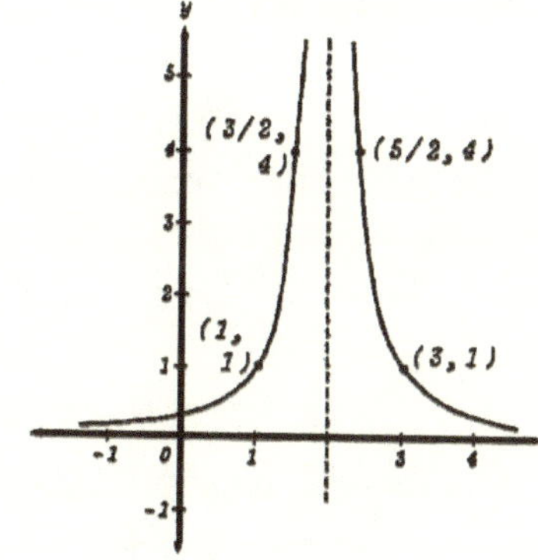

Figure 2.10

22.

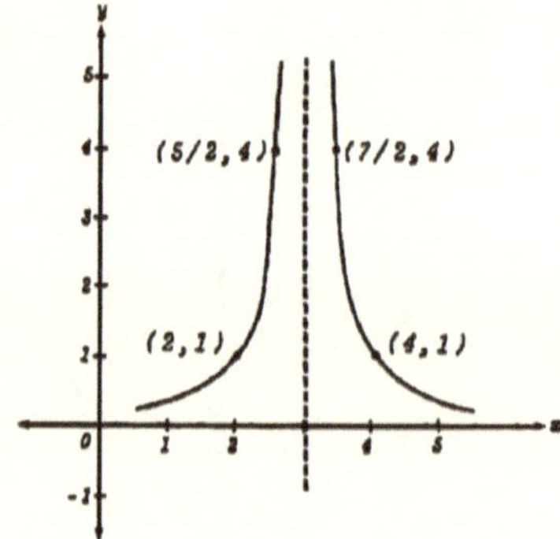

Figure 2.11

24.

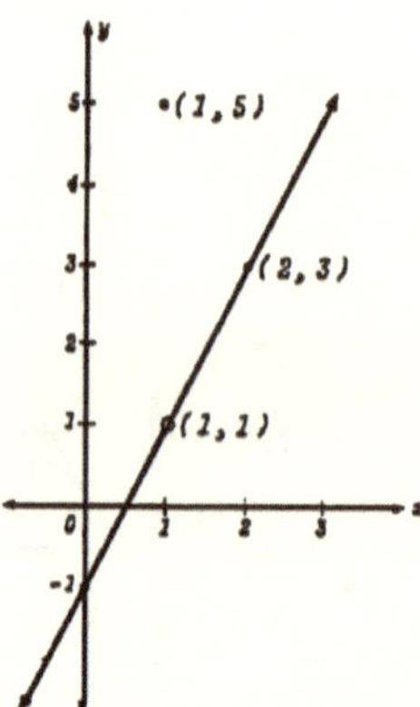

Figure 2.12

26.

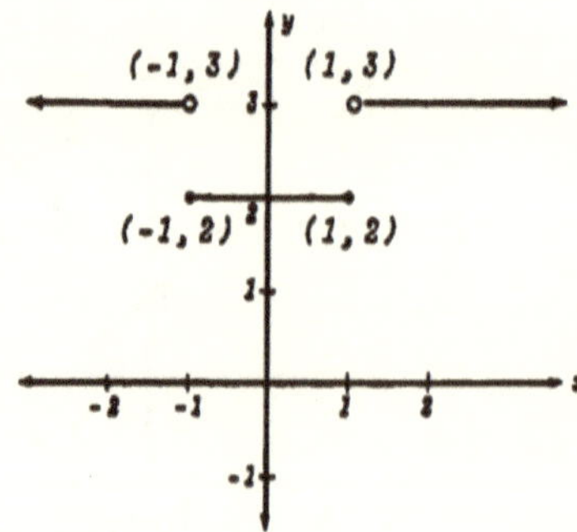

Section 2.3 (p. 83)

2. $I(x) = (200 - 10x)(100 + x)$, where x is the number of taxis added to the fleet. The domain of definitive of I(x) is 0, 1, 2, ..., 19.

4. $y = 2(2^t) = 2^{t+1}$; for t = 12, $y = 2^{13} = 8192$

6. $C(x) = \begin{cases} 32, \text{ where } 0 < x \leq 1 \\ 64, \text{ where } 2 < x \leq 2, \\ 96, \text{ where } 2 < x \leq 3 \end{cases}$ subject to the next increase in rates, where x is the weight of the letter in ounces.

Chapter 3. Limits of Functions

Section 3.1 (p. 93)

2. Values of $f(x) = 4x - 3$ for values of x getting closer and closer to 2 are getting closer and closer to 5. Thus $\lim_{x\to 2} (4x - 3) = 5$.

4. From Table 3.1 we see that values of $f(x) = x^2$ for values of x getting closer and closer to 3 are getting closer and closer to 9. Thus $\lim_{x\to 3} x^2 = 9$.

Table 3.2

x	2.9	3.1	2.99	3.01	2.999	3.001
f(x)	8.41	9.61	8.94	9.06	8.994	9.006

6. $f(x) = \dfrac{x^2 - 1}{x - 1} = \dfrac{(x-1)(x+1)}{x-1} = x + 1$, where $x \neq 1$. As x approaches 1, f(x) approaches 2. Thus $\lim_{x\to 1} f(x) = 2$. The fact that f(1) is not defined is irrelevant to the limit behavior of f(x) as x approaches 1. The graph of f(x) is shown in Figure 3.1.

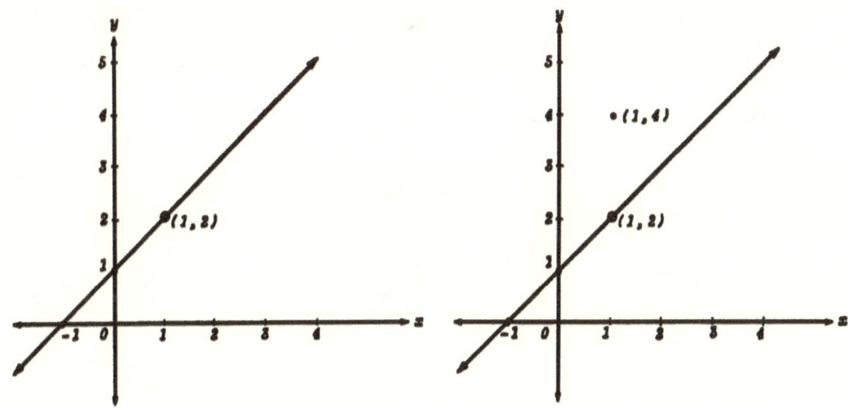

Figure 3.1 *Figure 3.2*

8. The graph of $f(x)=\begin{cases}\dfrac{x^2-1}{x-1}, \text{for } x \neq 1\\ 4, \text{for } x=1\end{cases}$ is shown in Figure 3.2. The fact that f(1) is defined in this case is irrelevant to the limit behavior of f(x) as x approaches 1. $\underset{x\to 1}{\text{limit}}\, f(x)=2$, since as x approaches 1, f(x) in turn approaches 2.

10. Values of $f(x)=\begin{cases}1, \text{ for } x \geq 0\\ -1, \text{ for } x<0\end{cases}$ for values of x that are getting closer and closer to 1 are shown in Table 3.2, from which we see that as x takes on values closer and closer to 1, f(x) remains constant at 1. Thus $\underset{x\to 1}{\text{limit}}\, f(x)=1$.

Table 3.2

x	.9	1.1	.99	1.01	.999	1.001
f(x)	1	1	1	1	1	1

12. Values of $f(x)=\begin{cases}x^2, \text{ for } x \neq 2\\ 5, \text{ for } x=2\end{cases}$ for values of x that are getting closer and closer to 2 are shown in Table 3.3, from which we see that as x takes on values closer and closer to 2, f(x) approaches 4. Thus $\underset{x\to 2}{\text{limit}}\, f(x)=4$. The behavior of f(x) at 2 (f(2) = 5; see Figure 3.3) is irrelevant to the limit behavior of f(x) as x approaches 2.

Table 3.3

x	1.9	2.1	1.99	2.01	1.999	2.001
f(x)	3.61	4.41	3.96	4.04	3.996	4.004

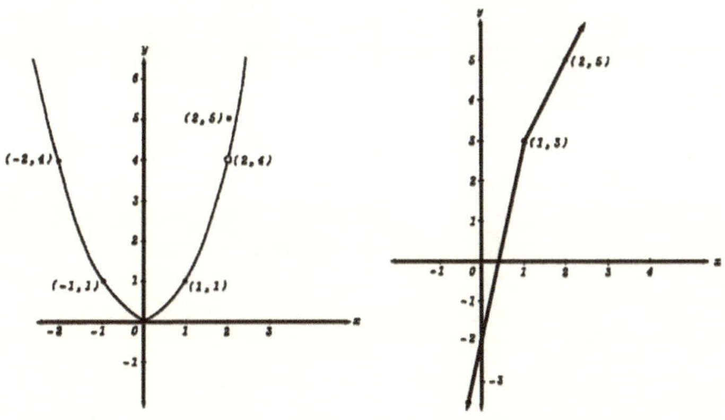

Figure 3.3 *Figure 3.4*

14. Values of $f(x)=\begin{cases}3, & \text{for } x \geq 1\\ 2, & \text{for } x<1\end{cases}$ for values of x that are getting closer and closer to 1 are shown in Table 3.4, from which we see that as x takes on values that are closer and closer to 1, f(x) is unstable in that it fluctuates between 3 and 2. Thus $\underset{x\to 1}{\text{limit}}\, f(x)$ does not exist.

Table 3.4

x	.9	1.1	.99	1.01	.999	1.001
f(x)	2	3	2	3	2	3

16. Values of $f(x)=\begin{cases}2x+1, & \text{for } x \geq 1\\ 5x-2, & \text{for } x<1\end{cases}$ for values of x that are getting closer and closer to 1 are shown in Table 3.5, from which we see that as x takes on values closer and closer to 1, f(x) gets closer and closer to 3. Thus $\underset{x\to 1}{\text{limit}}\, f(x)=3$. The graph of f(x) is shown in Figure 3.4.

Table 3.5

x	.9	1.1	.99	1.01	.999	1.001
f(x)	2.5	3.2	2.95	3.02	2.99	3.002

18. $\lim_{x \to 3} f(x) = 10$. As x takes on values closer and closer to 3, f(x) gets closer and closer to 10.

20. Values of $f(x) = \dfrac{1}{x-1}$ for values of x that are getting closer and closer to 1 are shown in Table 3.6. The graph of f(x) is shown in Figure 3.5. From Table 3.7 and Figure 3.5 we see that

Table 3.6

x	$1-\frac{1}{10}$	$1+\frac{1}{10}$	$1-\frac{1}{100}$	$1+\frac{1}{100}$
f(x)	−10	10	−100	100

f(x) is unstable as x approaches 1. As x approaches 1 through values less than 1 (such as $1-\frac{1}{10}$, $1-\frac{1}{100}$, etc.) f(x) decreases without bound; as x approaches 1 through values greater than 1 (such as $1+\frac{1}{10}$, $1+\frac{1}{100}$, etc.), f(x) increases without bound. Thus $\lim_{x \to 1} \dfrac{1}{x-1}$ does not exist.

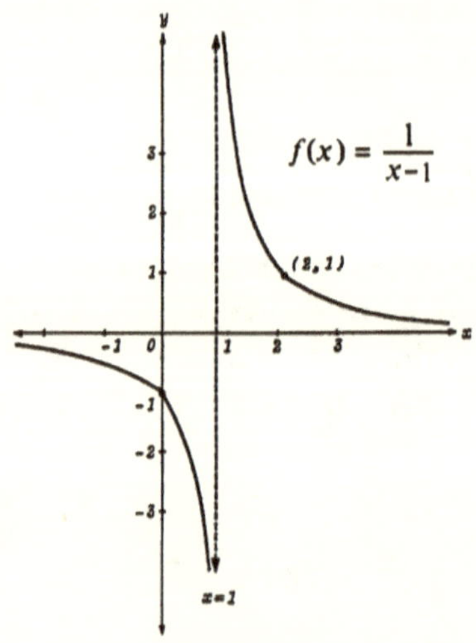

Figure 3.5

Section 3.2 (p. 101)

2.	48	4.	−7	6.	164/3
8.	30	10.	−182	12.	−1/3
14.	−69/31	16.	−14	18.	26/29

20. We first set up and simplify $g(h) = \frac{f(4+h) - f(4)}{h}$, where f(x) = 2x + 1.

$f(4 + h) = 2(4 + h) + 1 = 2h + 9$
$f(4) = 9$
$$\frac{f(4+h) - f(4)}{h} = \frac{2h}{h} = 2, \text{ where } h \neq 0.$$
Thus $\underset{h\to 0}{\text{limit}}\, g(h) = \underset{h\to 0}{\text{limit}}\, 2 = 2$.

22. $f(x + h) = (x + h)^2 = x^2 + 2xh + h^2$
$f(x) = x^2$
$f(x+h) - f(x) = x^2 + 2hx + h^2 - x^2 = 2hx + h^2$
Thus $g(h) = \frac{f(x+h) - f(x)}{h} = \frac{2hx + h^2}{h} = \frac{h(2x+h)}{h} = 2x + h$, where $h \neq 0$.
We therefore obtain $\underset{h\to 0}{\text{limit}}\, g(h) = \lim_{h\to 0} 2(x + h) = 2x$.

24. $f(x+h) = 2(x+h)^2 - 1 = 2x^2 + 4hx + 2h^2 - 1$
$f(x) = 2x^2 - 1$
$f(x+h) - f(x) = 4hx + 2h^2$
Thus $g(h) = \frac{f(x+h) - f(x)}{h} = \frac{4hx + 2h^2}{h} = 4x + 2h$, where $h \neq 0$..
We have, $\underset{h\to 0}{\text{limit}}\, g(h) = \underset{h\to 0}{\text{limit}}\, (4x + 2h) = 4x$.

Section 3.3 (p. 106)

2. $\underset{x\to 3}{\text{limit}}\, f(x) = 32$, f(3) = 32. Since $\underset{x\to 3}{\text{limit}}\, f(x) = f(3)$,
$f(x) = 3x^2 + 2x - 1$ is continuous at 3.

4. Since f(2) is not defined, $f(x) = \frac{x^2 - 4}{x - 2}$ is not continuous at 2.

6. Since $\lim_{x\to 0} f(x)$ does not exist, $f(x)=\begin{cases}1, & \text{for } x\geq 0\\ -1, & \text{for } x<0\end{cases}$ is not continuous at 0.

8. $\lim_{x\to 3} f(x)=259$, f(3) = 259. Since $\lim_{x\to 3} f(x)=f(3)$, $f(x)=3x-2)(4x^2+1)$ is continuous at 3.

10. Since $\lim_{x\to 2} f(x)$ does not exist, $f(x)=\begin{cases}2x-3, & \text{for } x\geq 2\\ 3x-1, & \text{for } x<2\end{cases}$ is not continuous at 2.

Section 3.4 (p. 114)

2. Let us observe that as *x* takes on values that decrease without bound, $-4/x$ goes to 0 (see Table 3.7). Thus $f(x)=-2-\frac{4}{x}$ gets closer and closer to -2 as $x\to\infty$. That is, $\lim_{x\to-\infty}\left(-2-\frac{4}{x}\right)=-2$.

Table 3.7

x	10	100	1000	10,000
−4/x	−0.4	−0.04	−0.004	−0.0004

4. As x takes on values that approach 0, f(x) = 1/x is unstable (see Table 3.8) in that f(x) takes on values that increase without bound and values that decrease without bound. Thus $\lim_{x\to 0}\frac{1}{x}$ does not exist. The fact that f(0) is not defined is irrelevant to the limit behavior of f(x) as x approaches 0.

Table 3.8

x	$-\frac{1}{10}$	$\frac{1}{10}$	$-\frac{1}{100}$	$\frac{1}{100}$	$-\frac{1}{1000}$	$\frac{1}{1000}$
1/x	−10	10	−100	100	−1000	1000

6. As x takes on values that approach 1, $f(x)=\frac{1}{x-1}$ is unstable in that f(x) takes on values that increase without bound and decrease without bound. Thus $\lim_{x\to 1}\frac{1}{x-1}$ does not exist. The fact that f(1) is undefined is irrelevant to the limit behavior of f(x) as x approaches 1.
8. As x takes on values that approach 0, $f(x)=2-1/x^2$ decreases without bound. Thus $\lim_{x\to 0} f(x)=-\infty$.
10. As x approaches 4, $f(x)=6+\frac{2}{(x-4)^2}$ increases without bound (see Table 3.9). Thus $\lim_{x\to 4} f(x)=\infty$.

Table 3.9

x	$4\pm\frac{1}{10}$	$4\pm\frac{1}{100}$	$4\pm\frac{1}{1000}$
f(x)	206	10,006	1,000,006

12. As

$$x\to\infty, \frac{x^2+3x-2}{3x^2+2x-1}=\frac{1\pm\frac{3}{x}-\frac{2}{x^2}}{3+\frac{2}{x}-\frac{1}{x^2}}\to\frac{1}{3} \text{ since } \frac{3}{x}\to 0,\ \frac{2}{x^2}\to 0,\ \frac{2}{x}\to 0,\ \frac{1}{x^2}\to 0$$

That is, $\lim_{x\to\infty}\frac{x^2+3x-2}{3x^2+2x-1}=\frac{1}{3}$.
14. $\lim_{x\to -1} f(x)=\infty$, since as x takes on values that approach −1, f(x) increases without bound.
16. As x takes on values that approach 5, f(x) is unstable. Thus $\lim_{x\to 5} f(x)$ does not exist.
18. As $x\to-\infty$, $\frac{2}{x}\to 0$ and $f(x)=5+\frac{2}{x}\to 5$. That is, $\lim_{x\to-\infty} f(x)=5$.
20. $\lim_{x\to\infty} f(x)=3e^{-5}$

Chapter 4. An Application to Finance

Section 4.1 (p. 122)

2. (a) \$2846.62
   (b) \$2851.52
   (c) \$2856.50

4. \$2768.40

6. 8.286%

8. For 12% per annum compounded 3 times a year, v = 12.486%; for 11% per annum compounded 11 times a year, v = 11.567%.

Chapter 5. Derivatives of Functions

Section 5.1 (p. 160)

2. By definition, $f'(3) = \underset{h\to 0}{\text{limit}} \dfrac{f(3+h)-f(3)}{h}$, where $f(x) = x^2 + 50$.

   Our first task is to set up and simplify $\dfrac{f(3+h)-f(3)}{h}$.

   $f(3+h) = (3+h)^2 + 50 = h^2 + 6h + 59$
   $f(3) = 59$
   $f(3+h) - f(3) = h^2 + 6h = h(h+6)$
   $\dfrac{f(3+h)-f(3)}{h} = \dfrac{h^2+6h}{h} = h + 6$, where $h \neq 0$

   Thus $f'(3) = \underset{h\to 0}{\text{limit}}\,(h+6) = 6$.

4. By definition, $f'(-2) = \underset{h\to 0}{\text{limit}} \dfrac{f(-2+h)-f(-2)}{h}$, where $f(x) = 3x^2 - 2x + 1$.

   We first set up and simplify $\dfrac{f(-2+h)-f(-2)}{h}$.

   $f(-2+h) = 3(-2+h)^2 - 2(-2+h) + 1 = 3h^2 - 14h + 17$
   $f(-2) = 17$

$$f(-2+h)-f(-2)=3h^2-14h=h(3h-14)$$
$$\frac{f(-2+h)-f(-2)}{h}=3h-14, \text{ where } h\neq 0$$

Thus $f'(-2)=\lim_{h\to 0}(3h-14)=-14$.

6. By definition, $f'(x)=\lim_{h\to 0}\frac{f(x+h)-f(x)}{h}$, where $f(x)=2x+5$.

   We first set up and simplify $\frac{f(x+h)-f(x)}{h}$.

   $$f(x+h)=2(x+h)+5=2x+2h+5$$
   $$f(x)=2x+5$$
   $$\frac{f(x+h)-f(x)}{h}=\frac{2h}{h}=2, \text{ where } h\neq 0$$

   Thus $f'(x)=\lim_{h\to 0}2=2$.

8. By definition, $f'(x)=\lim_{h\to 0}\frac{f(x+h)-f(x)}{h}$, where $f(x)=x^2-2x-1$.

   We first set up and simplify $\frac{f(x+h)-f(x)}{h}$.

   $$f(x+h)=(x+h)^2-2(x+h)-1=x^2+2xh+h^2-2x-2h-1$$
   $$f(x)=x^2-2x-1$$
   $$f(x+h)-f(x)=2xh+h^2-2h=h(2x+h-2)$$
   $$\frac{f(x+h)-f(x)}{h}=2x+h-2, \text{ where } h\neq 0$$

   Thus $f'(x)=\lim_{h\to 0}(2x+h-2)=2x-2$

10. (a) The tangent line to the graph of $f(x) = x^2 + 3x + 100$ at P(2, 110) has slope 7.
    (b) If $f(x) = x^2 + 3x + 100$ represents the time-distance function of some object in motion, then 7 is the instantaneous velocity of the object at time 2.

(c) If $f(x) = x^2 + 3x + 100$ represents the cost function of a firm, then 7 is the marginal cost for an output of 2 units.

(d) If $f(x) = x^2 + 3x + 100$ represents the revenue function of a firm, then 7 is the marginal revenue for an output of 2 units.

Section 5.2 (p. 162)

2. The graph of $f(x) = \begin{cases} x, & \text{for } x \geq 0 \\ -x, & \text{for } x < 0 \end{cases}$ is shown in Figure 5.1.

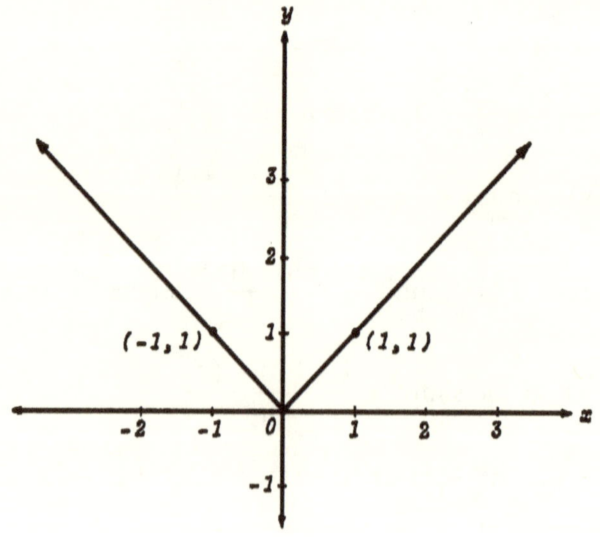

Figure 5.1

To determine if $f'(0)$ is defined, we use the definition of $f'(0)$. By definition

$$f'(0) = \underset{h \to 0}{\text{limit}} \frac{f(0+h) - f(0)}{h},$$

provided that this limit exists.

$$f(0+h)=f(h)=\begin{cases} h, & \text{for } h>0 \\ -h, & \text{for } h<0 \end{cases}$$

$$f(0)=0$$

$$\frac{f(0+h)-f(0)}{h}=\begin{cases} 1, & \text{for } h>0 \\ -1, & \text{for } h<0 \end{cases}$$

Since $\lim\limits_{h\to 0}\frac{f(0+h)-f(0)}{h}$ does not exist, $f'(0)$ is not defined.

To determine if f(x) is continuous at 0, we use the definition of continuity at 0.

(1) Let us observe that f(0) = 0.

(2) $\lim\limits_{x\to 0} f(x)=0$; as x approaches 0, f(x) approaches 0.

(3) Since $\lim\limits_{x\to 0} f(x)=f(0)$, f(x) is continuous at 0.

Geometrically speaking, the continuity of f(x) at 0 tells us that the graph of f(x) is in one piece at 0. Since $f'(0)$ does not exist, the graph of f(x) is not smooth enough to support a tangent line at P(0, 0).

4. f(x) = 2 is differentiable at 0 since $f'(0)$ exists $f'(0)=0$.) Since f(x) = 2 is differentiable at 0, f(x) = 2 is continuous at 0; differentiability at a value implies continuity at the value.

Chapter 6. Tools for Computing Derivatives

Section 6.1 (p. 166)

2. $f'(x) = 7x^6$; $f'(2) = 7(2)^6$

4. $f'(x)=\frac{-2}{x^3}$; $f'(1)=-2$

6. $f'(x) = 0$; $f'(\pi) = 0$

8. $f'(x)=\frac{-1}{4(\sqrt[4]{x})^3}$; $f'(16)=\frac{1}{32}$

10. $f'(x)=\frac{2}{3}x^{-1/3}$; $f'(-1)=-\frac{2}{3}$

12. $f'(x)=\frac{-1}{2(\sqrt{x})^3}$; $f'(4)=\frac{-1}{16}$

Section 6.2 (p. 171)

2. $f'(x) = 20x^9$

4. $f'(x) = 18x^5$

6. $f'(x) = -x^{-2/3}$

8. $f'(x) = -\frac{4}{3}x^{-4/3}$

10. $f'(x) = -2x^{-5/3}$

12. $f'(x) = 12x^2 - 10x$

14. $f'(x) = 18x^5 + 5 + 2x^{-3}$

16. $f'(x) = \frac{3}{4}x^2 - x$

18. $f'(x) = -3x^{-4} + x^{-2} - \frac{1}{2}x^{-1/2}$

20. $f'(x) = (x^2 - 1)2x + (x^2 + 2)2x$

22. $f'(x) = x^3(-2) + (1 - 2x)3x^2$

24. $f'(x) = (3x - 2)12x^2 + (4x^3 + 5)3$

26. $f'(x) = (3x^4 - 2x)28x^6 + (4x^7 + 13)(12x^3 - 2)$

28. $f'(x) = \dfrac{(1 - x^2)\,6x - (3x^2 + 2)\,(-2x)}{(1 - x^2)^2}$

30. $f'(x) = \dfrac{(3x + 2)\,3x^2 - (1 + x^3)\,3}{(3x + 2)^2}$

32. $f'(x) = \dfrac{(200 + x^4)\,2700x^2 - 900x^3\,(4x^3)}{(200 + x^4)^2}$

34. $f'(x) = \dfrac{(500 + 2x^4)\,4 - (4x - 500)\,(8x^3)}{(500 + 2x^4)^2}$

Section 6.3 (p. 178)

2. $6(2x^4 + x)^5\,(8x^3 + 1)$

4. $3(2x^3 - 5x^2 + 1)^2(6x^2 - 10)$

6. $\frac{1}{2}(5x^4 - 7x^2 + 13)^{-1/2}(20x^3 - 14x)$

8. $4(3x^5 - 4x^4 + 2)^3(15x^4 - 16x^3)$

10. $\dfrac{5\,(3x^2 - 4x)^4 - 4\,(5x + 1)\,(3x^2 - 4x)^3\,(6x - 4)}{(3x^4 - 4x)^8}$

12. $\frac{3}{2}x^2(4x^2 + 1)(x^3 + 6)^{-1/2} + 8x(x^3 + 6)^{1/2}$

14. $18x^2(3x + 7)^5 + 2x(3x + 7)^6$

16. $\dfrac{2x^4(4x+7)^{-1/2} - 4x^3(4x+7)^{1/2}}{x^8}$

Section 6.4 (p. 181)

2. $xD_xy + yD_x(x) - D_x(1) = D_x(0)$

$xD_xy + y = 0$

$D_xy = -\dfrac{y}{x}$

4. $x^3D_xy + yD_x(x^3) + D_x(12) = D_x(0)$

$x^3D_xy + 3x^2y = 0$

$D_xy = -\dfrac{3y}{x}$

6. $D_x(x^2) + D_x(y^2) = D_x(1)$

$2x + 2yD_xy = 0$

$D_xy = -\dfrac{x}{y}$

8. $D_x(4x^2) + D_x(9y^2) = D_x(25)$

$8x + 18yD_xy = 0$

$D_xy = -\dfrac{4x}{9y}$

10. $xD_xy + yD_xx = D_x4$

$xD_xy + y = 0$

$D_xy = -\dfrac{y}{x}$

12. $D_x(x^3) - D_xx = 3y^2D_xy - 2yD_xy + D_x(10)$

$3x^2 - 1 = (3y^2 - 2y)D_xy$

$$D_x y = \frac{3x^2 - 1}{3y^2 - 2y}$$

Section 6.5 (p. 183)

2. $x = \frac{1}{5}y - 24$

4. $y = \frac{x+1}{x}, \; y = \frac{x}{x} + \frac{1}{x}, \; y = 1 + \frac{1}{x}, \; y - 1 = \frac{1}{x}, \; x(y-1) = 1, \; x = \frac{1}{y-1}$

6. $y = \sqrt{x+4}, \; y^2 = x + 4, \; x = y^2 - 4$

8. $D_y x = \frac{1}{D_x y} = \frac{1}{3}$

10. $D_y x = \frac{1}{D_x y} = \frac{1}{3(x+1)^2 D_x (x+1)} = \frac{1}{3(x+1)^2}$

12. $D_x y = \frac{1}{D_y x} = \frac{1}{81y^2}$

14. We first rewrite $y = \frac{x+1}{x}$ so as to simplify the task of finding its derivative.

$$y = \frac{x+1}{x}, \; y = \frac{x}{x} + \frac{1}{x}, \; y = 1 + x^{-1}. \text{ Now } D_y x = \frac{1}{D_x y} = \frac{1}{-x^{-2}} = -x^2$$

Section 6.6 (p. 191)

4. $D_x y = 4e^x$

6. $D_x y = \frac{x^2 e^x - 2xe^x}{x^4}$

8. $D_x y = 2x^3 D_x(e^x) + e^x D_x(2x^3) = 2x^3 e^x + 6x^2 e^x$

10. $D_x y = -2e^{-2x+1}$

12. $D_x y = 2x^3 e^x + 6x^2 e^x$

14. $D_x y = 3e^x$

16. $D_x y = \dfrac{e^x D_x (x^2+1) + (x^2+1) D_x (e^x)}{(e^x)^2} = \dfrac{2xe^x + (x^2+1)(e^x)}{(e^{2x})}$

18. $D_x y = e^{3x+8} D_x (3x+8) = 3e^{3x+8}$

20. $D_x y = 5x^2 e^{5x} + 2xe^{5x}$

22. $D_x y = 7e^{-x} D_x (-x) = -7e^{-x}$

24. $D_x y = \dfrac{12e^x - 3(4x+3)e^x}{(9e^{2x})}$

26. $D_x y = 3x^2 D_x \left(e^{-x^2}\right) + \left(e^{-x^2}\right) D_x (3x^2)$

$= 3x^2 e^{-x^2} D_x (-x^2) + 6xe^{-x^2}$

$= -6x^3 e^{-x^2} + 6xe^{-x^2}$

Section 6.7 (pp. 193, 200)

2. $\log_{10} 1000 = 3$
4. $\log_2 8 = 3$
6. $\log_{10} 0.01 = -2$
8. $\log_{32}(1/2) = -1/5$
10. $\log_5 1 = 0$
12. $\log_{10} 0.001 = -3$
14. $3^0 = 1$
16. $12^2 = 144$
18. $8^{2/3} = 4$
20. $10^4 = 10{,}000$
22. 3
24. −3
26. 3/2

30. $D_x y = x^2 D_x (\ln x) + (\ln x) D_x (x^2) = x^2 \cdot \dfrac{1}{x} + 2x\ln(x) = x + 2x\ln(x)$

32. $D_x y = \dfrac{(\ln x) D_x (4x+3) - (4x+3) D_x (\ln x)}{(\ln x)^2} = \dfrac{4\ln x - \dfrac{(4x+3)}{x}}{(\ln x)^2}$

34. $D_x y = 8x + 2xe^{x^2} + \dfrac{1}{x}$

36. $D_x y = \dfrac{D_x (4x^2+5)}{4x^2+5} = \dfrac{8x}{4x^2+5}$

38. $y = \ln \dfrac{50}{x} = \ln 50 - \ln x, \quad D_x y = 0 - \dfrac{1}{x} = -\dfrac{1}{x}$

40. $D_x y = \dfrac{6x^2}{x^2+1} + 3\ln(x^2+1)$

42. $D_x y = e^{x^2} D_x(\ln x) + (\ln x) D_x\left(e^{x^2}\right) = \dfrac{e^{x^2}}{x} + (\ln x)e^{x^2} D_x(x^2)$

$= \dfrac{e^{x^2}}{x} + 2xe^{x^2} \ln x$

44. $D_x y = \dfrac{(\ln x) D_x (3x^2 + e^{-x}) - (3x^2 + e^{-x}) D_x (\ln x)}{(\ln x)^2} = \dfrac{(6x - e^{-x}) \ln x - \dfrac{(3x^2 + e^{-x})}{x}}{(\ln x)^2}$

Section 6.8 (p. 209)

2. For $f(x) = 2x^2 - x + 4$, $f'(x) = 4x - 1$. The slope of the tangent line T to the graph of $f(x) = 2x^2 - x + 4$ at $P(-1, 7)$ is $f'(-1) = -5$. The equation of T is $y = -5x + 2$.
4. $c'(x) = \dfrac{1}{2}x + 2$, $c'(50) = 27$
6. Before taking the derivative of c(x) to obtain the marginal cost function $c'(x)$ it is useful to put c(x) in a more convenient form for this purpose by multiplying out the numerator and denominator components. This yields

$$c(x) = \frac{x^3 + 50x^2}{50x + 10{,}000} + 75$$

$$c'(x) = \frac{(50x + 10{,}000)(3x^3 + 100x) - \left(x^3 + 50x^2\right)50}{(50x + 10{,}000)^2}$$

By multiplying and combining terms we obtain

$$c'(x) = \frac{100x^3 + 32{,}500x^2) + 1{,}000{,}000x}{(50x + 10{,}000)^2}$$

$$c'(10) = \frac{13{,}350{,}000}{110{,}250{,}000} = 0.121$$

8. $$R'(x) = \frac{(x+40)\,2000 - 2000x}{(x+40)^2} - 10 = \frac{80,000}{(x+40)^2} - 10$$

$$R'(10) = \frac{80,000}{2500} - 10 = 22$$

10. We first express R(x) in an equivalent form that is more convenient for differentiation.

$$\begin{aligned} R(x) &= 20x \ln \frac{70,000}{x} = 20x(\ln 70,000 - \ln x) \\ &= 20x(\ln 70,000) - 20x\ln x \end{aligned}$$

$$\begin{aligned} R'(x) &= 20(\ln 70,000) - [20xD_x(\ln x) + (\ln x)D_x(20x)] \\ &= 20(\ln 70,000) - 20x\frac{1}{x} - 20(\ln x) \\ &= 20(\ln 70,000) - 20 - 20(\ln x) \end{aligned}$$

12. $D_t^2 y = -100ke^{-kt}$

14. The marginal revenue function $R'(x)$ is, by the product theorem,

$$R'(x) = 30x \frac{d\left(\ln \frac{90,000}{x}\right)}{dx} + \left[\ln\left(\frac{90,000}{x}\right)\right]\frac{d\,(30x)}{dx} \qquad (6.1)$$

Now $\frac{d\,(30x)}{dx} = 30$. By using the chain rule we obtain

$$\frac{d\left(\ln \frac{90,000}{x}\right)}{dx} = -\frac{1}{x}$$

Substituting these results into equation (6.1) yields

$$R'(x) = -30 + 30\left(\ln \frac{90,000}{x}\right)$$

for the marginal revenue function.

16. $D_t d = 9t^2 + 5$; $D_t^2 d = 18t$

18. The point elasticity of demand at price p is defined by

$$\eta = -\frac{p}{f(p)} f'(p)$$

where x = f(p) expresses x, the quantity purchased per unit time, as a function of p, the unit price. For p = 64,

$$\eta = -\frac{64}{f(64)} f'(64)$$

Since $f(p) = 25 - \frac{1}{2}\sqrt{p}$, $f'(p) = -\frac{1}{4}p^{-1/2} = -\frac{1}{4\sqrt{p}}$.

Thus $f'(64) = -\frac{1}{4\sqrt{64}} = -\frac{1}{32}$.

Therefore,

$$\eta = -\frac{64}{21}\left(-\frac{1}{32}\right) = \frac{2}{21}.$$

The total revenue function R(x) is obtained by multiplying output x by price $p = (50 - 2x)^2$.

$$\begin{aligned} R(x) &= x(50 - 2x)^2 \\ &= x(2500 - 200x + 4x^2) \\ &= 2500x - 200x^2 + 4x^3 \end{aligned}$$

The marginal revenue function $R'(x)$ is

$$R'(x) = 2500 - 400x + 12x^2.$$

When p = 64, x = 21. The marginal revenue corresponding to a price of 64¢ a gallon is thus

$$R'(21) = 2500 - 400(21) + 12(21)^2 = -608.$$

20. Proof. Since R = xp, where p = g(x), we have by the product theorem

$$\frac{dR}{dx} = x\frac{dp}{dx} + p\frac{dx}{dx} = x\frac{dp}{dx} + p = p + x\frac{dp}{dx} \tag{6.2}$$

By factoring p from the right member of equation (6.2) we obtain

$$\frac{dR}{dx} = p\left(1 + \frac{x}{p}\cdot\frac{dp}{dx}\right).$$

To complete our proof we must show that

$$\frac{x}{p}\cdot\frac{dp}{dx} = -\frac{1}{\eta}$$

Since

$$\eta = -\frac{p}{x}\cdot\frac{dx}{dp}, \text{ then } -\frac{1}{\eta} = \frac{x}{p}\cdot\frac{1}{\frac{dx}{dp}} = \frac{x}{p}\cdot\frac{dp}{dx}$$

Section 6.9 (p. 213)

2. $f'(x) = 12x^2 - 4x, \quad f''(x) = 24x - 4, \quad f'''(x) = 24$
   $f'(-2) = 56, \quad f''(-2) = -52, \quad f'''(-2) = 24$

4. $f'(x) = \frac{1}{5}x - 2, \quad f''(x) = \frac{1}{5}, \quad f'''(x) = 0$

   $f'(-2) = -\frac{62}{5}, \quad f''(-2) = \frac{1}{5}, \quad f'''(-2) = 0$

6. $f'(x) = 3x + 4 + \frac{1}{x^2}, \quad f''(x) = 3 - \frac{2}{x^3}, \quad f'''(x) = \frac{6}{x^4}$

   $f'(-2) = -\frac{7}{4}, \quad f''(-2) = \frac{13}{4}, \quad f'''(-2) = \frac{3}{8}$

Chapter 7. Optimization Problems

Section 7.1 (pp. 224, 232)

2. $f(x) = 3x^2 + 12x + 2$. To find type 1 critical values we determine $f'(x)$, set $f'(x)$ equal to zero and solve for x.

$$f'(x) = 6x + 12$$
$$6x + 12 = 0$$
$$x = -2$$

To determine if $f(-2)$ is an extreme value we employ the first derivative test and examine the behavior of $f'(x) = 6(x + 2)$ for $x < -2$ and $x > -2$ in a small interval about -2.

For $x < -2$, $f'(x) = 6(x + 2)$ is negative.

For $x > -2$, $f'(x) = 6(x + 2)$ is positive.

Since $f'(-2) = 0$ and $f'(x)$ goes from negative to positive, $f(-2) = -10$ is a minimum value. There are no end point critical values and there are no values at which $f'(x)$ does not exist.

4. $f(x) = -3x^2 - 6x + 8, f'(x) = -6x - 6 = -6(x+1)$. Setting $f'(x)$ equal to zero and solving for x yields

$$-6(x+1) = 0, x = -1$$

For $x < -1, f'(x) = -6(x+1)$ is positive.

For $x > -1, f'(x) = -6(x+1)$ is negative.

From the first derivative test it follows that $f(-1)$ is a maximum value.

There are no end point critical values and there are no values at which $f'(x)$ does not exist.

6. $f(x) = 2x^3 + 3x^2$, $f'(x) = 6x^2 + 6x = 6x(x + 1)$. From $6x(x + 1) = 0$ we obtain the critical values $x = 0$ and $x = -1$.

   For $-1 < x < 0, f'(x) = 6x(x+1)$ is negative.

   For $x > 0$, $f'(x) = 6x\,(x + 1)$ is positive.

   Thus $f(0) = 0$ is a minimum value.

   For $x < -1$, $f'(x) = 6x(x+1)$ is positive.

   For $x > -1\,(-1 < x < 0)$, $f'(x) = 6x(x+1)$ is negative.

   Thus $f(-1) = 1$ is a maximum value.

   There are no end point critical values and there are no values at which $f'(x)$ does not exist.

8. $f(x) = \frac{1}{3}x^3 + x^2 - 3x + 4$, $f'(x) = x^2 + 2x - 3$. From $x^2 + 2x - 3 = 0$ we obtain the critical values $x = -3$ and $x = 1$. $f(-3) = 13$ is a maximum value; $f(1) = 7/3$ is a minimum value.

10. $f(x) = x^3 - 2x^2 - 4x + 3$, $f'(x) = 3x^2 - 4x - 4$. From $3x^2 - 4x - 4$ we obtain $(3x+2)(x-2) = 0$, which yields $x = -2/3$ and $x = 2$. $f(-2/3) = 4.48$ is a maximum value and $f(2) = 5$ is a minimum value.

12. $f(x) = x^3 - 7x^2 - 24x + 2$, $f'(x) = 3x^2 - 14x - 24$, $f''(x) = 6x - 14$. From $3x^2 - 14x - 24 = 0$ we obtain $(3x+4)(x-6) = 0$, which yields the critical values $x = -4/3$ and $x = 6$. Since $f''(-4/3) = -22 < 0, f(-4/3) = 19.19$ is a maximum value; since $f''(6) > 0, f(6) = -178$ is a minimum value.

14. $f(x) = x^4 + \frac{7}{3}x^3 - x^2 + 4$, $f'(x) = 4x^3 + 7x^2 - 2x$, $f''(x) = 12x^2 + 14x - 2$. From $f'(x) = 0$ we obtain $x(4x-4)(x+2) = 0$, which yields the

critical values $x = 0$, $x = 1/4$ and $x = -2$. Since $f''(0) = -2 < 0, f(0) = 4$ is a maximum value; $f'(1/4) > 0$, and thus $f(1/4) = 3.98$ is a minimum value; $f''(-2) > 0$, and thus $f(-2) = -2.67$ is a minimum value.

16. $f(x) = x + 10 + \frac{100}{x}$, $f'(x) = 1 - \frac{100}{x^2}$, $f''(x) = \frac{200}{x^3}$. From $f'(x) = 0$ we obtain the critical values $x = -10$ and $x = 10$. $f(-10) = -10$ is a maximum value; $f(10) = 30$ is a minimum value.

18. $f(x) = \frac{1}{4}x + 5 + \frac{1}{x}$, $f'(x) = \frac{1}{4} - \frac{1}{x^2}$, $f''(x) = \frac{2}{x^3}$. From $f'(x) = 0$ we obtain the critical values $x = -2$ and $x = 2$. $f(-2) = 4$ is a maximum value; $f(2) = 6$ is a minimum value.

20. $f(x) = x(\ln x)$, $f'(x) = 1 + \ln x$, $f''(x) = 1/x$. From $f'(x) = 0$ we obtain $1 + \ln x = 0$, $\log_e x = -1$, $e^{-1} = x$ or $x = 1/e$ as a critical value. Since $f''(1/e) > 0$, $f(1/e) = (1/e)(\ln 1/e) = 1/e(\ln 1 - \ln e) = 1/e(0 - 1) = -1/e$ is a minimum value.

22. $f(x) = \frac{1}{\sqrt{2\pi}} e^{-x^2/2}$. See Section 8.2, Example 3 (p. 265) of the text for discussion of the extreme values of this function.

23. The selling price is \$25 per lamp, the profit is \$3000, and the number of lamps made and solid is 200 per week.

25. Let x and y denote the length and width of the type plate (see Figure 7.1). The length of the poster is x + 1.5 inches and its width is y +1 inches Thus the area of the poster is expressed by the product

$$(x + 1.5)(y + 1) =$$
$$xy + 1.5y + x + 1.5$$

Since the area of the type plate is to be 24 square inches, we have $xy = 24$, $y = 24/x$. Substituting $24/x$ for y in the above expression for the area of the poster yields the area function

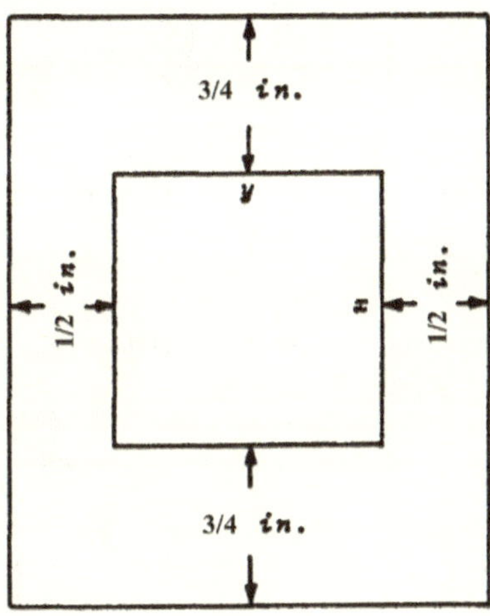

Figure 7.1

$$A(x) = 25.5 + x + \frac{36}{x}, x > 0$$

to be minimized.

$$A'(x) = 1 - \frac{36}{x^2},\ A''(x) = \frac{72}{x^3}$$

From $A'(x) = 0$ we obtain the critical value $x = 6$. Since $A''(6) > 0$, $x = 6$ minimizes the area function $A(x)$.
The type plate that gives the poster with least area has length $x = 6$ inches and width $y = 4$ inches.

27. Let x and y denote the length and width in feet of the building (see Figure 7.2). Let k denote the cost of a linear foot of building material. The cost of the four walls is expressed by $2ky + kx + ky + kx = 3ky + 2kx$ Since the building is to have 60,000 square feet of floor space, $xy = 60{,}000$, which yields $y = \frac{60{,}000}{x}$. By substituting $\frac{60{,}000}{x}$ for y in $3ky + 2kx$ we obtain the cost function

$$C(x) = 2kx + \frac{180{,}000k}{x}, x > 0$$

to be minimized.

$$C'(x) = 2k - \frac{180,000k}{x^2}, \quad C''(x) = \frac{360,000k}{x^3}$$

From $C'(x) = 0$ we obtain $x = 300$. Since $C''(300) > 0$, $x = 300$ minimizes the cost function $C(x)$. From the condition $xy = 60,000$ we obtain $y = 200$.
For the cost of material to be minimized the building should have length 300 feet and width 200 feet.

Figure 7.2

29. The volume function to be maximized is

$$\begin{aligned} v(x) &= x(20-2x)(32-2x) \\ &= 640x - 104x^2 + 4x^3, \quad 0 < x < 10 \end{aligned}$$

Setting $v'(x)$ equal to 0 and solving for x yields $x = 4$ and $x = \frac{40}{3}$, which is extraneous (Why?).

Since $v''(4) < 0$, $x = 4$ yields a maximum value.
To maximize the volume of the box take $x = 4$ inches.

Section 7.2 (p. 243)

2. $R(x) = 310x - 2x^2$ $\qquad$ $c(x) = \frac{1}{2}x^2 + 10x + 1500$

$R'(x) = 310 - 4x$ $\qquad$ $c'(x) = x + 10$

(a) The producer's optimal output level is determined by the condition $R'(x) = c'(x)$. We have

$$310 - 4x = x + 10$$

$$5x = 300$$

$$x = 60$$

The profit derived from this output level is $R(60) - c(60) = 11,400 - 3900 = \$7500$.

(b) Since the imposition of a lump-sum tax does not affect the optimal output level, we have x = 60 from part (a). The profit after taxes is $\$7500 - \$1200 = \$6300$.

(c) Since the imposition of a percent of profit tax does not affect the optimal output level, we have x = 60 from part (a). Since the producer's profit after taxes is 60% of the profit before the tax is computed, the profit after taxes is 0.60(7500) = \$4500.

(d) The producer's optimal output level is determined by the condition $R'(x) = c'(x) + 60$. This yields x = 48 as the optimal output level. The tax is 60(48) = \$2880 and the profit after taxes is $R(48) - c(48) - \text{tax} = \$10,272 - \$3132 - \$2880 = \$4260$.

(e) The lump-sum equivalent of the \$60 per ton output tax is \$2880, the total tax revenue obtained from this tax mechanism, imposed as a lump-sum tax (see part (d)). Since the imposition of a lump-sum tax does not affect the optimal output level, we have x = 60 from part (a). The profit after taxes is $R(60) - c(60) - \text{tax} = \$11,400 - \$3900 - \$2880 = \$4620$.

(f) The basic problem is to set up the tax revenue function that expresses tax revenue T(x) as a function of the output tax t to be imposed on each ton of output. We have:

Tax revenue = Tax per ton of output × producer's output

$$= t \cdot x$$

The producer's output, x, after the tax has been imposed is determined by the condition $R'(x) = c'(x) + t$ (see p. 240 of the text). This gives us

$$310 - 4x = x + 10 + t$$

which yields

$$x = 60 - \frac{1}{5}t$$

Thus the tax revenue function T(t) is defined by

$$T(t) = t(60 - \frac{1}{5}t)$$

$$= 60t - \frac{1}{5}t^2, \ t > 0$$

and this function is to be maximized. Doing so yields t = 150 as the optimal output tax.

(g) For t = 150 we have $x = 60 - \frac{1}{5}(150) = 30$ tons per day. Tax revenue is 150(30) = \$500; the profit after taxes is $R(30) - c(30) - \text{tax} = \$7500 - \$2250 - \$4500 = \$750$.

(h) The lump-sum equivalent tax is \$4500. The optimal output level, which is not affected by this tax, is x = 60 tons per day. Profit after taxes is $R(60) - c(60) - \text{tax} = \$11,400 - \$3900 - \$4500 = \$3000$.

Section 7.3 (p. 249)

2. $y = Se^{-0.10x} = 2000e^{\sqrt{x}-0.10x}\ e^{-0.10x} = 2000e^{\sqrt{x}-0.10x}$ expresses the dollar value of the timber-bearing land x years after its acquisition when the given interest conditions have been taken into account.

By using the chain rule we obtain:

$$\frac{dy}{dx} = 2000e^{\sqrt{x}-0.20x}\left(\frac{1}{\sqrt{x}} - 0.20\right)$$

Setting $\frac{dy}{dx}$ equal to zero and solving for x yields

$$2000e^{\sqrt{x}-0.20x}\left(\frac{1}{\sqrt{x}} - 0.20\right) = 0$$

$$\frac{1}{\sqrt{x}} - 0.20 = 0$$

$$\frac{1}{\sqrt{x}} = \frac{2}{10}$$

$$\sqrt{x} = 5$$

Thus x = 25 is a critical value. To determine if it yields an extreme value we use the first derivative test.

If $x < 25$, $2000\underbrace{e^{\sqrt{x}-0.20x}}_{\text{positive}}\left(\underbrace{\frac{1}{\sqrt{x}} - 0.20}_{\text{positive}}\right)$ is positive.

If $x > 25$, $2000\underbrace{e^{\sqrt{x}-0.20x}}_{\text{positive}}\left(\underbrace{\frac{1}{\sqrt{x}} - 0.20}_{\text{negative}}\right)$ is negative.

Thus, from the first derivative test, x = 25 yields a maximum value. With respect to the given conditions, the optimal time to sell the land is 25 years after its acquisition.

Chapter 8. Curve Sketching

Section 8.1 (p. 258)

2. $f(x)=\sqrt{x},\ f'(x)=\frac{1}{2\sqrt{x}},\ f''(x)=-\frac{1}{4(\sqrt{x})^3}$

Increasing and Decreasing Behavior: f(x) is defined for $x \geq 0$ and $f'(x) > 0$. Thus f(x) is increasing for $x \geq 0$.

Points of inflection and Concavity Behavior: Since $f''(x) \neq 0$ for $x > 0$, there are no points of inflection. $f''(x) < 0$ for $x > 0$, and thus the graph of f(x) is concave downward for $x > 0$.

4. $f(x)=\frac{1}{x},\ f'(x)=-\frac{1}{x^2},\ f''(x)=\frac{2}{x^3}$

Increasing and Decreasing Behavior: f(x) is not defined at 0 and $f'(x) < 0$ for $x < 0$ and $f'(x) < 0$ for $x > 0$; thus f(x) is increasing for $x < 0$ and for $x > 0$.

Points of inflection and Concavity Behavior: Since $f''(x) \neq 0$, there are no points of inflection. $f''(x) > 0$ for $x > 0$, and $f''(x) < 0$ for $x < 0$; thus the graph of f(x) is concave upward for $x > 0$ and concave downward for $x < 0$.

6. $f(x) = 2x^3 + 3x^2,\ f'(x) = 6x^2 + 6x,\ f''(x) = 12x + 6.$

Increasing and Decreasing Behavior: $f'(x) = 6x(x + 1)$; the condition $f'(x) = 0$ yields $x = 0$ and $x = -1$.

For $x < -1,\ f'(x) = 6x(x+1)$ is positive.

For $-1<x<0$, $f'(x)=6x(x+1)$ is negative.

For $x > 0$, $f'(x) = 6x(x + 1)$ is positive.

Thus $f(-1)=1$ is a maximum value, $f(0) = 0$ is a minimum value, $f(x)$ is increasing for $x \leq -1$, decreasing for $-1 \leq x \leq 0$, and increasing for $x \geq 0$.

Points of infection and Concavity Behavior: From $f''(x) = 0$ we obtain $x=-1/2$.

For $x<-\frac{1}{2}$, $f''(x)=12(x+\frac{1}{2})$ is negative.

For $x>-\frac{1}{2}$, $f''(x)=12(x+\frac{1}{2})$ is positive.

Thus, $(-\frac{1}{2},\frac{1}{2})$ is a point of inflection and the graph of f(x) is concave downward for $x<-\frac{1}{2}$ and concave upward for $x>-\frac{1}{2}$.

8. $f(x)=3x^4-4x^3$, $f'(x)=12x^3-12x^2$, $f''(x)=36x^2-24x$

   f(x) is decreasing for $x \leq 1$ and increasing for $x \geq 1$. (0, 0) and $(2/3,-0.59)$ are inflection points; the graph of f(x) is concave upward for $x < 0$, concave downward for $0 < x < 2/3$, and concave upward for $x > 2/3$.

10. $f(x)=e^{-x^2}$, $f'(x)=-2xe^{-x^2}$, $f''(x)=(4x^2-2)e^{-x^2}$

    f(x) is increasing for $x \leq 0$ and decreasing for $x \geq 0$. $\left(-\frac{1}{\sqrt{2}},\frac{1}{\sqrt{e}}\right)$ and $\left(\frac{1}{\sqrt{2}},\frac{1}{\sqrt{e}}\right)$ are inflection points. The graph of f(x) is concave upward for $x<-\frac{1}{\sqrt{2}}$, concave downward for $-\frac{1}{\sqrt{2}}<x<\frac{1}{\sqrt{2}}$, and concave upward for $x>\frac{1}{\sqrt{2}}$. For a detailed analysis of a similar function see Section 8.2, Example 3 (p. 265) of the text.

12. $f(x) = x + 10 + \frac{100}{x}$, $f'(x) = 1 - \frac{100}{x^2}$, $f''(x) = \frac{200}{x^3}$

f(x) is not defined at 0. $f(-10) = -10$ is a minimum value; f(x) is increasing for $x \le -10$ and decreasing for $-10 \le x < 0$. f(10) = 30 is a minimum value; f(x) is decreasing for $0 < x \le 10$ and increasing for $x > 10$. There are no inflection points and the graph of f(x) is concave downward for $x < 0$ and concave upward for $x > 0$.

Section 8.2 (p. 268)

4. $f(x) = 2x^3 - 3x^2 - 12x + 4$, $f'(x) = 6x^2 - 6x - 12$, $f''(x) = 12x - 6$

   1. Since f′(x) exists for all x, the graph of f(x) has no gaps, jumps or breaks and is smooth.
   2. $f(-1) = 11$ is a maximum value and $f(2) = -16$ is a minimum value. f(x) is increasing for $x \le -1$, decreasing for $-1 \le x \le 2$, and increasing for $x \ge 2$.
   3. $(1/2, -5/2)$ is an inflection point and the graph of f(x) is concave downward for $x < 1/2$ and concave upward for $x > 1/2$.
   4. Table 8.1 gives values of f(x) for selected values of x. The graph of f(x) is shown in Figure 8.1.

Table 8.1

x	−3	−2	−1	0	1/2	1	2	3	4
f(x)	−41	0	11	4	−5/2	−9	−16	−5	36

6. $f(x) = 3x^4 - 4x^3$, $f'(x) = 12x^3 - 12x^2$, $f''(x) = 36x^2 - 24x$

   1. Since f′(x) exists for all x, the graph of f(x) has no gaps, jumps or breaks and is smooth.

   2. $f(1) = -1$ is a minimum value. f(x) is decreasing for $x \le 1$ and increasing for $x \ge 1$.

3. (0,0) and (2/3, − 0.59) are inflection points. The graph of f(x) is concave upward for x < 0 and concave downward for 0 < x < 2/3 and concave upward for x > 2/3.

4. Table 8.2 gives values of f(x) for selected values of x. The graph of f(x) is shown in Figure 8.2

Table 8.2

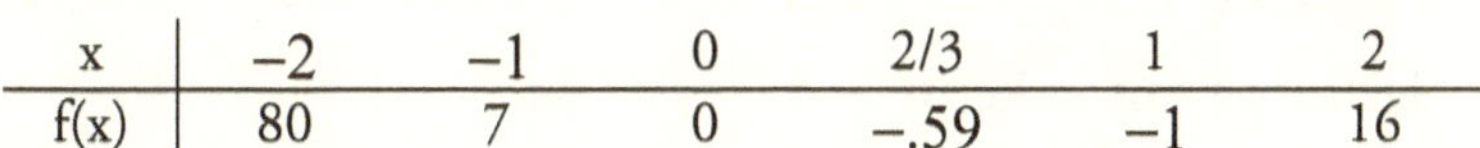

x	−2	−1	0	2/3	1	2
f(x)	80	7	0	−.59	−1	16

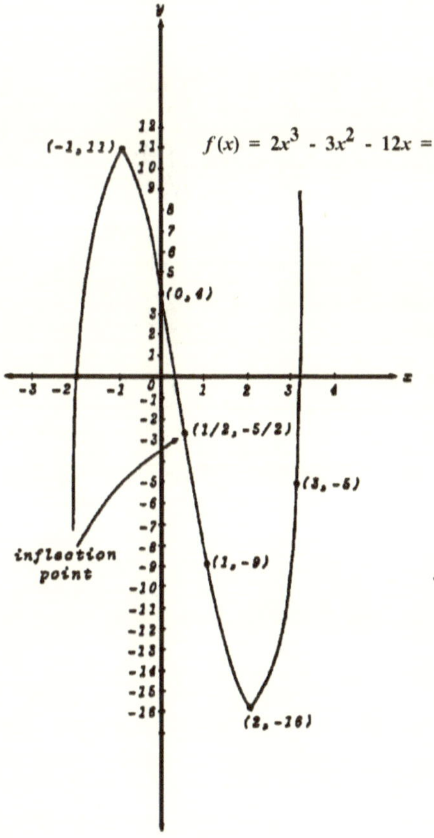

Figure 8.1

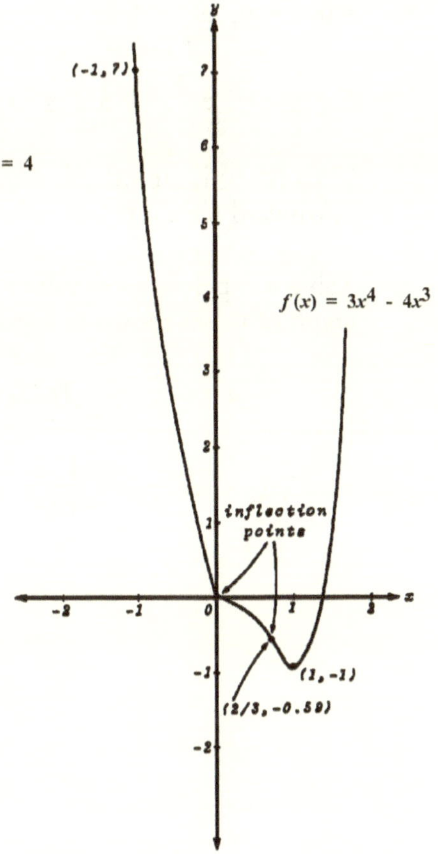

Figure 8.2

12. $f(x) = e^x - e^{-x}$, $f'(x) = e^x + e^{-x}$, $f''(x) = e^x - e^{-x}$

1. Since $f'(x)$ exists for all x, the graph of f(x) has no gaps, jumps or breaks and is smooth.
2. $f'(x) = e^x + e^{-x}$ is the sum of two positive quantities for all x, which cannot be zero. Thus there are no critical values and f(x) is increasing for all x.
3. From $f'(x) = 0$ we have $e^x - e^{-x} = 0$, x = 0. By taking e^{-x} as a factor we can write $e^x - e^{-x} = e^{-x}(e^{2x} - 1)$.

   For $x < 0$, $e^{2x} < 1$ and $f''(x) = e^{-x}(e^{2x} - 1)$ is negative.

   For $x > 0$, $e^{2x} > 1$ and $f''(x) = e^{-x}(e^{2x} - 1)$ is positive.

   Thus (0, 0) is an inflection point and the graph of f(x) is concave downward for x < 0 and concave upward for x > 0.

4. Table 8.3 below gives values of f(x) for selected values of x. The graph of f(x) is shown in Figure 8.3.

Table 8.3

x	−2	−1	−1/2	0	1/2	1	2
f(x)	−7.25	−2.35	−1.04	0	1.04	2.35	7.25

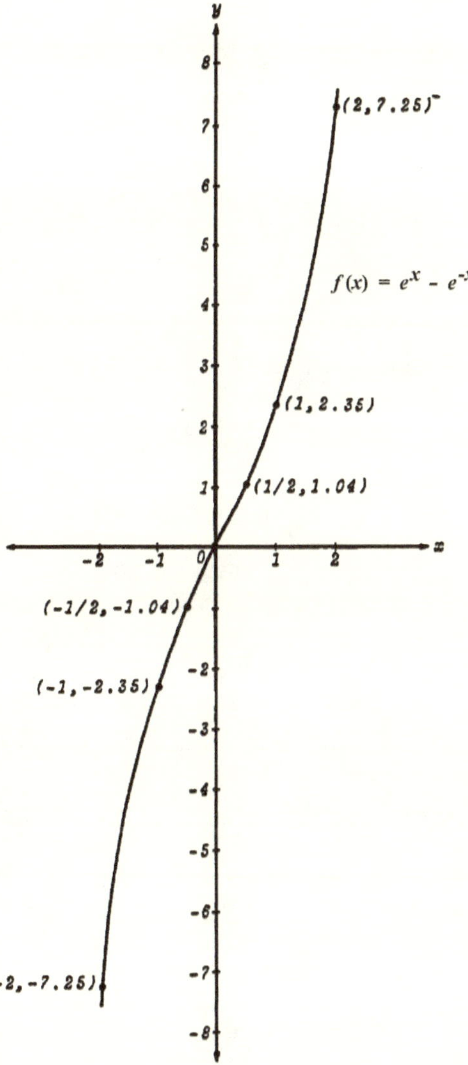

Figure 8.3

14. $f(x) = xe^{-x^2}$, $f'(x) = x(-2xe^{-x^2}) + e^{-x^2}(1 - 2x^2)$

1. The graph of f(x) has no gaps, jumps or breaks and is smooth.
2. From $f'(x) = 0$ we have $2x^2 - 1 = 0$, $x^2 = \frac{1}{2}$, $x = \pm\frac{1}{\sqrt{2}} = \pm\frac{\sqrt{2}}{2}$

$\approx \pm 0.707$. By factoring $(1-2x^2)$ we can write $f'(x)$ in the useful, but unusual, form $f'(x) = e^{-x^2}(1-\sqrt{2}x)(1+\sqrt{2}x)$.

For $x < -\frac{\sqrt{2}}{2}$, $f'(x) = \underbrace{e^{-x^2}}_{\text{pos.}}\underbrace{(1-\sqrt{2}x)}_{\text{pos.}}\underbrace{(1+\sqrt{2}x)}_{\text{neg.}}$ is negative.

For $-\frac{\sqrt{2}}{2} < x < \frac{\sqrt{2}}{2}$, $f'(x) = \underbrace{e^{-x^2}}_{\text{pos.}}\underbrace{(1-\sqrt{2}x)}_{\text{pos.}}\underbrace{(1+\sqrt{2}x)}_{\text{pos.}}$ is positive.

For $x > \frac{\sqrt{2}}{2}$, $f'(x) = \underbrace{e^{-x^2}}_{\text{pos.}}\underbrace{(1-\sqrt{2}x)}_{\text{neg.}}\underbrace{(1+\sqrt{2}x)}_{\text{pos.}}$ is negative.

Thus f(x) is decreasing for $x \le -\frac{\sqrt{2}}{2}$, increasing for $-\frac{\sqrt{2}}{2} \le x \le \frac{\sqrt{2}}{2}$, and decreasing for $x \ge \frac{\sqrt{2}}{2}$. $f(-\frac{\sqrt{2}}{2}) = -0.43$ is a minimum value and $f(\frac{\sqrt{2}}{2}) = 0.43$ is a maximum value.

3. $$\begin{aligned} f''(x) &= e^{-x^2}(-4x) + (1-2x^2)e^{-x^2} D_x(-x^2) \\ &= -4xe^{-x^2} - 2x(1-2x^2)e^{-x^2} \\ &= (-4x - 2x + 4x^3)e^{-x^2} \\ &= [(4x(x^2 - \frac{3}{2})]e^{-x^2} \end{aligned}$$

From $f''(x) = 0$ we obtain x = 0 and $x = \pm\sqrt{1.5} \approx 1.23$.

For $x < -\sqrt{1.5}$, $f''(x) = [\underbrace{4x}_{-}\underbrace{(x^2 - \frac{3}{2})}_{+}]\underbrace{e^{-x^2}}_{+}$ is negative.

$$\text{For } -\sqrt{1.5} < x < 0,\ f''(x) = [\underbrace{4x}_{-}\underbrace{(x^2 - \tfrac{3}{2})}_{-}]\underbrace{e^{-x^2}}_{+} \text{ is positive.}$$

$$\text{For } 0 < x < \sqrt{1.5},\ f''(x) = [\underbrace{4x}_{+}\underbrace{(x^2 - \tfrac{3}{2})}_{-}]\underbrace{e^{-x^2}}_{+} \text{ is negative.}$$

$$\text{For } x > \sqrt{1.5},\ f''(x) = [\underbrace{4x}_{+}\underbrace{(x^2 - \tfrac{3}{2})}_{+}]\underbrace{e^{-x^2}}_{+} \text{ is positive.}$$

Thus $f(-\sqrt{1.5}) = -0.27$, f(0) = 0, and $f(\sqrt{1.5}) = 0.27$ are inflection point values, the graph of f(x) is concave downward for $x < -\sqrt{1.5}$, concave upward for $-\sqrt{1.5} < x < 0$, concave downward for $0 < x < \sqrt{1.5}$, and concave upward for $x > \sqrt{1.5}$. See Figure 8.4.

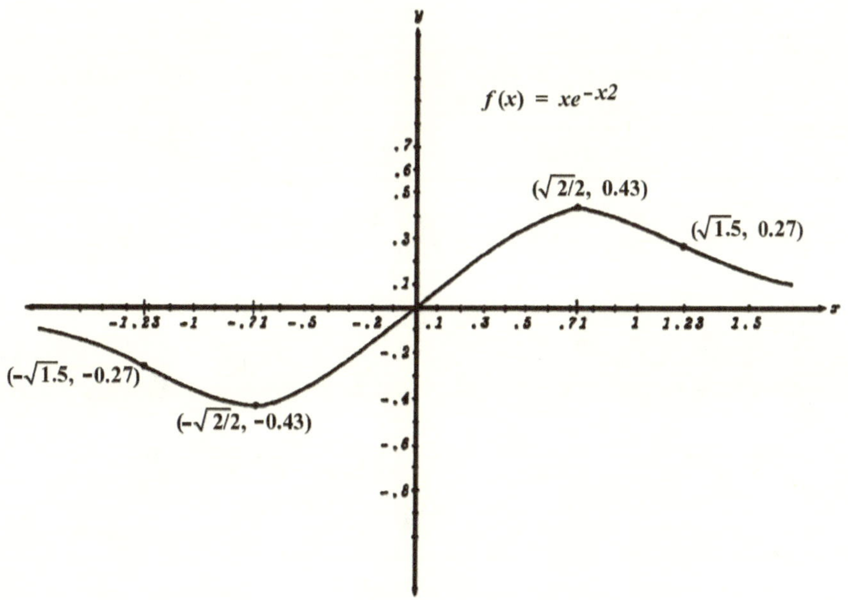

Figure 8.4

Chapter 9. Integral Calculus: Indefinite Integrals

Section 9.1 (p. 280)

2. $\frac{1}{9}x^9 + C$

4. $-\frac{1}{x} + C$

6. $-\frac{1}{7}x^{-7} + C$

8. $\frac{3}{5}x^{5/3} + C$

10. $4x^{1/2} + C$

12. $\frac{4}{7}x^7 + \frac{3}{2}x^2 + C$

14. $\frac{3}{2}x^{4/3} - \frac{1}{2}x^2 + 4\ln x + C$

16. $3e^x - \frac{2}{3}x^{3/2} + C$

18. $\frac{8}{3}x^{3/2} - 3\ln x + C$

20. $\frac{15}{2}x^{2/3} - 3x + C$

22. We first write $f(x) = \frac{1+x}{x^2} = \frac{1}{x^2} + \frac{x}{x^2} = x^{-2} + x^{-1}$. Integration yields $-\frac{1}{x} + \ln x + C$.

Section 9.2 (p. 287)

2. $\frac{1}{6}(5x^2 - 1)^6 + C$

4. $\frac{1}{3}(5x^3 - 2)^3 + C$

6. $\frac{1}{24}(x^4 - 1)^6 + C$

8. $\frac{1}{3}(2x^3 - 7)^{3/2} + C$

10. $\frac{1}{8}\ln(8x - 1) + C$

12. $-\frac{1}{4x^2 + 1} + C$

14. $2(x^2 + 3x)^{1/2} + C$

16. $\frac{3}{2}e^{2x} + C$

18. $\ln(x^3 + 4x + 1) + C$

20. $\frac{2}{5}(5x^2 + 1)^{1/2} + C$

Section 9.3 (p. 290)

2. Take $u = x$, $dv = e^{3x}$; then $du = dx$,
$v = \frac{1}{3}e^{3x} \cdot \int xe^{3x}dx = \frac{1}{3}xe^{3x} - \frac{1}{9}e^{3x} + C$.

4. Take $u = 1000 - 20t$, $dv = e^{-0.05t}$; then $du = -20dt$, $v = \frac{-1}{0.05}e^{-0.05t}$.
$\int (1000 - 20t)\, e^{-0.05t}\, dt = -12000e^{-0.05t} + 400te^{-0.05t} + C$.

6. Take $u = 100 + 3t$, $dv = e^{0.1(5-t)}$; then $du = 3dt$, $v = -\frac{1}{0.1}e^{0.1(5-t)}$
$\int (1000 - 20t)\, e^{0.1(5-t)}\, dt = -1300e^{0.1(5-t)} - 30te^{0.1(5-t)} + C$.

8. Take $u = x^2$, $dv = x(1 + x^2)^{-1/2}$; then $du = 2xdv$, $v = (1 + x^2)^{1/2}$.
$\int \frac{x^3}{\sqrt{1 + x^2}}dx = x^2(1 + x^2)^{1/2} - \frac{2}{3}(1 + x^2)^{3/2} + C$

10. Take $u = x^2$, $dv = e^{-x}$; then $du = 2xdx$, $v = -e^{-x}$.

$$\begin{aligned}\int x^2e^{-x}dx &= -x^2e^{-x} + 2\int xe^{-x}dx \\ &= -x^2e^{-x} + 2[-xe^{-x} - e^{-x}] + C \\ &= -x^2e^{-x} - 2xe^{-x} + C.\end{aligned}$$

Chapter 10. Introduction to Mathematical Modeling

Section 10.1 (p. 296)

1. (a) $t = \frac{300}{60} = 5$; 5 hours travel time.
   (b) No; it's irrelevant to the issue of validity of the conclusion. The validity of the conclusion is established by the mathematical argument (division) given in answer to (a).
   (c) The discrepancy establishes that the valid conclusion reached in (a) is unrealistic (false, if you will).

Section 10.2 (p. 301)

2. Disagree; precise mathematical reasoning in whatever context, geometric or other, can only establish the validity of the conclusions in question based on the assumptions or postulates set up as a starting point. The truth of the conclusions is another issue altogether.

Chapter 11. Differential Equation Models

Section 11.1 (p. 308)

2. $f(x)=\frac{4}{3}x^3+x^2-x-12$

4. $f(x)=-\ln x+\frac{1}{2}x^2+3x+\ln 3-\frac{27}{2}$

6. $y=x^3-x+5$

8. $y=\frac{1}{2}\ln(x^2+1)+5$

10. $R(x)=500x-2x^2$

Section 11.3 (p. 319)

1. The function $f(t)=N_0e^{-0.0001209t}$ describes the amount of carbon-14 present in the organic remains, f(t), after t years. Since the artifacts found contain 60% of the original amount of carbon-14, we have $f(t) = 0.6N_0$, and thus
$0.6N_0=N_0e^{-0.0001209t}$, $0.6=e^{-0.0001209t}$, $\ln 0.6=-0.0001209t$
Thus

$$t=\frac{\ln 0.6}{-0.0001209}=\frac{-0.5108}{-0.0001209}=4225$$

The artifacts are approximately 4225 years old.

2. $N(t)=N_0e^{kt}$; from $N_0/2$ = 1600 we obtain $N(t)=N_0e^{-0.0004t}$ where t is time in years and N_0 is the amount of radium (in grams) initially present in the given portion of matter.

3. $f(t)=N_0e^{-0.0001209t}$; $0.80N_0=N_0e^{-0.0001209t}$, $\ln 0.8=-0.0001209t$,

$$t=\frac{-0.22314}{-0.0001209}=1846 \text{ years}.$$

Chapter 12. Integral Calculus: Definite Integrals

Section 12.1 (p. 343)

2. (a) $1^3+2^3+3^3+4^3+5^3$
   (b) $x_1(x_2-3)+x_2(x_3-3)+x_3(x_4-3)+x_4(x_5-3)$
   (c) $x_1x_0^{\,2}+x_2x_1^{\,2}+x_3x_2^{\,2}+x_4x_3^{\,2}$
   (d) $3+3^22^2+3^33^2+3^44^2+3^55^2$

4. (a) $378\le A(S)\le 1791$
   (b) $612\le A(S)\le 1497$
   (c) $660\le A(S)\le 1371$
   (d) A(S) = 990

6. Region S is shown in Figure 12.1.

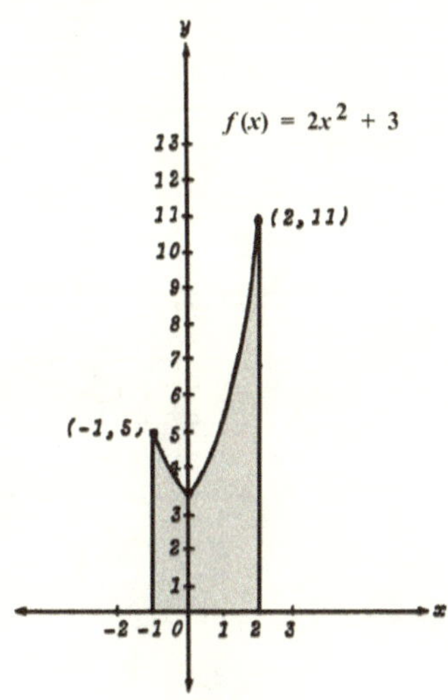

Figure 12.1

$F(x) = \frac{2}{3}x^3 + 3x$ is an indefinite integral of $f(x) = 2x^2 + 3$. Thus

$$A(S) = F(2) - F(-1)$$
$$= \frac{34}{3} - \left(\frac{-11}{3}\right) = 15 .$$

8. Region S is shown in Figure 12.2.

$$F(x) = \frac{1}{2}x^4 + 4x .$$

$$A(S) = F(2) - F(-1) = 19.5$$

10. Region S is shown in Figure 12.3.

$$F(x) = \frac{2}{3}x^{2/3} + x$$

$$A(S) = F(9) - F(1) = 27 - \frac{2}{3} = 25.3 .$$

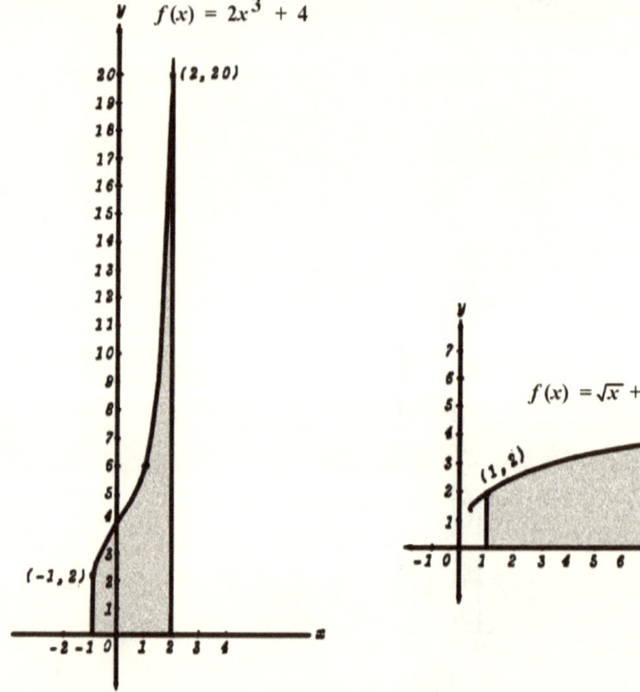

Figure 12.2

Figure 12.3

Section 12.2 (p. 353)

2. $(\frac{5}{2}x^2+3x]_1^4=47.5$

4. $(x^4+2x]_{-2}^{2}=8$

6. $(\frac{1}{2}x^4-\frac{1}{2}x^2+2x]_{-1}^4=130$

8. $(x^4-\frac{3}{2}x^2-x]_1^3=66$

10. $(\frac{2}{3}(4+x)^{3/2}]_0^{12}=37.3$

12. $(\frac{2}{3}(1+x^3)^{3/2}]_0^2=17.3$

14. $(25e^{0.2x}]_0^{10}=159.7$

16. $(\frac{2}{9}(x^3+3)^{3/2}]_1^3=34.7$

18. $(\frac{2}{3}(5+x)^{3/2}]_{-1}^4=12.7$

20. $(-100e^{-0.1x}]_0^5=39.3$

22. $(\frac{600}{0.09}e^{0.09x}]_0^4=\frac{600}{0.09}(e^{0.45}-1)=3789$; note $e^{0.45}=1.56831$.

24. $(-\frac{600}{0.09}e^{0.06(2-t)}]_0^2=-16{,}666.7(1-e^{0.12})=2125$

Section 12.3 (p. 356)

2. $\int_{-\infty}^{1}\frac{dx}{x^4}=\lim_{u\to\infty}\left(\frac{1}{3}+\frac{1}{3u^3}\right)=\frac{1}{3}$

4. $\int_0^{\infty}e^{-x}dx=\lim_{u\to\infty}\left(-\frac{1}{e^u}+1\right)=1$

6. $\int_{-\infty}^{3}\frac{dx}{(4-x)^2}=\lim_{u\to-\infty}\left(1-\frac{1}{4-u}\right)=1$

8. $\int_{-\infty}^{0}e^x dx=\lim_{u\to-\infty}(1-e^u)=1$

Section 12. 4 (p. 358)

2. 9; 9 is an upper bound and no number smaller than 9 is an upper bound.

4. Denote the set by A. LUB A = 1; 1 is an upper bound of A and no number smaller than 1 is an upper bound of A. GLB A = 0. 0 is a lower bound of A and no number larger than 0 is a lower bound of A.

Chapter 13. Concepts with a Definite Integral Structure

Section 13.1 (p. 366)

2. Region S consists of three components, as shown in Figure 13.1 The computation of A(S) thus requires the computation of the areas of three components - the region above the x-axis between $x = -2$ and $x = -1/2$, the region below the x-axis between $x = -1/2$ and x = 1/2, and the region above the x-axis between x = 1/2 and x = 4.

$$F(x) = \frac{4}{3}x^3 - x$$

$$A(S) = F(-\frac{1}{2}) - F(-2)) - (F(\frac{1}{2}) - F(-\left(\frac{1}{2}\right)))$$

$$+ (F(4) - F(\frac{1}{2}))$$

$$= 9 - (-\frac{2}{3}) + \frac{245}{3} = 91.3$$

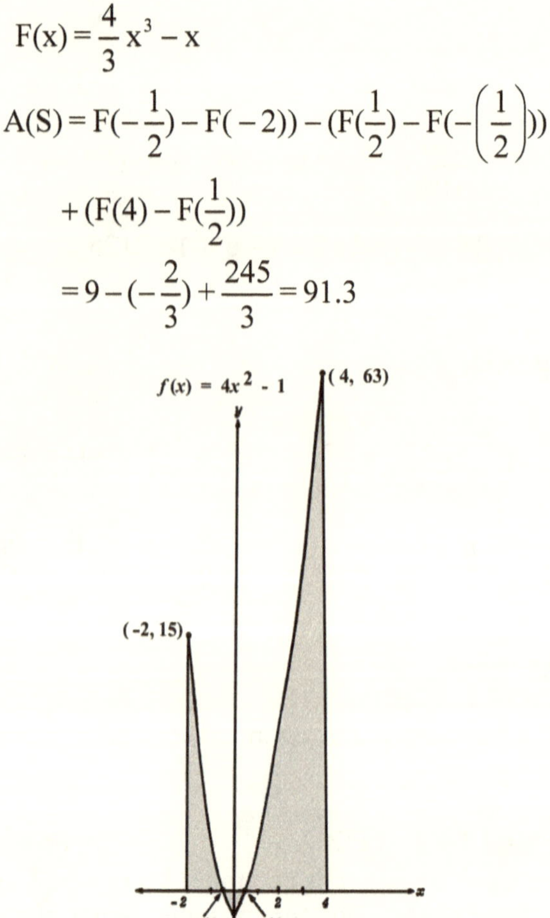

Figure 13.1

4. Region S is shown in Figure 13.2.

$$F(x) = \frac{2}{3}x^3 + 2x^2 + x$$

$$A(S) = F(3) - F(1) = 35.3$$

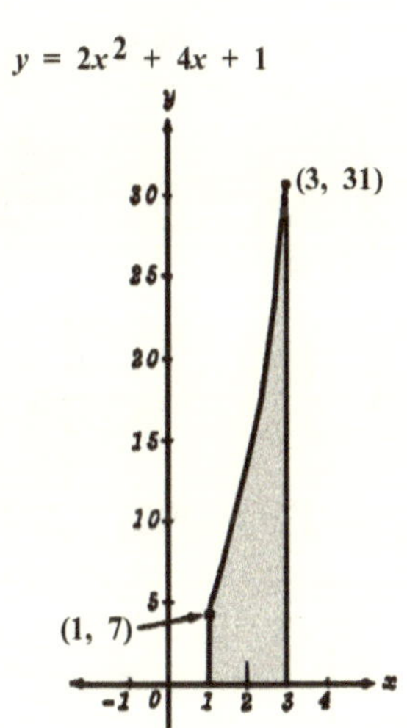

Figure 13.2

6. Region S is shown in Figure 13.3.

$$f(x) = x + 10,\ g(x) = 2x^2 + 1$$

$$f(x) - g(x) = 2x^2 + x + 9$$

$F(x) = -\frac{2}{3}x^3 + \frac{1}{2}x^2 + 9x$ is an indefinite integral of f(x) - g(x). $A(S) = F(1) - F(-1) = 16.7$

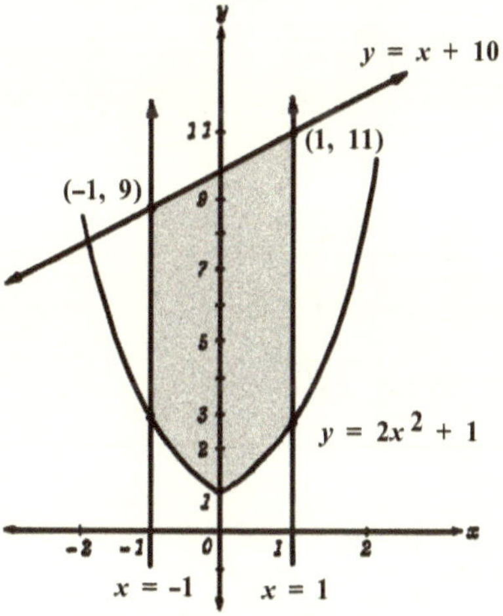

Figure 13.3

8. Region S is shown in Figure 13.4

$$F(x) = \frac{1}{4}x^4 - \frac{3}{2}x^2 - 3x$$
$$A(S) = -[F(2) - F(-2)] = 12$$

10. Region S is shown in Figure 13.5.

$$F(x) = -\frac{1}{3}(4 - x^2)^{3/2}$$
$$A(S) = F(2) - F(0) = 2.7$$

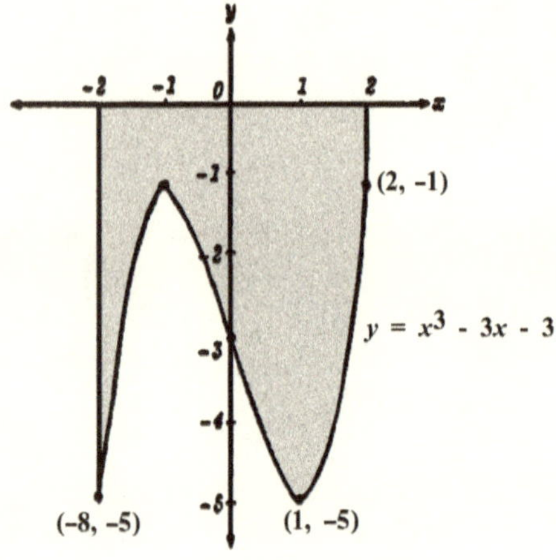

Figure 13.4

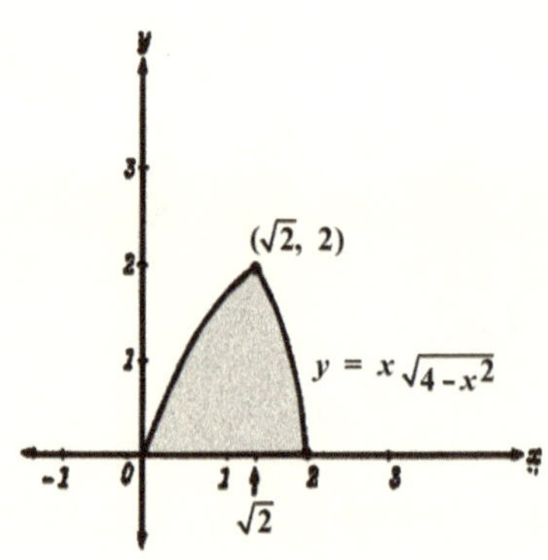

Figure 13.5

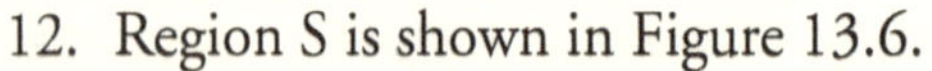

12. Region S is shown in Figure 13.6.

$$F(x) = \frac{2}{3}x^{3/2} - \frac{1}{4}x^4$$

$$A(S) = F(1) - F(0) = \frac{5}{12}$$

14. Region S is shown in Figure 13.7.

$$F(x) = \frac{1}{4}x^4 - 2x^3 + \frac{9}{2}x^2$$

$$A(S) = F(3) - F(0) = 22.8$$

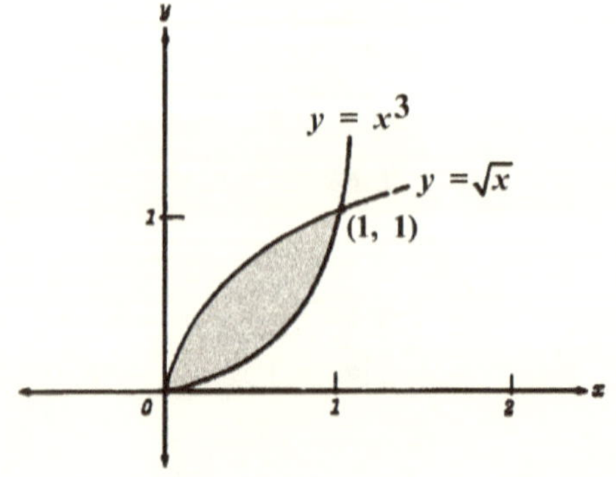

Figure 13.6

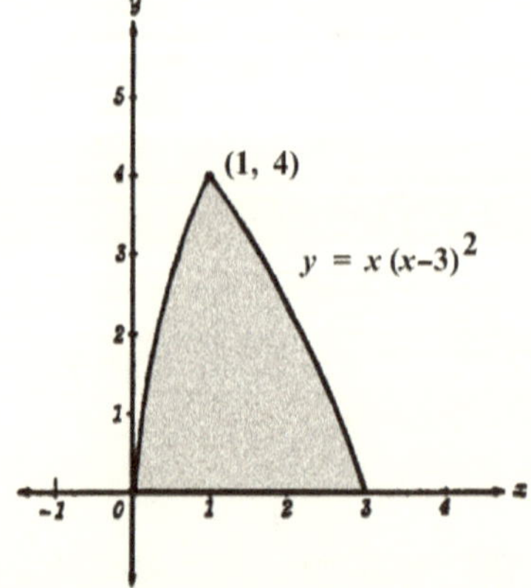

Figure 13.7

Section 13.2 (p. 370)

1. (a) The demand functions are $p = 2000 - 2x^2$ and $x = \sqrt{(2000-p)/2}$. For p = 200, x = 30. The consumers' surplus for p = 200 is the area of the region shown in Figure 13.8. $F(x) = 200x - \frac{2}{3}x^3$ is an indefinite integral of $p = 2000 - 2x^2$. The consumers' surplus for p = 200 is

$$(2000x - \frac{2}{3}x^3]_0^{30} - 200(30)$$
$$= 42,000 - 6000 = 36,000$$

(b) For p = 200, x = 30; for p = 78, x = 31. The change in consumers' surplus for a price change from $p_1 = 78$ to $p_2 = 200$ is the sum of the areas of regions R and S shown in Figure 13.9, which is given by

$$(2000x - \frac{2}{3}x^3]_{30}^{31} - (78(31) - 200(30)) = 3721$$

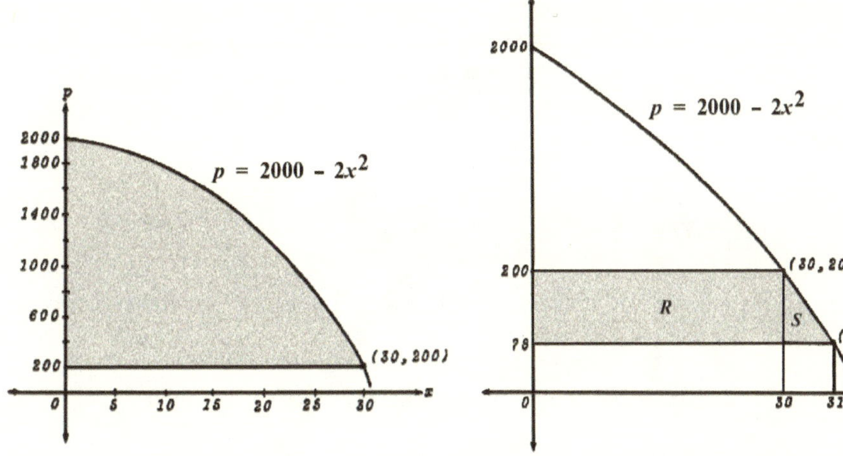

Figure 13.8 *Figure 13.9*

(c) For p = 200, x = 30; for p = 318, x = 29. The change in consumers' surplus for price change from p_1 = 200 to p_2 = 318 is the sum of the areas of regions R and S shown in Figure 3.10, which is given by

$$(2000x - \frac{2}{3}x^3]_{29}^{30} - (200(30) - 318(29)) = 3481$$

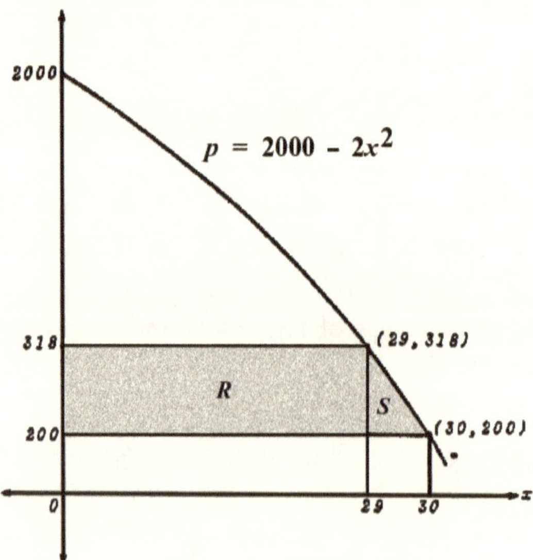

Figure 13.10

2. $4333.33

Section 13.3 (p. 372)

1. (a) When p = 200, we obtain 200 = $2x^2$, x = 10. Thus the producers' surplus for p=200 is

$$200(10) - \int_0^{10} 2x^2dx = 2000 - (\frac{2}{3}x^3]_0^{10} = 0020$$

$$2000 - \frac{2000}{3} = \frac{4000}{3}.$$

(b) When p = 162, we obtain $162 = x^2$, $x = 9$. Thus the producers' surplus for p = 162 is $162(9) - \int_0^9 2x^2 dx = 1458 - (\frac{2}{3}x^3]_0^9 = 1458 - 486 = 972$.

2. (a) When p = 16, we obtain $16 = \frac{1}{4}x^3, x = 4$. Thus the producers' surplus for p = 16 is $16(4) - \int_0^4 \frac{1}{4}x^3 dx = 64 - (\frac{1}{16}x^4]_0^4 = 64 - 16 = 48$.

(b) When p = 54, we obtain $54 = \frac{1}{4}x^3, x = 6$. Thus the producers' surplus for p = 54 is $54(6) - \int_0^6 \frac{1}{4}x^3 dx = 324 - (\frac{1}{16}x^4]_0^6 = 324 - 81 = 243$.

Section 13.4 (p.377)

1. $\int_0^5 3000e^{-0.09t}dt =$
$(\frac{-3000}{0.09}e^{-0.09t}]_0^5 = 12,079$

Note: $e^{-0.45} = 0.63763$

2. \$27,534

Section 13.5 (p. 380)

1. (a) $\Pi = \int_0^{10}(1800 - 50t)e^{-0.1t}dt$
$= (-13000)e^{-0.1t} + 500te^{-0.1t}]_0^{10}$
$= -800e^{-1} + 13,000 = \$10,057$

(b) $\Pi = 10,057 + 100e^{-1} - 800 = \$9294$

2. (a) $\Pi = \int_0^{20} (1000 - 20t)e^{-0.05t}dt$

$= -12000e^{-0.05t} + 400te^{-0.05t} \Big]_0^{20}$

$= -4000e^{-1} + 12,000$

$= -4000(0.36788) + 12,000 = \$10,528$

(b) $\Pi = 10,528 + 50e^{-1} - 500 = \$10,046$

Section 13.6 (p. 382)

1. $\int_0^u 4000e^{-0.1t}dx = (-40,000e^{-0.1t}\Big]_0^u$

$= -40,000e^{-0.1u} - (-40,000)e^0$

$= \dfrac{-40,000}{e^{0.1u}} + 40,000$

Thus $\Pi = \int_0^u 4000e^{-0.1t}dt = \lim_{u\to\infty}\left(\dfrac{-40,000}{e^{0.1u}} + 40,000\right)$

$= \$40,000$.

2. $\int_0^u 1200e^{-0.12t}dt = (-10,000e^{-0.12t}\Big]_0^u$

$= -10,000e^{-0.12u} - (-10,000)e^0$

$= -\dfrac{-10,000}{e^{0.12u}} + 10,000$

Thus $\Pi = \int_0^\infty 1200e^{-0.12t}dt = \lim_{u\to\infty}\left(\dfrac{-10,000}{e^{0.12u}} + 10,000\right)$

$= \$10,000$.

Section 13.7 (p. 384)

1. $S = \int_0^{10} 2000e^{0.1(10-t)}dt = (-20,000e^{-0.1(10-t)}]_0^{10}$

$$= -20,000e^0 - (-20,000)e^1$$
$$= 20,000(e-1) = 20,000(1.71828)$$
$$= \$34,365.60.$$

2. $S = \int_0^{8} 3000e^{0.12(8-t)}dt = (-25,000e^{0.12(8-t)}]_0^{8}$

$$= -25,000e^0 - (-25,000e^{0.96})$$
$$= 25,000(e^{0.96} - 1) = 25,000(2.6117 - 1)$$
$$= \$40,292.50.$$

3. $S = \int_0^{11} 5000e^{0.08(11-t)}dt = (-62,500e^{-0.08(11-t)}]_0^{11}$

$$= 62,000(1.4109) = \$88,181.25$$

4. $S = \int_0^{x} Ae^{r(x-t)}dt = \left(-\frac{A}{r}e^{r(x-t)}\right]_0^x$

$$= \frac{A}{r} + \frac{A}{r}e^{rx}$$

5. $S = \int_0^{5} (100+3t)e^{0.1(5-t)}dt = (-1300e^{0.1(5-t)} - 30te^{0.1(5-t)}]_0^{5}$

$$= -1450 + 1300e^{0.5}$$
$$= -1450 + 2143.31 = \$693.31$$

6. $S = \int_0^{8} (500-5t)e^{0.12(8-t)}dt = \left(-\frac{500-5t}{0.12}e^{0.12(8-t)} + \frac{5}{(0.12)^2}e^{0.12(8-t)}\right]_0^8$

$= \$6489$

Section 13.8 (p. 391)

1. $W = \int_1^{11} (x^2 + 2x + 1)\, dx = (S = \int_0^8 (500 - 5t)e^{0.12(8-t)} dt = (\frac{1}{3}x^3 + x^2 + x]_1^{11}$

$= \frac{1727}{3} - \frac{7}{3} = \frac{1720}{3}$ foot - pounds

2. 27 foot-pounds

Chapter 14. Topics in Multivariable Calculus

Section 14.1 (p. 398)

2. $f(2,3,1) = 21,\ f(-1,2,4) = 12,\ f(3,2,-1) = 8$

Section 14.2 (p. 404)

2. $f_x(x,y) = 12x^3y - y,\ f_x(1,3) = 33,\ f_y(x,y) = 3x^4 - x,\ f_y(2,4) = 46$

4. $f_x(x,y) = D_x \ln xy = \frac{D_x(xy)}{xy} = \frac{y}{xy} = \frac{1}{x},\ f_x(1,3) = 1,\ f_y(x,y) = \frac{1}{y},\ f_y(2,4) = \frac{1}{4}$

6. $f_x(x,y) = D_x(x^2 + y)^{1/2} = \frac{1}{2}(x^2 + y)^{-1/2} D_x(x^2 + y) = \frac{x}{\sqrt{x^2 + y}}$

$f_y(x,y) = D_y(x^2 + y)^{1/2} = \frac{1}{2\sqrt{x^2 + y}}$

$f_x(1,3) = \frac{1}{2},\ f_y(2,4) = \frac{1}{2\sqrt{8}}$

8. $f_x(x,y) = D_x(e^x \ln y) = (\ln y)D_x e^x = e^x \ln y,\ f_y(x,y) = \frac{e^x}{y},\ f_x(1,3) = e \ln e,$

$f_y(2,4) = e^2/2$

10. $f_y(x,y,z) = xD_y e^{yz} = x(e^{yz} D_y{}^{(yz)}) = x(e^{yz} z) = xze^{yz}$

$f_x(x,y,z) = e^{yz} D_x{}^{x} = e^{yz},\ f_z(x,y,z) = xye^{yz}$

12. $f_x(x, y, z) = x^2\ D_x(x^2 + z^2)^{1/2} + (x^2 + z^2)^{1/2}\ D_x x^2$

$$= x^2 \cdot \frac{1}{2}(x^2 + z^2)^{-1/2}\ D_x(x^2 + z^2) + 2x(x^2 + z^2)^{1/2}$$

$$= \frac{x^3}{\sqrt{x^2 + z^2}} + 2x\sqrt{x^2 + z^2}$$

$f_y(x, y, z) = 0$

$$f_z(x, y, z) = x^2 D_z(x^2 + z^2)^{1/2} = x^2 \cdot \frac{1}{2}(x^2 + z^2)^{-1/2}\ D_z(x^2 + z^2) = \frac{x^2 y}{\sqrt{x^2 + z^2}}$$

14. $f_x(x, y, z) = \dfrac{D_x(x^3y + z)}{y^2 + z^2} = \dfrac{3x^2y}{y^2 + z^2}$

$$f_y(x, y, z) = \frac{(y^2 + z^2)\,D_y(x^3y + z) - (x^3y + z)\,D_y(y^2 + z^2)}{(y^2 + z^2)^2}$$

$$= \frac{(y^2 + z^2)\,x^3 - 2y\,(x^3y + z)}{(y^2 + z^2)^2}$$

$$f_z(x, y, z) = \frac{(y^2 + z^2)\ - 2z\,(x^3y + z)}{(y^2 + z^2)^2}$$

Section 14.3 (p. 409)

2. $f_x(x, y) = 40x^{-1/3y^{1/3}}$, $f_y(x, y) = 20x^{2/3}y^{-2/3}$

Section 14.4 (p. 412)

2. $f_x(x, y) = y^2 - 6xy$, $f_y(x, y) = 2xy - 3x^2$, $f_{xy}(x, y) = f_{yx}(x, y) = 2y - 6x$, $f_{xx}(x, y) = -6y$, $f_x(x, y) = y^2 - 6xy$, $f_{yy}(x, y) = 2x$, $f_{xx}(-1, 4) = -24$, $f_{yy}(-3, 1) = -6$, $f_{xy}(3, 4) = -10$

4. $f_x(x, y, z) = 3y^2z^3D_xx = 3y^2z^3$, $f_y(x, y, z) = 3xz^3D_yy^2 = 6xyz^3$, $f_z(x, y, z) = 3xy^2D_zz^3 = 9xy^2z^2$, $f_{xy}(x, y, z) = f_{yx}(x, y, z) = 6yz^3$, $f_{xx}(x, y, z) = 0$, $f_{yy}(x, y, z) = 6xz3$, $f_{zz}(x, y, z) = 18xy^2z$, $f_{xz}(x, y, z) = f_{zx}(x, y, z) = 9y^2z^2$, $f_{yz}(x, y, z) = f_{zy}(x, y, z) = 18xyz^2$

Section 14.5 (p. 425)

2. $f_x(x, y) = x^2 - xy + 2y^2 + 3$, $f_x(x, y) = 2x - y$, $f_y(x, y) = -x + 4y$, $f_{xx}(x, y) = 2$, $f_{yy}(x, y) = 4$, $f_{yx}(x, y) = -1$

   Setting $f_x(x, y)$ and $f_y(x, y)$ equal to zero and solving for x and y yields

   $2x - y = 0$
   $-x + 4y = 0,$
   $x = 0, y = 0$

   $A(x, y) = [f_{xx}(x, y)][f_{yy}(x, y)] - [f_{yx}(x, y)]^2$
   $= 2(4) - (-1)^2 = 7$
   $A(0, 0) = 7$

   Since $A(0, 0) > 0$ and $f_{xx}(0, 0) = 2 > 0$, $f(0, 0) = 3$ is a local minimum value.

4. $f(x, y) = x^3 - 3x - y^2 + 10$, $f_x(x, y) = 3x^2 - 3$, $f_y(x, y) = -2y$
   $f_{xx}(x, y) = 6x$, $f_{yy}(x, y) = -2$, $f_{yx}(x, y) = 0$

   From

   $3x^2 - 3 = 0$
   $-2y = 0$

   we obtain the critical points $(1, 0)$ and $(-1, 0)$.

$A(x, y) = [f_{xx}(x, y)][f_{yy}(x, y)] - [f_{yx}(x, y)]^2$
$= -12x$
$A(1, 0) = -12 < 0$, and thus f(1, 0) is not an extreme value.
$A(-1, 0) = 12 > 0$, $f_{xx}(-1, 0) = -6 < 0$, and thus $f(-1, 0) = 12$ is a local maximum value.

6. $f(x, y) = 3x^2 - xy + 2y^2$, f(0, 0) = 0 is a local minimum value.

8. $f(x, y) = x^3 - 2xy^2 + x^2$, $f_x(x, y) = 3x^2 - 2y^2 + 2x$, $f_y(x, y) = -4xy$,
$f_{xx}(x, y) = 6x + 2$, $f_{yy}(x, y) = -4x$, $f_{yx}(x, y) = -4y$
Setting $f_x(x, y)$ and $f_y(x, y)$ equal to zero and solving for x and y yields

$$3x^2 - 2y^2 + 2x = 0 \qquad (1)$$
$$-4xy = 0 \qquad (2)$$

From (2) we have x =0 or y = 0. Substitution of 0 for x in (1) yields the critical point (0, 0). Substitution of 0 for y in (1) yields the critical point $(-2/3, 0)$.

$$A(x, y) = [f_{xx}(x, y)][f_{yy}(x, y)] - [f_{yx}(x, y)]^2$$
$$A(x, y) = -4x(6x + 2) - 16y^2$$

A(0, 0) = 0, and thus the test is inconclusive and no conclusion can be drawn about the nature of f(0, 0). It can be shown by other means, however, that f(0, 0) = 0 is not an extreme value.

If we approach (0, 0) along the positive x-axis (where y = 0), f(x, y) is positive; if we approach (0, 0) along the arc $y^2 = x$ f(x, y) is negative for x < 1/2.
$A(-2/3, 0) = (-16/3) < 0$, and thus $f(-2/3, 0)$ is not an extreme value.

10. The Lagrangian function is

$$H(x, y, L) = x^2 + 4y + 12 - L(x + y - 6) = x^2 + 4y + 12 - Lx - Ly + 6L$$
$$H_x(x, y, L) = 2x - L$$

$$H_y(x, y, L) = 4 - L$$

From

$$\begin{aligned} 2x - L &= 0 \\ 4 - L &= 0 \\ x + y - 6 &= 0 \end{aligned}$$

we obtain $x = \frac{1}{2}$, $y = \frac{11}{2}$.

12. The Lagrangian function is

$$\begin{aligned} H(x, y, L) &= 8x - 2x^2 + 8y - y^2 - L(x + y - 9) \\ &= 8x - 2x^2 + 8y - y^2 - Lx - Ly + 9L \end{aligned}$$

$$H_x(x, y, L) = 8 - 4x - L$$

$$H_y(x, y, L) = 8 - 2y - L$$

From

$$\begin{aligned} 8 - 4x - L &= 0 \\ 8 - 2y - L &= 0 \\ x + y - 9 &= 0 \end{aligned}$$

we obtain $x = \frac{9}{4}$, $y = \frac{25}{4}$.

14. The Lagrangian function is

$$\begin{aligned} H(x, y, L) &= x^2 + y^2 - L(xy - 4) \\ &= x^2 + y^2 - Lxy + 4L \end{aligned}$$

$$H_x(x, y, L) = 2x - Ly$$

$$H_y(x, y, L) = 2y - Lx$$

From

$$\begin{aligned} 2x - Ly &= 0 \\ 2y - Lx &= 0 \\ xy - 4 &= 0 \end{aligned}$$

we obtain $(2, 2)$ and $(-2, -2)$.